Signals and Communication Technology

The series "Signals and Communications Technology" is devoted to fundamentals and applications of modern methods of signal processing and cutting-edge communication technologies. The main topics are information and signal theory, acoustical signal processing, image processing and multimedia systems, mobile and wireless communications, and computer and communication networks. Volumes in the series address researchers in academia and industrial R&D departments. The series is application-oriented. The level of presentation of each individual volume, however, depends on the subject and can range from practical to scientific.

More information about this series at http://www.springer.com/series/4748

Lucian Trifina · Daniela Tarniceriu

Permutation Polynomial Interleavers for Turbo Codes

Springer

Lucian Trifina
Faculty of Electronics,Telecommunications and Information Technology
Gheorghe Asachi Technical University
Iaşi, Romania

Daniela Tarniceriu
Faculty of Electronics, Telecommunications and Information Technology
Gheorghe Asachi Technical University
Iaşi, Romania

ISSN 1860-4862 ISSN 1860-4870 (electronic)
Signals and Communication Technology
ISBN 978-981-13-2624-0 ISBN 978-981-13-2625-7 (eBook)
https://doi.org/10.1007/978-981-13-2625-7

Library of Congress Control Number: 2018955452

This Springer imprint is published by the registered company Springer Nature Singapore Pte Ltd.
The registered company address is: 152 Beach Road, #21-01/04 Gateway East, Singapore 189721, Singapore

Contents

Abbreviations

4-PP	Permutation polynomial of fourth degree
5-PP	Permutation polynomial of fifth degree
ARP	Almost regular permutation
AWGN	Additive white Gaussian noise
BCJR	Bahl–Cocke–Jelinek–Raviv
BER	Bit error rate
CF	Contention-free
CNP	Cubic null polynomial
CPP	Cubic permutation polynomial
DRP	Dithered relative prime
DVB	Digital Video Broadcasting
FER	Frame error rate
IWR	Interwindow randomization
LNP	Linear null polynomial
LPP	Linear permutation polynomial
LSB	Least significant bit
LTE	Long-Term Evolution
MCF	Maximum contention-free
NP	Null polynomial
NRNS	Non-recursive non-systematic
PLPP	Parallel linear permutation polynomial
PP	Permutation polynomial
QNP	Quadratic null polynomial
QPP	Quadratic permutation polynomial
RS	Recursive systematic
RSC	Recursive systematic convolutional
SISO	Soft-input soft-output
SNR	Signal-to-noise ratio
TUB	Truncated upper bound
UB	Upper bound

Symbols

$\mathbb{N}$	set of natural numbers
$\mathbb{N}^*$	set of natural numbers greater than zero
$\mathcal{P}$	set of prime numbers
$\mathbb{R}$	set of real numbers
$\mathbb{Z}$	set of integer numbers
$\mathbb{Z}^*$	set of integer numbers different by zero
$\mathbb{Z}_L$	set of integers modulo L, i.e. the set $\{0, 1, \ldots, L-1\}$
$\mathbb{Z}_L^*$	set of integers modulo L greater than zero, i.e. the set $\{1, 2, \ldots, L-1\}$
$\emptyset$	the empty set
$\mathcal{C}$	corner merit of an interleaver
ζ	nonlinearity degree of a PP interleaver
ζ'	refined nonlinearity degree of a PP interleaver
ϵ	degree of shift invariance of a PP interleaver
$\ln(x)$	natural logarithm (i.e. logarithm to base e) of x ($x \in \mathbb{R}$, $x > 0$)
$\Phi(\cdot)$	Euler function
$w_H(\mathbf{v})$	Hamming weight of the bit sequence $\mathbf{v}$
$\gcd(a, b)$	greatest common divisor of numbers a and b ($a, b \in \mathbb{N}$)
$\max(a, b)$	maximum of numbers a and b ($a, b \in \mathbb{R}$)
$\min(a, b)$	minimum of numbers a and b ($a, b \in \mathbb{R}$)
$\lvert x\rvert$	absolute value of x ($x \in \mathbb{R}$)
$[x]$	integer part of x ($x \in \mathbb{R}$)
$\lceil x \rceil$	ceiling function of x ($x \in \mathbb{R}$), i.e. smallest integer x_1 so that $x_1 \geq x$
$\lfloor x \rfloor$	floor function of x ($x \in \mathbb{R}$), i.e. the greatest integer x_1 so that $x_1 \leq x$
(mod L)	modulo L operation ($L \in \mathbb{N}^*$)
$\oplus$	modulo 2 sum
$(f \circ g)(\cdot)$	composing operation of functions $f(\cdot)$ and $g(\cdot)$, i.e. the function $f(g(\cdot))$
$\overline{m, n}$	the set $\{m, m+1, m+2, \ldots, n\}$ ($m, n \in \mathbb{N}, m < n$)

C_n^k	n choose k (binomial coefficient)
$k!$	factorial of k ($k \in \mathbb{N}^*$)
$a \mid b$	a divides b
$a \nmid b$	a does not divide b
$a \,\vdots\, b$	a is divisible by b
$a \equiv b \pmod{L}$	a is equivalent to b modulo L, i.e. $a \pmod{L} = b \pmod{L}$ ($a, b \in \mathbb{N}, L \in \mathbb{N}^*$)
$[a, b)$	the interval of real numbers x, so that $a \leq x < b$ ($a, b \in \mathbb{R}$, $a < b$)

List of Figures

List of Tables

Chapter 1
Introduction

Current communications systems cannot be conceived without error-correcting codes in their composition due to different propagation environments affected by disturbances (noise, fading, interference, etc.).

The bit error probability (or bit error rate (BER)) is a widely used measure which assesses the performance of error-correcting codes. It is the ratio between the number of erroneous bits remaining after decoding and the total number of information bits transmitted.

Another measure used in assessing the performance of error correcting codes is the frame error probability (or frame error rate (FER)) representing the ratio between the number of erroneous frames remaining after decoding and the total number of frames transmitted. BER and FER values are expressed by curves depending on the signal to noise ratio (SNR), usually expressed in deciBells (dB).

In Shannon (1948a, b), Claude E. Shannon demonstrated that for a certain channel there is a minimum value for SNR required to achieve a given BER. This value of SNR is called *Shannon limit*.

The *coding gain* of a coded system is the difference between the SNR required to achieve a given BER or FER for a non-coded system and the SNR for the coded system.

Error-correcting codes fall into two categories: block codes and convolutional codes.

Block codes carry out a bijective correspondence between the set of messages to be encoded or information words and the so-called code words. Each information word and code word has a fixed length. The symbols from a code word depend only on the symbols from the information word, i.e. the encoding is done on blocks of symbols, hence the name of these codes. The most popular error correcting block codes are Hamming codes (Hamming 1950), BCH codes (the name comes from the three researchers who discovered them, Bose, Chauduri and Hocquenghem) Bose and Ray-Chaudhuri (1960a, b), Hocquenghem (1959), and Reed–Solomon codes (Reed and Solomon 1960).

L. Trifina and D. Tarniceriu, *Permutation Polynomial Interleavers for Turbo Codes*, Signals and Communication Technology,
https://doi.org/10.1007/978-981-13-2625-7_1

In the case of convolutional codes (Elias 1955), the coding is no longer performed on symbol blocks. The symbols at the encoder output at a given time does not depend only on the input symbols from that time instant, but also on a number of previous input symbols. Therefore, convolutional codes have memory.

Shannon limit may be reached by increasing the length of information sequence to be coded, in the case of block codes, or by increasing the memory of the convolutional encoder, in the case of convolutional codes. However, in both cases, the decoding complexity becomes too large to be implemented, so that for a long time the performance of error correcting codes was quite far from the theoretical Shannon limit. Approaching this limit was made once the "turbo revolution" came into being, through the discovery of turbo codes in 1993 by three French researchers (Berrou et al. 1993).

The core concept of turbo codes is based on two main elements: parallel concatenated encoding of two or more recursive convolutional codes with one or more interleavers and the iterative decoding at reception. Iterative decoding is suboptimal, but it has a reasonable complexity for practical implementation and leads to very good performance. In (Berrou et al. 1993), the performance achieved by a turbo code with the information block length of 65536 bits and 18 iterations of turbo decoding was 0.5 dB far from the Shannon limit, which was an outstanding result.

The crucial component of turbo codes is the interleaver, a device that interlaces a block of symbols. The interleaver role in decoding is to decorrelate the inputs in the component decoders of the turbo decoder, which proves essential for turbo code suboptimal decoding. During encoding, the interleaver combines the low weight code words provided by a component encoder with those of high weight provided by the other component encoder, so that the codeword from the turbo encoder output has an overall large weight.

Since 1993 when turbo codes were discovered, many efforts have been made in finding performative interleavers. Research has been conducted in two main directions: *generic interleavers*, which do not take into account the component codes and are especially focused on its random behaviour and *code matched interleavers*, which consider the component codes and improve significantly the performance of turbo codes.

The search in a class of interleavers aims at reducing the number of interleavers. This is made using some metrics, out of which the best known are those that measure the spread and the randomization.

The main direction in code-matched interleaver design consists in considering the distance spectrum.

The interleavers can also be classified as: *deterministic interleavers*, which are described by mathematical laws, *random interleavers* and *combined interleavers* (described both by mathematical laws and random permutations). Deterministic interleavers are of particular interest, as they can be mathematically analyzed and designed. Out of the deterministic interleavers, the PP ones are among the best known and the most used, due to their outstanding performance and simple, practical implementation with high-speed, low-power consumption and little memory requirements (Takeshita 2007).

In addition to some general aspects regarding turbo codes, described in Chap. 2, this book attempts to present PP interleavers, focusing on the following issues:

- conditions on the coefficients of a polynomial so that it is PP (in Chap. 3),
- determining the number of true different PP interleavers of a certain degree for a certain length (in Chap. 4),
- results regarding the minimum distance of turbo codes with PP interleavers (in Chap. 5),
- the contention–free property of PP interleavers (in Chap. 6),
- reduced complexity methods to search PP interleavers (in Chap. 7) and
- presenting PP interleaver performances on channels with additive white Gaussian noise (AWGN) (in Chap. 7).

References

C. Berrou, A. Glavieux, P. Thitimajshima, Near Shannon limit error-correcting coding and decoding: turbo-codes, in *IEEE International Conference Communication (ICC)*, vol. 2 (Geneva, Switzerland, 1993), pp. 1064–1070

R.C. Bose, D.K. Ray-Chaudhuri, Further results on error correcting binary group codes. Inf. Control **3**(3), 279–290 (1960a)

R.C. Bose, D.K. Ray-Chaudhuri, On a class of error correcting binary group codes. Inf. Control **3**(1), 68–79 (1960b)

P. Elias, Coding for noisy channels. IRE Int. Conv. Rec. **3**, 37–46 (1955)

R.W. Hamming, Error detecting and correcting codes. Bell Syst. Tech. J. **29**(2), 147–160 (1950)

A. Hocquenghem, Codes correcteurs d'Erreurs. Chiffres **2**, 147–160 (1959)

I.S. Reed, G. Solomon, Polynomial codes over certain finite fields. SIAM J. Appl. Math. **8**, 300–304 (1960)

C.E. Shannon, A mathematical theory of communications. Bell Syst. Tech. J. **27**(3), 379–423 (1948a)

C.E. Shannon, A mathematical theory of communications. Bell Syst. Tech. J. **27**(4), 623–656 (1948b)

O.Y. Takeshita, Permutation polynomial interleavers: an algebraic-geometric perspective. IEEE Trans. Inf. Theory **53**(6), 2116–2132 (2007)

Chapter 2
Fundamentals of Turbo Codes

2.1 Convolutional Codes

As mentioned in the Introduction section, for *block codes* the bits from a code word depend only on the bits from the information word. If k is the number of information bits and n is the number of bits from the corresponding code word, then we have a (n, k) block code. The ratio $R_c = k/n$ is named the *coding rate* of the block code and it is a measure of the redundancy introduced by the code.

Unlike block codes, where the output bits depend only on the input bits at the current time, *convolutional codes* are non-block codes, where the output bits depend both on the input bits at the current time, and on a number of bits from earlier times. Non-recursive convolutional codes are structures without feedback, while the recursive ones are structures with feedback. Turbo codes use recursive convolutional codes, because of the interleaver gain. It means that the coding gain increases along with interleaver length (Benedetto and Montorsi 1996).

A binary convolutional encoder is a finite memory system, which provides n_0 output bits for k_0 input bits. It is composed of a shift register with $N \cdot k_0$ delay elements, n_0 modulo 2 adders and another shift register of n_0 elements for the output bits, as shown in Fig. 2.1.

The *coding rate* of a convolutional code, denoted by R_c, is the ratio between the number of information bits k_0 and the number of output bits n_0:

$$R_c = \frac{k_0}{n_0} \tag{2.1}$$

The current n_0 output bits are linear combinations of the present k_0 input bits and the previous $(N-1) \cdot k_0$ input bits. Beside the coding rate, the convolutional codes are characterized by the *memory order of the encoder*,

$$m = (N-1) \cdot k_0 \tag{2.2}$$

L. Trifina and D. Tarniceriu, *Permutation Polynomial Interleavers for Turbo Codes*, Signals and Communication Technology,
https://doi.org/10.1007/978-981-13-2625-7_2

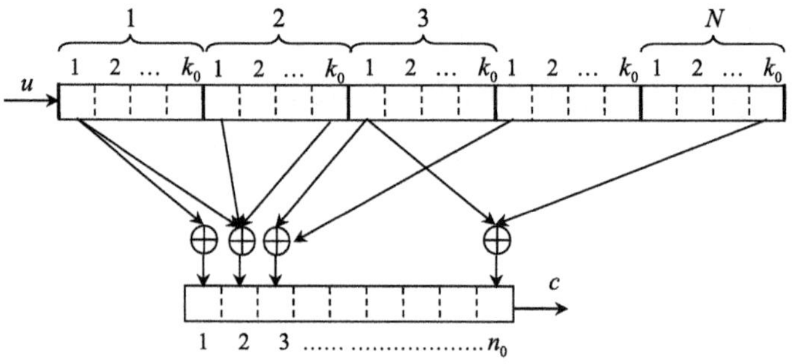

Fig. 2.1 The structure of a convolutional encoder

and *the constraint length* N. A convolutional code (n_0, k_0, N) provides n_0 output bits for k_0 information (input) bits and its constraint length is N.

The *generator matrix* of a convolutional code can be composed of N submatrices $G_1, G_2, \ldots, G_N$, each of them with k_0 rows and n_0 columns. Submatrix G_i, of size $k_0 \times n_0$, describes the connections of the "*i*th" structure of k_0 elements to the n_0 modulo 2 adders. The semi-infinite generator matrix of the convolutional code is of the form:

$$G_\infty = \begin{bmatrix} G_1 \; G_2 \ldots G_N & & & \\ & G_1 \; G_2 \ldots G_N & & \\ & & G_1 \; G_2 \ldots G_N & \\ & & & G_1 \; G_2 \ldots G_N \\ & & & \ldots \; \ldots \; \ldots \; \ldots \end{bmatrix}. \tag{2.3}$$

The coded output sequence is given by:

$$c = u \cdot G_\infty, \tag{2.4}$$

where u is the input (information) sequence and the sums in the matrix multiplication are modulo 2.

Alternatively, the structure of the encoder can be given by n_0 *generator vectors* g_i, $i = \overline{1, n_0}$, of size $N \cdot k_0$, describing the connections between the delay elements of the shift register from the input to each of the n_0 modulo 2 adders (they are usually given as polynomials or in octal form).

Example 2.1 For the convolutional code (3, 1, 3) (i.e. $n_0 = 3$, $k_0 = 1$, $N = 3$), equivalent representations of the encoder are given in Fig. 2.2.

In this case, the generator submatrices are of the form:

$$\begin{aligned} G_1 &= [1\ 0\ 1] \\ G_2 &= [1\ 1\ 1] \\ G_3 &= [0\ 1\ 1] \end{aligned} \tag{2.5}$$

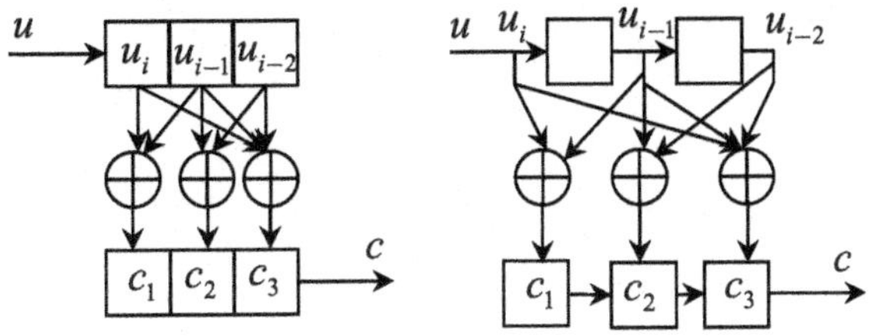

Fig. 2.2 Equivalent representations of the encoder in Example 2.1

For this particular case, the semi-infinite generator matrix becomes:

$$G_\infty = \begin{bmatrix} 101 & 111 & 011 & 000 \ldots & \ldots & \ldots \\ 000 & 101 & 111 & 011\ 000 & & \ldots \\ 000 & 000 & 101 & 111\ 011\ 000 & \ldots & \\ \ldots & \ldots & \ldots & \ldots\ \ldots & \ldots & \ldots \end{bmatrix}. \tag{2.6}$$

For the input sequence $u = (1011)$, the output sequence $c = (101\ 111\ 110\ 010)$ results.

The generator vectors for the encoder in this example are:

$$\begin{aligned} g_1 &= [1\ 1\ 0] \\ g_2 &= [0\ 1\ 1] \\ g_3 &= [1\ 1\ 1] \end{aligned} \tag{2.7}$$

In polynomial form, we can write:

$$\begin{aligned} g_1(D) &= 1 \oplus D \\ g_2(D) &= D \oplus D^2 \\ g_3(D) &= 1 \oplus D \oplus D^2, \end{aligned} \tag{2.8}$$

where D is the delay operator.

In octal representation, with the least significant bit (LSB) in the right side, we have:

$$\begin{aligned} g_1 &= 6 \\ g_2 &= 3 \\ g_3 &= 7 \end{aligned} \tag{2.9}$$

■

A convolutional code can be equivalently described by:

- generator submatrices G_i, or generator vectors g_i;
- state diagram;
- trellis diagram.

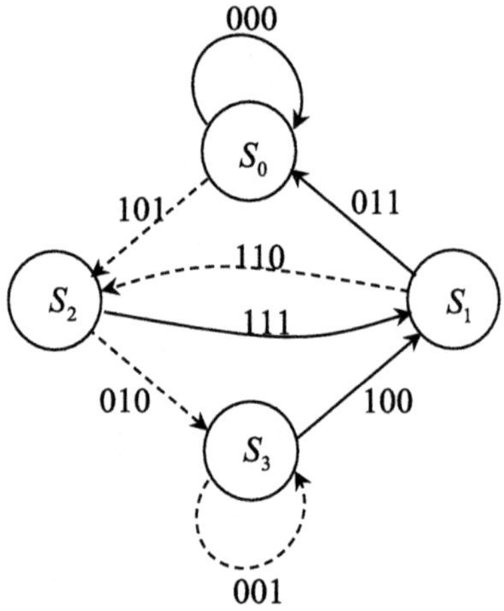

Fig. 2.3 State diagram of the encoder in Example 2.1

Table 2.1 Table describing the state diagram of the encoder in Example 2.1

u_l	σ_l	σ_{l+1}	c_l
0	00 (S_0)	00 (S_0)	000
1	00 (S_0)	10 (S_2)	101
0	10 (S_2)	01 (S_1)	111
1	10 (S_2)	11 (S_3)	010
0	01 (S_1)	00 (S_0)	011
1	01 (S_1)	10 (S_2)	110
0	11 (S_3)	01 (S_1)	100
1	11 (S_3)	11 (S_3)	001

The description by generator submatrices G_i or generator vectors g_i was given above.

A convolutional encoder is a finite memory system, whose output depends on the input sequence and its state. The *state of the encoder* is defined as the content of its shift register at a certain time. The operations of a convolutional encoder can be described by a *state diagram*. The number of the encoder states is $N_S = 2^m$, where m is the encoder memory order. There are 2^{k_0} branches leaving any state, corresponding to all entries at a time. 2^{k_0} branches enter the encoder states. The transition from one state to the other is through a branch that is labelled with the output sequence and the input bits at that time.

For the encoder in Example 2.1, we denote by $\sigma_l = (u_{l-1}, u_{l-2})$ the state at time l. There are $N_S = 2^2 = 4$ states, denoted by $S_0 = (00)$, $S_1 = (01)$, $S_2 = (10)$, $S_3 = (11)$. The corresponding state diagram is given in Fig. 2.3, as well as in Table 2.1. The branches with solid line indicate transitions due to input bit "0" and those with dashed line indicate transitions corresponding to "1". The size of diagram grows exponentially with memory order m.

The *trellis diagram* describes the time evolution of the state diagram. For the code presented in Example 2.1, the trellis is given in Fig. 2.4. The four nodes on each

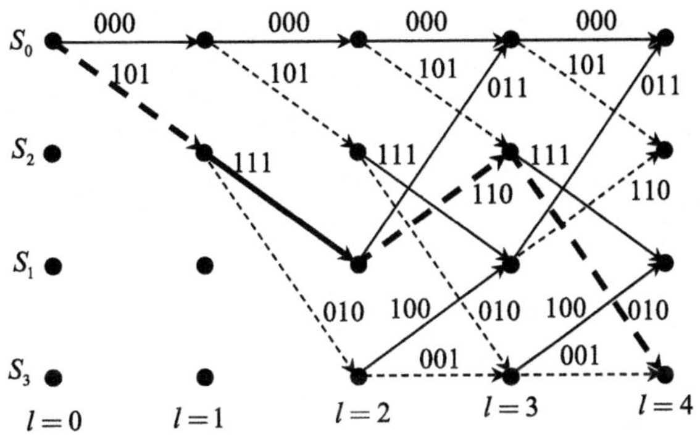

Fig. 2.4 Trellis diagram of the encoder in Example 2.1

vertical direction represent the four states at the discrete time l, which is called the *depth* of the trellis. The representations with solid and dashed lines have the same meaning as for the state diagram. The path with bold line corresponds to the input sequence $u = (1011)$.

2.1.1 Systematic Recursive Convolutional Encoders

These types of encoders present feedback branches and were introduced by Costello (1969) and Forney Jr. (1970).

For example, two encoders with memory of order 2 (two delay elements required for implementation) are given in Fig. 2.5, both in non-recursive and recursive versions. The coding rate of the two encoders is 1/2. The second encoder is *systematic* because the input sequence appears in the codeword, whereas the first encoder is non-systematic.

Let $g_{1,1}(D)$ and $g_{1,2}(D)$ be the generator polynomials of a convolutional encoder of rate 1/2. These polynomials describe the connections of delay elements of the encoder at each of modulo 2 adders (this information can also be given as generator vectors or in octal form), as shown in Example 2.1.

When the entry in polynomial form is $u(D)$, the output sequences are:

$$\begin{aligned} c_1(D) &= u(D) \cdot g_{1,1}(D) \\ c_2(D) &= u(D) \cdot g_{1,2}(D) \end{aligned} \tag{2.10}$$

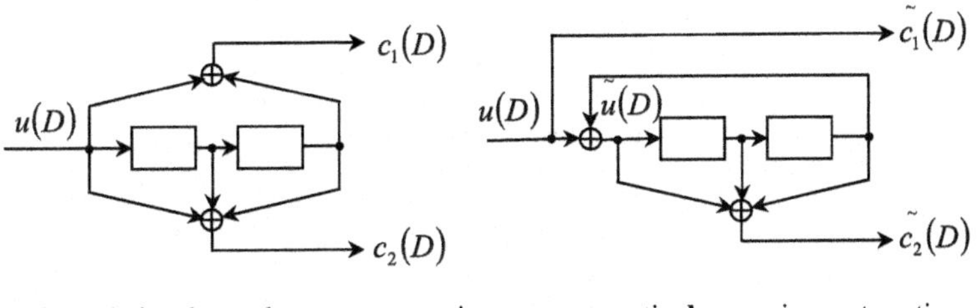

Fig. 2.5 Convolutional encoder: **a** non-recursive non-systematic; **b** recursive systematic

To get a systematic code with $c_1(D) = u(D)$, we divide (2.10) by $g_{1,1}(D)$, obtaining:

$$\begin{aligned}\tilde{c}_1(D) &= \frac{c_1(D)}{g_{1,1}(D)} = u(D)\\ \tilde{c}_2(D) &= \frac{c_2(D)}{g_{1,1}(D)} = \frac{u(D)}{g_{1,1}(D)} \cdot g_{1,2}(D)\end{aligned} \tag{2.11}$$

Considering the new input sequence

$$\tilde{u}(D) = \frac{u(D)}{g_{1,1}(D)} \tag{2.12}$$

relations (2.11) become:

$$\begin{aligned}\tilde{c}_1(D) &= \tilde{u}(D) \cdot g_{1,1}(D)\\ \tilde{c}_2(D) &= \tilde{u}(D) \cdot g_{1,2}(D)\end{aligned} \tag{2.13}$$

As $\tilde{u}(D)$ is the reordered sequence $u(D)$ (according to a recursive digital filter), from (2.10) and (2.13) it results that the set of code sequences $\tilde{c}(D)$ is the same as the set $c(D)$ and thus the two codes have the same weight enumeration function (Proakis 1995).

For the non-recursive non-systematic (NRNS) encoder in Fig. 2.5, the generators can be written in polynomial form as:

$$\begin{aligned}g_1^{NRNS}(D) &= 1 \oplus D^2\\ g_2^{NRNS}(D) &= 1 \oplus D \oplus D^2\end{aligned} \tag{2.14}$$

and for the recursive systematic (RS) encoder, equivalent to the former, the generators are:

$$\begin{aligned}g_1^{RS}(D) &= 1\\ g_2^{RS}(D) &= \frac{1 \oplus D \oplus D^2}{1 \oplus D^2}\end{aligned} \tag{2.15}$$

The generator vectors are:

$$\begin{aligned}g_1^{NRNS} &= [1\ 0\ 1]\\ g_2^{NRNS} &= [1\ 1\ 1]\end{aligned} \tag{2.16}$$

and the octal representation for the non-recursive non-systematic encoder is:

$$\begin{aligned}g_1^{NRNS} &= 5\\ g_2^{NRNS} &= 7\end{aligned} \tag{2.17}$$

and for the recursive systematic encoder:

$$\begin{aligned} g_1^{RS} &= 1 \\ g_2^{RS} &= \frac{7}{5}. \end{aligned} \tag{2.18}$$

2.2 Turbo Code Structure

Turbo codes are composed by the parallel concatenation of two or more convolutional codes, where the encoder input sequences are interleaved versions of the information sequence. They are obtained with the so-called interleaver devices. The structure of a turbo encoder of coding rate 1/3 is given in Fig. 2.6.

To increase the coding rate, some output bits can be removed (usually from the parity check bits c_{1k}^p and c_{2k}^p). This operation is known as *puncturing*.

In general, the number of component codes can be greater than two, with an appropriate number of interleavers, resulting the so-called multiple turbo codes.

The interleavers and their complementary devices, de-interleavers, were previously used to correct error bursts by concatenating with an independent error correction code. The role of these devices in the turbo code structure is to obtain statistically independent encoded sequences at the output of the component encoders, which is essential for the iterative decoding. The interleavers take symbol blocks of size L from the input and provide at output the same block, but in a different order, performing the so-called entry permutation. De-interleavers perform the inverse operation, restoring the original order of symbols.

A block interleaver is described by an invertible function:

$$\pi : \mathbb{Z}_L \to \mathbb{Z}_L \tag{2.19}$$

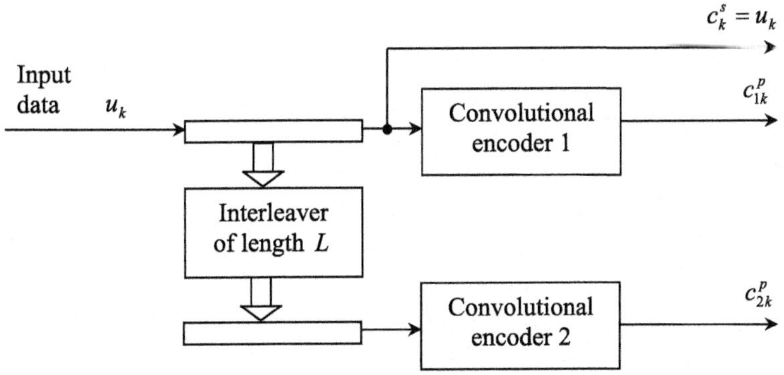

Fig. 2.6 Structure of a turbo encoder of coding rate 1/3

which is a permutation of the integers from the set $\mathbb{Z}_L = \{0, 1, 2, \ldots, L-1\}$, so that the symbol on position i, with $i \in \mathbb{Z}_L$, at the output is the symbol on position $\pi(i)$ from the input.

The inverse device, the de-interleaver, acts on the output interleaver symbols and places them back in the original order. The permutation describing the de-interleaver is $\mu = \pi^{-1}$.

Thus, the turbo code becomes a block code $(3L, L)$ (for the case shown in Fig. 2.6), where L is the length of the information bit block, and $3L$ is the length of the codeword. Convolutional encoders are systematic and their coding rate is 1/2; the input sequence is transmitted once. Besides block interleavers, there is another class called convolutional interleavers, working with continuous data flow, as code name.

2.3 Trellis Termination

The interleaver size and type influence the turbo code performance and for a good decoding it is required that the trellises of the two codes start from the same state (usually the null state) for every block of information bits. This imposes the operation named *trellis termination*.

If the trellis is truncated in an unknown state, the decoding performance is weak towards its end.

Assume the convolutional encoders have the memory of order m. If only the first trellis is terminated with m bits included in the interleaver, the equivalent block code corresponding to the turbo-code in Fig. 2.6 is $(3L, L-m)$. If both trellises are terminated with m bits after the interleaver, the equivalent block code corresponding to the turbo-code in Fig. 2.6 is $(3(L+m), L)$.

There are five main classes of trellis termination. Assuming that the two encoders have m_1 and m_2 memory elements, respectively, the trellis termination methods can be briefly described as follows (Hokfelt et al. 2001):

I. *Without trellis termination*

In this case, none of the trellises is finished and the decoding performance is the weakest. The coding rate is $R_c = \dfrac{1}{3}$.

II. *Termination of the first encoder*

The first encoder closes the trellis while the second does not. The termination is carried out by adding m_1 tail bits to the input sequence, so that the first encoder reaches state 0. These bits are included in the sequence entering the interleaver and therefore after permutation they no longer correspond to the termination bits of the second encoder. For $m_1 = m_2 = m$ the coding rate is $R_c = \dfrac{L-m}{3L}$.

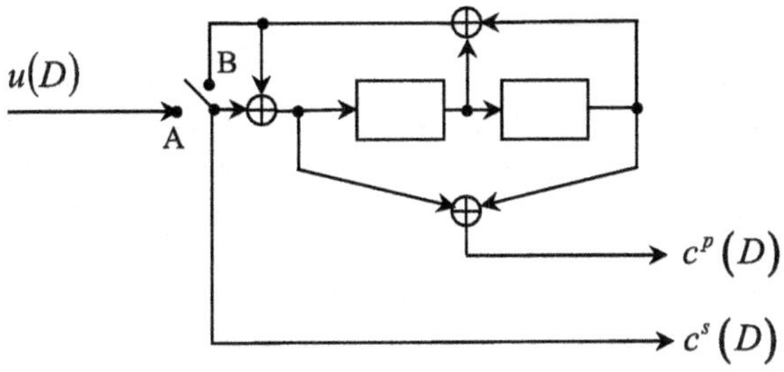

Fig. 2.7 Scheme for generating "flush" bits by a recursive convolutional encoder

III. *Termination of both encoders (dual termination)*
This can be accomplished in at least two ways:

(1) Imposing constraints to the interleaver, so that the second encoder can be forced in the same state as the first encoder, namely:

$$\pi(i) = i \mod p, i = \overline{0, L-1}, \tag{2.20}$$

where p is the period of the impulse response of the encoder and $\pi(\cdot)$ is the permutation describing the interleaver (Breiling et al. 1999). Then, a single sequence of "tail bits" is required to end both encoders.

For $m_1 = m_2 = m$, the coding rate is $R_c = \dfrac{L-m}{3L}$.

(2) Identifying specific input positions, dependent on the interleaver, in order to bring the encoders in state 0, independently of one another. This is achieved without constraints on the interleaver, but with a slight increase in the number of input bits required to close the trellis (m_t bits, where $\max(m_1, m_2) \leq m_t \leq m_1 + m_2$). The method is called *dual termination* (Guinand and Lodge 1994).

For $m_t = m_1 + m_2$ and $m_1 = m_2 = m$, the corresponding coding rate is $R_c = \dfrac{L-2m}{3L}$.

IV. *Post-interleaver flushing*

With this method both encoders are brought into the same state independently of each other, after encoding the input bit sequence of length L. A scheme for generating the "flush" bits is given in Fig. 2.7 (Divsalar and Pollara 1995).

The switch is in position A for L tact periods and in position B for m additional tact periods ($m = 2$ for the case shown in Fig. 2.7), thereby determining the input bits that bring the encoder in state 0.

In this case, for $m_1 = m_2 = m$ the coding rate is $R_c = \dfrac{L}{3L+3m}$ if the ending bits of the second trellis are not transmitted and $R_c = \dfrac{L}{3L+4m}$ if these bits are transmitted.

V. *Tailbiting termination*

In this method both encoders start and end in the same state (Berrou et al. 1999). The drawback of this method is that the data sequence has to be encoded twice: firstly from the state "all-zero" and secondly from a state S_c, determined according to the block of input bits and the final states of the encoders in the first coding step. However, in most cases, the double encoding operation is done at a much higher frequency than the data rate, so that the delay effects are reduced.

This termination method is called *circular termination*. In this termination method the coding rate is $R_c = \frac{1}{3}$.

2.4 Calculation of Turbo Code Distance Spectrum

The *distance spectrum* of a turbo code represents the values of triplets (d_i, N_i, w_i), with $d_1 < d_2 < d_3 < \cdots$. d_i is the Hamming distance of code words to the "all-zero" word (consisting of all zeros) or the weight of the codeword (because the turbo code is linear), i.e. the number of bit 1 in the word. N_i represents the multiplicity of codewords of weight d_i, i.e. the number of such codewords. w_i represents the overall weight of information words generating codewords of weight d_i, i.e. the sum of the weights of all the information words leading to codewords of weight d_i. Sometimes index 1 is replaced by "min", specifying the minimum distance, or with *free*, specifying the free distance (similar to convolutional codes) (Proakis 1995).

Knowing the exact first few values of the distance spectrum for a particular interleaver is useful because we can assess the performance of turbo code (BER, FER) at medium to large SNR. It is known that for this range of SNR the "error-floor" phenomenon appears. It consists in flattening the curve of error probability according to SNR. We can also design turbo codes requiring that the upper bounds of BER/FER be as small as possible for a certain SNR. Especially in the case of large interleaver lengths, it has been shown by simulations that, even if the decoding of turbo codes is suboptimal, the curves BER/FER converge for medium to high SNR to the upper bounds (UBs) of BER/FER.

The truncated upper bounds (TUB) of the bit and frame error rate on AWGN channel, to the first M terms of distance spectrum, are given by the relationships below

$$TUB(BER) \lesssim 0.5 \cdot \sum_{i=1}^{M} \frac{w_i}{L} \cdot erfc\left(\sqrt{d_i \cdot R_c \cdot E_b/N_0}\right), \tag{2.21}$$

$$TUB(FER) \lesssim 0.5 \cdot \sum_{i=1}^{M} N_i \cdot erfc\left(\sqrt{d_i \cdot R_c \cdot E_b/N_0}\right), \tag{2.22}$$

where $SNR = E_b/N_0$, E_b is the uncoded bit energy, and N_0 is the power spectral density of AWGN noise, equal to twice its variance, R_c is the coding rate, and the error function is defined by

$$erfc(x) = \frac{2}{\sqrt{\pi}} \cdot \int_{x}^{\infty} e^{-t^2} dt \tag{2.23}$$

For the independent Rayleigh fading channel with known channel state information, the upper bounds of the BER and FER are given by the following relationships, respectively:

$$TUB(BER) \lesssim 0.5 \cdot \sum_{i=1}^{M} \frac{w_i}{L} \cdot \left(\frac{1}{1 + R_c \cdot SNR} \right)^{d_i}, \tag{2.24}$$

$$TUB(FER) \lesssim 0.5 \cdot \sum_{i=1}^{M} N_i \cdot \left(\frac{1}{1 + R_c \cdot SNR} \right)^{d_i}, \tag{2.25}$$

where SNR is the signal to noise ratio for the above mentioned channel.

An efficient method for the exact calculation of the first terms of the distance spectrum was given by Garello et al. (2001). This method was implemented in C programming language and it was used to determine the distance spectra for PP interleavers presented in this book. Previous inaccurate methods for determining the distance spectrum of a turbo code were given in Perez et al. (1996) and Daneshgaran and Mondin (1997).

Garello's algorithm is based on the computation of the minimum distance (or distance spectrum) of a constrained subcode, i.e., a subset of a code defined via constraints on the edges of its trellis. Further, we introduce notations and concepts necessary for the algorithm presentation.

We denote by $\boldsymbol{u}^{(i)} = (u_0, u_1, \ldots, u_{i-1})$ a binary input sequence of length $i \le L$. The main idea of the algorithm is the computation of the minimum Hamming weight of a turbo codeword generated by a L-bit information frame $\boldsymbol{u} = (u_0, u_1, \ldots, u_{L-1})$, whose first i bits coincide with $\boldsymbol{u}^{(i)}$. This minimum weight, denoted by $v(\boldsymbol{u}^{(i)})$, consists of two parts, one for each constituent code: $v(\boldsymbol{u}^{(i)}) = v_1(\boldsymbol{u}^{(i)}) + v_2(\boldsymbol{u}^{(i)})$. To evaluate $v_1(\boldsymbol{u}^{(i)})$, we firstly encode $\boldsymbol{u}^{(i)}$ by the first encoder to produce $\boldsymbol{c}_1^{(i)} = (\boldsymbol{u}^{(i)}, \boldsymbol{p}_1^{(i)})$, where $\boldsymbol{p}_1^{(i)} = (p_{1,0}, p_{1,1}, \ldots, p_{1,i-1})$ denotes the parity check bits added by the first encoder, and then we compute its Hamming weight $w_H(\boldsymbol{c}_1^{(i)})$.

Let $\sigma_1^{(i)}$ be the state reached by the first encoder at time i after encoding $\boldsymbol{u}^{(i)}$. We add to $w_H(\boldsymbol{c}_1^{(i)})$ the minimum weight of the parity check sequence of the code path through the trellis of the first encoder, starting from $\sigma_1^{(i)}$ and reaching the all-zero state, denoted by $v(\sigma_1^{(i)})$.

Therefore:

$$v_1(\boldsymbol{u}^{(i)}) = w_H(\boldsymbol{c}_1^{(i)}) + v(\sigma_1^{(i)}). \tag{2.26}$$

The bit $u_k \in \{0, 1\}$ forces a multiple constraint in the second encoder trellis on position $j = \pi^{-1}(k)$, where $\pi(\cdot)$ is the permutation that describes the interleaver. Thereby, according to Garello et al. (2001), the input sequence $\boldsymbol{u}^{(i)}$ induces a constraint set $V(\boldsymbol{u}^{(i)})$ of cardinality i. Then

$$v_2\left(\boldsymbol{u}^{(i)}\right) = \min\left\{w_H\left(\boldsymbol{p}_2\right) + w_H\left(\boldsymbol{u}_2^{(j\neq\pi^{-1}(k))}\right)\right\}, \tag{2.27}$$

where $\boldsymbol{p}_2$ is the parity check sequence at the output of the second encoder, and $\boldsymbol{u}_2^{(j\neq\pi^{-1}(k))}$ is the sequence of information bits on the positions $j \neq \pi^{-1}(k)$, $k = 0, 1, \ldots, i-1$, of the interleaver output, for which the minimum in the previous relation is achieved.

In the following, we present Garello's algorithm approach derived from (Garello (C program) 2001).

Algorithm Description

- Set the value d^* for which all distances in the final computed spectrum are less than or equal to d^*, that is:
  $d_1 < d_2 \cdots < d_M \leq d^*$, where M is the number of distances in the spectrum to be calculated and $d_1, d_2, \ldots, d_M$ are the computed distances.
- Initialize the distances with a very large value and the multiplicities with 0, i.e.
  $d_1 = d_2 = \cdots = d_M = 10,000$,
  $N_1 = N_2 = \cdots = N_M = 0$,
  $w_1 = w_2 = \cdots = w_M = 0$.
- Start with the sequences $\boldsymbol{u}^{(L)} = \Big(\underbrace{0, 0, \ldots, 0}_{L-1 \text{ zeroes}}, 1\Big)$, then

  $\Big(\underbrace{0, 0, \ldots, 0}_{L-2 \text{ zeroes}}, 1, \times\Big) = \left(\boldsymbol{u}^{(L-1)}, \times\right), \Big(\underbrace{0, 0, \ldots, 0}_{L-3 \text{ zeroes}}, 1, \times, \times\Big) =$
  $\left(\boldsymbol{u}^{(L-2)}, \times, \times\right), \ldots, \Big(0, 1, \underbrace{\times, \times, \ldots, \times}_{L-2 \text{ of } \times}\Big) = \Big(\boldsymbol{u}^{(2)}, \underbrace{\times, \times, \ldots, \times}_{L-2 \text{ of } \times}\Big)$,
  $\Big(1, \underbrace{\times, \times, \ldots, \times}_{L-1 \text{ of } \times}\Big) = \Big(\boldsymbol{u}^{(1)}, \underbrace{\times, \times, \ldots, \times}_{L-1 \text{ of } \times}\Big)$, where $\times$ denotes the unknown bits, determined by the encoding of the sequence at the second encoder input, so that $v_2\left(\boldsymbol{u}^{(i)}\right)$ is minimum.
- Initialize $j = L$

(1) Set $i = j$
(2) For the sequences $\boldsymbol{u}^{(i)}$ and those formed from these in step (3), compute $v\left(\boldsymbol{u}^{(i)}\right) = v_1\left(\boldsymbol{u}^{(i)}\right) + v_2\left(\boldsymbol{u}^{(i)}\right)$. Then

- If $v\left(\boldsymbol{u}^{(i)}\right) > d_M$, eliminate the sequence $\boldsymbol{u}^{(i)}$.
- If $v\left(\boldsymbol{u}^{(i)}\right) \leq d_M$, keep the sequence $\boldsymbol{u}^{(i)}$. Let $\boldsymbol{c}$ be the turbo-coded sequence resulted by encoding the information sequence consisting of $\boldsymbol{u}^{(i)}$ together with $\boldsymbol{u}_2^{(j\neq\pi^{-1}(k))}$, $k = 0, 1, \ldots, i-1$. Denote by w the Hamming weight of the formed information sequence and by $w_H(\boldsymbol{c})$, the Hamming weight of the turbo-coded sequence.
- If $w_H(\boldsymbol{c}) = d_l$, for an $l \in \{1, 2, \ldots, M\}$, update the multiplicities for the distance d_l, thus $N_l \to N_l + 1$ and $w_l \to w_l + w$.
- If $d_{l-1} < w_H(\boldsymbol{c}) < d_l$, for an $l \in \{1, 2, \ldots, M\}$, update the distance spectrum as follows:

- o Let unmodified the distances $d_1, d_2, \ldots, d_{l-1}$ with afferent multiplicities, and
- o Set
$$d_M = d_{M-1}, d_{M-1} = d_{M-2}, \ldots, d_{l+1} = d_l, d_l = w_H(\mathbf{c}),$$
$$N_M = N_{M-1}, N_{M-1} = N_{M-2}, \ldots, N_{l+1} = N_l, N_l = 1,$$
$$w_M = w_{M-1}, w_{M-1} = w_{M-2}, \ldots, w_{l+1} = w_l, w_l = w.$$

(3) Add to the set of sequences $\boldsymbol{u}^{(i)}$, of length $i < L$, to be tested, a 0 and a 1, respectively, forming sequences $\boldsymbol{u}^{(i+1)}$ with $u_i = 0$ or $u_i = 1$. Then $i \to i + 1$. If $i \leq L$, go to step (2), otherwise, go to step (4).
(4) Set $j \to j - 1$. If $j \geq 1$, go to step (1), otherwise, go to step (5).
(5) For $j = 0$ the final distance spectrum is obtained.

Garello's algorithm has been improved in Rosnes and Ytrehus (2005) by reducing the processing average cost of each constrained set (changing the method for determining the weight $v_2(\boldsymbol{u}^{(i)})$ in (2.27) and reducing the total number of the investigated constraint set (i.e. the number of iterations in steps 3 and 4).

References

2001, http://www.tlc.polito.it/garello/turbodistance/turbodistance.html

S. Benedetto, G. Montorsi, Unveiling turbo codes: some results on parallel concatenated coding schemes. IEEE Trans. Inf. Theory **42**(2), 409–428 (1996)

C. Berrou, C. Douillard, M. Jezequel, Multiple parallel concatenation of circular recursive systematic convolutional (CRSC) codes. Ann. Telecommun. **54**(3–4), 166–172 (1999)

M. Breiling, S. Peeters, J. Huber, Class of double terminating turbo code interleavers. Electron. Lett. **35**(5), 389–391 (1999)

D.J. Costello, *Construction of convolutional codes for sequential decoding* (University of Notre Dame, Notre Dame, Indiana, USA, 1969)

F. Daneshgaran, M. Mondin, An efficient algorithm for obtaining the distance spectrum of turbo codes. In: 1st International Symposium on Turbo Codes. Brest, France, 3–5 September 1997, pp. 251–254

D. Divsalar, F. Pollara, Turbo codes for PCS applications. In: IEEE International Conference on Communications (ICC). Seattle, WA, USA, 18–22 June 1995, pp. 54–59

G.D. Forney Jr., Convolutional codes I: algebraic structure. IEEE Trans. Inf. Theory **16**(6), 720–738 (1970)

R. Garello, P. Pierleoni, S. Benedetto, Computing the free distance of turbo codes and serially concatenated codes with interleavers: algorithms and applications. IEEE J. Sel. Areas Commun. **19**(5), 800–812 (2001)

P. Guinand, J. Lodge, Trellis termination for turbo encoders. In: 17th Biennial Symposium on Communications. Queen's University, Kingston, Canada, May 30–June 1 1994, pp. 389–392

J. Hokfelt, O. Edvors, T. Maseng, On the theory and performance of trellis termination methods of turbo codes. IEEE J. Sel. Areas Commun. **19**(5), 838–847 (2001)

C.L. Perez, J. Seghers, D.J. Costello Jr., A distance spectrum interpretation of turbo codes. IEEE Trans. Inf. Theory **42**(6), 1698–1709 (1996)

J.G. Proakis, *Digital Communications*, 3rd edn. (McGraw-Hill, New York, 1995)

E. Rosnes, Ø. Ytrehus, Improved algorithms for the determination of turbo-code weight distributions. IEEE Trans. Commun. **53**(1), 20–26 (2005)

L. Trifina, V. Munteanu, *Coduri Turbo* (Politehnium, Iasi, 2008)

Chapter 3
Permutation Polynomial Based Interleavers. Conditions on Coefficients

3.1 Definition of a Permutation Polynomial Interleaver

The PP based interleavers for turbo codes were introduced by Jing Sun and Oscar Y. Takeshita in 2003 (Sun et al. 2003; Sun and Takeshita 2005). They are preferred because of several advantages: outstanding performance, complete analytical structure and simple, practical implementation with high-speed, low-power consumption and low memory requirements (Takeshita 2007).

A PP-based interleaver of degree d and length L is defined as:

$$\pi(x) = q_0 + q_1 \cdot x + q_2 \cdot x^2 + \cdots + q_d \cdot x^d \pmod{L}, x = \overline{0, L-1}, \qquad (3.1)$$

where the coefficients $q_1, q_2, \ldots, q_d$ are chosen so that the polynomial $\pi(x)$ of degree d from (3.1), for $x = \overline{0, L-1}$, performs a permutation of the set $\mathbb{Z}_L$ and q_0 determines only a shift of the permutation elements. As q_0 has no influence on whether $\pi(x)$ in (3.1) is or is not a permutation polynomial modulo L, we will consider $q_0 = 0$.

3.2 Necessary and Sufficient Conditions on the Coefficients of a Polynomial of Any Degree so that It Is PP Modulo a Number Equal to a Power of 2

For the particular case $L = 2^n$, with $n \in \mathbb{N}^*$, the necessary and sufficient conditions on the coefficients of a polynomial of any degree so that it is PP were given by Ronald L. Rivest in (2001). Before giving the main theorem with this result, we present some helpful lemmas also given in Rivest (2001). The authors have tried to give detailed proofs of lemmas and theorems for their easy understanding.

Case $L = 2$ ($n = 1$) is trivial. It is given in Lemma 3.1.

L. Trifina and D. Tarniceriu, *Permutation Polynomial Interleavers for Turbo Codes*, Signals and Communication Technology,
https://doi.org/10.1007/978-981-13-2625-7_3

Lemma 3.1 *$\pi(x)$ from (3.1) is PP modulo 2 if and only if $(q_1 + q_2 + q_3 + \cdots + q_d)$ is odd.*

Proof We have $\pi(0) = 0$ and $\pi(1) = q_1 + q_2 + q_3 + \cdots + q_d \pmod 2$. For $\pi(x)$ to be PP modulo 2 we need $\pi(1) = 1$, i.e. $(q_1 + q_2 + q_3 + \cdots + q_d)$ must be odd. ■

Lemma 3.2 *Let L be $L = 4 \cdot r$, with $r \in \mathbb{N}^*$. If $\pi(x)$ in (3.1) is PP then coefficient q_1 is odd.*

Proof We have $\pi(0) = 0$. If q_1 were even, we could write

$$\pi(2 \cdot r) = \left(\sum_{i=1}^{d} q_i \cdot (2 \cdot r)^i \right) \pmod{(4 \cdot r)} =$$
$$= \left(q_1 \cdot (2 \cdot r) + \sum_{i=2}^{d} q_i \cdot (2 \cdot r)^i \right) \pmod{(4 \cdot r)} =$$
$$= q_1 \cdot (2 \cdot r) \pmod{(4 \cdot r)} +$$
$$+ \left(\sum_{i=2}^{d} q_i \cdot (4 \cdot r) \cdot (2^{i-2} \cdot r^{i-1}) \right) \pmod{(4 \cdot r)} = 0 + 0 = 0. \quad (3.2)$$

Therefore, we would have $\pi(0) = \pi(2 \cdot r)$, which means that $\pi(x)$ in (3.1) is not PP. This is why q_1 must be odd. ■

Lemma 3.3 *Let L and L_1 be $L = 2^n$, with $n \in \mathbb{N}^*$, and $L_1 = L/2 = 2^{n-1}$, respectively. If $\pi(x)$ in (3.1) is PP modulo L, then $\pi(x)$ is PP modulo L_1.*

Proof As for $i \in \mathbb{N}^*$, $(x + L_1)^i \pmod{L_1} = \left(\sum_{k=0}^{i-1} C_i^k \cdot x^k \cdot L_1^{i-k} + x^i \right) \pmod{L_1} = x^i \pmod{L_1}$, $\forall x \in \mathbb{Z}_L$, obviously, we have $\pi(x + L_1) \pmod{L_1} = \pi(x) \pmod{L_1}$, $\forall x \in \mathbb{Z}_L$. We note that in the set $\mathbb{Z}_L$, there are only two values equal modulo L_1. We assume that $\pi(x)$ in (3.1) is PP modulo L. If $\pi(x)$ were not PP modulo L_1, then there are two distinct values $x, x' \in \mathbb{Z}_{L_1}$, so that $\pi(x) \pmod{L_1} = \pi(x') \pmod{L_1} = y \pmod{L_1}$.

Taking into account the considerations above, we have $\pi(x) \pmod{L_1} = \pi(x + L_1) \pmod{L_1} = \pi(x') \pmod{L_1} = \pi(x' + L_1) \pmod{L_1}$. Therefore, as $\pi(x)$ is PP modulo L, to distinct values modulo L, $x, x + L_1, x', x' + L_1$, there correspond the values $\pi(x), \pi(x + L_1), \pi(x'), \pi(x' + L_1)$, also distinct modulo L. The four values $\pi(x) \pmod L$, $\pi(x + L_1) \pmod L$, $\pi(x') \pmod L$, $\pi(x' + L_1) \pmod L$ are distinct in the set $\mathbb{Z}_L$. But as we have already shown, only two values equal modulo L_1 can be in the set $\mathbb{Z}_L$. This is inconsistent with the fact that the four values are equal modulo L_1 and consequently with the fact that $\pi(x)$ would not be PP modulo L_1. Therefore, if $\pi(x)$ is PP modulo L, then $\pi(x)$ is PP modulo $L_1 = L/2$. ■

Lemma 3.4 *Let L be $L = 2^n = 2 \cdot L_1$, with $n \in \mathbb{N}^*$. If $\pi(x)$ in (3.1) is PP modulo L, then $\pi(x + L_1) \pmod L = (\pi(x) + L_1) \pmod L$, $\forall x \in \mathbb{Z}_L$.*

Proof As shown in the proof of Lemma 3.3, we have $\pi(x + L_1) \pmod{L_1} = \pi(x) \pmod{L_1}$, $\forall x \in \mathbb{Z}_L$. We note that $\pi(x) \pmod{L}$ and $(\pi(x) + L_1) \pmod{L}$ are two distinct values in the set $\mathbb{Z}_L$. The values $\pi(x) \pmod{L}$ and $\pi(x + L_1) \pmod{L}$ are also distinct in the set $\mathbb{Z}_L$, $\pi(x)$ being PP modulo L. As in the set $\mathbb{Z}_L$ there are only two distinct values equal modulo L_1 and $\pi(x + L_1) \pmod{L_1} = \pi(x) \pmod{L_1} = (\pi(x)+L_1) \pmod{L_1}$, it is required that $\pi(x+L_1) \pmod{L} = (\pi(x)+L_1) \pmod{L}$. ■

Lemma 3.5 *Let L and L_1 be $L = 4 \cdot r$, with $r \in \mathbb{N}^*$ and $L_1 = L/2 = 2 \cdot r$, respectively. If $\pi(x)$ in (3.1) is PP modulo L_1, then $\pi(x)$ in (3.1) is PP modulo L if and only if $(q_3 + q_5 + q_7 + \cdots)$ is even.*

Proof "$\Rightarrow$" Consider that $\pi(x)$ in (3.1) is PP both modulo L_1 and L, and prove that $(q_3 + q_5 + q_7 + \cdots)$ is even. From Lemma 3.2 we have that q_1 is odd. From Lemma 3.4 we have $\pi(x + L_1) \pmod{L} = (\pi(x) + L_1) \pmod{L}$. Using Newton's binomial theorem, we have:

$$(x + L_1)^i = \sum_{k=0}^{i} C_i^k \cdot x^k \cdot L_1^{i-k}, \text{ where } i \in \mathbb{N}^*. \tag{3.3}$$

Because

$$\begin{aligned} C_i^k \cdot x^k \cdot L_1^{i-k} = C_i^k \cdot x^k \cdot (2 \cdot r)^{i-k} = \\ = 2 \cdot 2 \cdot r \cdot C_i^k \cdot x^k \cdot 2^{-1} \cdot (2 \cdot r)^{i-k-1} = L \cdot C_i^k \cdot x^k \cdot 2^{i-k-2} \cdot r^{i-k-1} \end{aligned} \tag{3.4}$$

and C_i^k is a positive integer, it follows that

$$\left(C_i^k \cdot x^k \cdot L_1^{i-k}\right) \pmod{L} = 0, \tag{3.5}$$

for any $k \le i - 2$.

If $k = i - 1$, we have:

$$C_i^k \cdot x^k \cdot L_1^{i-k} = L \cdot C_i^{i-1} \cdot x^{i-1} \cdot 2^{-1} \cdot r^0 = i \cdot L_1 \cdot x^{i-1}. \tag{3.6}$$

If $k = i$, we have:

$$C_i^k \cdot x^k \cdot L_1^{i-k} = C_i^i \cdot x^i \cdot L_1^0 = x^i. \tag{3.7}$$

Therefore

$$\left(x + L_1\right)^i \pmod{L} = \left(x^i + i \cdot L_1 \cdot x^{i-1}\right) \pmod{L}. \tag{3.8}$$

Then

$$\pi(x+L_1) \pmod{L} = \left(\sum_{i=1}^{d} q_i \cdot (x+L_1)^i\right) \pmod{L} =$$
$$= \left(\sum_{i=1}^{d} q_i \cdot \left(x^i + i \cdot L_1 \cdot x^{i-1}\right)\right) \pmod{L} =$$
$$= \left(\sum_{i=1}^{d} q_i \cdot x^i\right) \pmod{L} + \left(\sum_{i=1}^{d} q_i \cdot \left(i \cdot L_1 \cdot x^{i-1}\right)\right) \pmod{L} =$$
$$= \left(\pi(x) + \sum_{i=1}^{d} q_i \cdot \left(i \cdot L_1 \cdot x^{i-1}\right)\right) \pmod{L}. \tag{3.9}$$

But

$$\left(\sum_{i=1}^{d} q_i \cdot \left(i \cdot L_1 \cdot x^{i-1}\right)\right) \pmod{L} =$$
$$= \left(q_2 \cdot \left(2 \cdot L_1 \cdot x^1\right) + q_4 \cdot \left(4 \cdot L_1 \cdot x^3\right) + q_6 \cdot \left(6 \cdot L_1 \cdot x^5\right) + \cdots\right)$$
$$\pmod{L} + \left(q_1 \cdot \left(L_1 \cdot x^0\right) + q_3 \cdot \left(3 \cdot L_1 \cdot x^2\right) +\right.$$
$$\left.+ q_5 \cdot \left(5 \cdot L_1 \cdot x^4\right) + \cdots\right) \pmod{L} = \left(q_2 \cdot \left(L \cdot x^1\right) + q_4 \cdot \left(2 \cdot L \cdot x^3\right) +\right.$$
$$\left.+ q_6 \cdot \left(3 \cdot L \cdot x^5\right) + \cdots\right) \pmod{L} +$$
$$+ \left(q_1 \cdot L_1 + q_3 \cdot L_1 \cdot x^2 + q_5 \cdot L_1 \cdot x^4 + \cdots\right) \pmod{L} =$$
$$= 0 + \left(q_1 \cdot L_1 + q_3 \cdot L_1 \cdot x^2 + q_5 \cdot L_1 \cdot x^4 + \cdots\right) \pmod{L} =$$
$$= \left(L_1 + q_3 \cdot L_1 \cdot x^2 + q_5 \cdot L_1 \cdot x^4 + \cdots\right) \pmod{L}. \tag{3.10}$$

In the last equality we used that q_1 is odd and an odd number multiplied by L_1, evaluated modulo L, is equal to L_1.

For any x even, we obviously have

$$\left(L_1 + q_3 \cdot L_1 \cdot x^2 + q_5 \cdot L_1 \cdot x^4 + \cdots\right) \pmod{L} =$$
$$= L_1 \pmod{L} \tag{3.11}$$

and thus

$$\pi(x+L_1) \pmod{L} = (\pi(x) + L_1) \pmod{L}, \tag{3.12}$$

and for x odd, as x^i with $i \in \mathbb{N}^*$ is also odd, we have

$$\left(L_1 + q_3 \cdot L_1 \cdot x^2 + q_5 \cdot L_1 \cdot x^4 + \cdots\right) \pmod{L} =$$

$$= \left(L_1 + q_3 \cdot L_1 + q_5 \cdot L_1 + \cdots\right) \pmod{L} =$$

$$= L_1 \cdot \left(1 + q_3 + q_5 + q_7 + \cdots\right) \pmod{L} \tag{3.13}$$

and thus

$$\pi(x + L_1) \pmod{L} = \left(\pi(x) + L_1 \cdot \left(1 + q_3 + q_5 + q_7 + \cdots\right)\right) \pmod{L}, \tag{3.14}$$

Because the equality

$$\pi(x + L_1) \pmod{L} = (\pi(x) + L_1) \pmod{L} \tag{3.15}$$

must be true for any $x \in \mathbb{Z}_L$, it follows that $\left(1 + q_3 + q_5 + q_7 + \cdots\right)$ should be odd, i.e. $\left(q_3 + q_5 + q_7 + \cdots\right)$ is even. So, the first part of lemma is proved.

" $\Leftarrow$" Consider now that $\pi(x)$ in (3.1) is PP modulo L_1 and $\left(q_3 + q_5 + q_7 + \cdots\right)$ is even and prove that $\pi(x)$ is PP modulo L. We notice that equalities (3.3)–(3.14) also hold in this case. As $\left(q_3 + q_5 + q_7 + \cdots\right)$ is even, equality (3.15) is true for any $x \in \mathbb{Z}_L$.

Assume that $\pi(x)$ is no PP modulo L. This means that there are two distinct values $x, x' \in \mathbb{Z}_L$, so that

$$\pi(x) \pmod{L} = \pi(x') \pmod{L} \tag{3.16}$$

Because the equality (3.15) is true, we have

$$\pi(x + L_1) \pmod{L} = \pi(x' + L_1) \pmod{L} \tag{3.17}$$

The four numbers x, x', $(x + L_1) \pmod{L}$, $(x' + L_1) \pmod{L}$ are distinct in the set $\mathbb{Z}_L$. Obviously, two of these four numbers are in the set $\mathbb{Z}_{L_1}$. Let these be x_1 and x_1'. Evaluating the equality (3.15) modulo L_1, it follows that

$$\pi(x + L_1) \pmod{L_1} = \pi(x) \pmod{L_1} \tag{3.18}$$

for any $x \in \mathbb{Z}_L$. Considering equalities (3.16)–(3.18), it follows that

$$\pi(x) \pmod{L_1} = \pi(x') \pmod{L_1} =$$

$$= \pi(x + L_1) \pmod{L_1} = \pi(x' + L_1) \pmod{L_1}. \tag{3.19}$$

This means that to the two distinct values x_1 and x_1', in the set $\mathbb{Z}_{L_1}$, there corresponds through permutation $\pi(\cdot)$, the same modulo L_1 value, i.e.

$$\pi(x_1) \pmod{L_1} = \pi(x_1') \pmod{L_1}. \tag{3.20}$$

Equality (3.20) contradicts the hypothesis that $\pi(x)$ is PP modulo L_1. Therefore, the assumption that $\pi(x)$ is no PP modulo L is false. So, the second part of the lemma is proved. ■

In the following we give the main theorem regarding the particular case $L = 2^n$, with $n \in \mathbb{N}, n > 1$.

Theorem 3.6 *For $L = 2^n$, with $n \in \mathbb{N}$, $n > 1$, $\pi(x)$ in (3.1) is PP if and only if q_1 is odd, $(q_2 + q_4 + q_6 + \cdots)$ is even and $(q_3 + q_5 + q_7 + \cdots)$ is even.*

Proof "$\Rightarrow$" Assume first that $\pi(x)$ is PP modulo L. From Lemma 3.2 we have that q_1 is odd. From Lemma 3.3 we also have that $\pi(x)$ is PP modulo $L_1 = L/2$. Then, from Lemma 3.5 it results that $(q_3+q_5+q_7+\cdots)$ is even. By repeated application of Lemma 3.3 we have that $\pi(x)$ is PP modulo $L_1 = L/2$, modulo $L/4$, a. s. o. modulo 2. Then, from Lemma 3.1 we have that $(q_1+q_2+q_3+\cdots+q_d)$ is odd. As q_1 is odd and $(q_3 + q_5 + q_7 + \cdots)$ is even, it results that $(q_2 + q_4 + q_6 + \cdots)$ is even. Thus, the first part of the theorem is proved.

"$\Leftarrow$" Assume now that q_1 is odd, $(q_2+q_4+q_6+\cdots)$ is even and $(q_3+q_5+q_7+\cdots)$ is even. We use the mathematical induction technique to show that $\pi(x)$ is PP modulo L. As $(q_1+q_2+q_3+\cdots+q_d)$ is odd, from Lemma 3.1 we have that $\pi(x)$ is PP modulo 2. We assume that $\pi(x)$ is PP modulo 2^{n_1}, with $n_1 \in \mathbb{N}, n_1 \geq 2$. As $(q_3+q_5+q_7+\cdots)$ is even, from Lemma 3.5 we have that $\pi(x)$ is also PP modulo $2 \cdot 2^{n_1} = 2^{n_1+1}$. Thus, induction is complete and the second part of the theorem is proved. ■

3.3 Necessary and Sufficient Conditions on the Coefficients of a Polynomial so that It Is PP Modulo a Number Equal to a Power of a Prime Number

For the particular case $L = p^n$, with p a prime number and $n \in \mathbb{N}^*$, the necessary and sufficient conditions on polynomial coefficients so that it is PP are given in the following theorem. This theorem is a Nöbauer's result (Nöbauer 1965), but it is also given in Mullen and Stevens (1984); Sun and Takeshita (2005); Chen et al. (2006). In Mullen and Stevens (1984) it is mentioned that the result of this theorem is a direct consequence of Theorem 123 from Hardy and Wright (1975).

Theorem 3.7 *For $L = p^n$, with p a prime number and $n \in \mathbb{N}^*$, $\pi(x)$ in (3.1) is PP modulo L if and only if $\pi(x)$ is PP modulo p and $\pi'(x) \neq 0 \pmod{p}$, $\forall x \in \mathbb{Z}_{p^n}$, where $\pi'(x)$ is the derivative of the polynomial $\pi(x)$.*

We can check that when $p = 2$, the conditions from Theorem 3.6 can be obtained from Theorem 3.7. So, if $\pi(x)$ is PP modulo 2 it results that $(q_1+q_2+q_3+\cdots+q_d)$ is odd. Imposing the condition $\pi'(x) \neq 0 \pmod{2}$, $\forall x \in \mathbb{Z}_{2^n}$, we have

$$(q_1 + 2 \cdot q_2 \cdot x + 3 \cdot q_3 \cdot x^2 + \cdots + d \cdot q_d \cdot x^{d-1}) \pmod{2} \neq 0,$$

$$\forall x \in \mathbb{Z}_{2^n} \tag{3.21}$$

Taking into account that the even terms from the left hand side of (3.21) are equal to zero modulo 2, it results that (3.21) is equivalent to

$$(q_1 + q_3 \cdot x^2 + q_5 \cdot x^4 + \cdots) \pmod 2 \neq 0, \forall x \in \mathbb{Z}_{2^n} \tag{3.22}$$

For x an even number in (3.21) or (3.22), we have that q_1 is odd. For x an odd number in (3.22) we have that $(q_1 + q_3 + q_5 + \cdots)$ is odd and considering that q_1 is odd, we have that $(q_3 + q_5 + q_7 + \cdots)$ is even. As $(q_1 + q_2 + q_3 + \cdots + q_d)$ is odd, it results that $(q_2 + q_4 + q_6 + \cdots)$ is even.

3.4 Simplified Necessary and Sufficient Conditions on the Coefficients of a Polynomial so that It Is PP Modulo Any Positive Integer

For the general case when the interleaver length L is any positive integer, Jing Sun and Oscar Y. Takeshita gave in Sun and Takeshita (2005) a theorem that simplifies the conditions for a polynomial to be PP modulo L. This theorem is given below.

Theorem 3.8 *For any* $L = \prod_{i=1}^{n_L} p_i^{n_{L,p_i}}$, *where* $n_L \in \mathbb{N}^*$, p_i, *with* $i = \overline{1, n_L}$, *are distinct prime numbers and* $n_{L,p_i} \in \mathbb{N}^*$, $\forall i = \overline{1, n_L}$, $\pi(x)$ *in (3.1) is PP modulo L if and only if* $\pi(x)$ *also is PP modulo* $p_i^{n_{L,p_i}}$, $\forall i = \overline{1, n_L}$.

Proof "$\Rightarrow$" Assume first that $\pi(x)$ in (3.1) is PP modulo L. Using Newton's binomial formula, for any $l \in \mathbb{N}^*$ we have:

$$\begin{aligned} &\pi\left(x + l \cdot p_i^{n_{L,p_i}}\right)\left(\text{mod } p_i^{n_{L,p_i}}\right) = \\ &= \Big(q_0 + q_1 \cdot \left(x + l \cdot p_i^{n_{L,p_i}}\right) + q_2 \cdot \left(x + l \cdot p_i^{n_{L,p_i}}\right)^2 + \cdots \\ &\quad + \cdots + q_d \cdot \left(x + l \cdot p_i^{n_{L,p_i}}\right)^d\Big)\left(\text{mod } p_i^{n_{L,p_i}}\right) = \\ &= \left(q_0 + q_1 \cdot x + q_2 \cdot x^2 + \cdots + q_d \cdot x^d\right)\left(\text{mod } p_i^{n_{L,p_i}}\right) \end{aligned} \tag{3.23}$$

Thus

$$\pi\left(x + l \cdot p_i^{n_{L,p_i}}\right)\left(\text{mod } p_i^{n_{L,p_i}}\right) = \pi(x)\left(\text{mod } p_i^{n_{L,p_i}}\right) \tag{3.24}$$

Assume that $\pi(x)$ is no PP modulo $p_i^{n_{L,p_i}}$. Then there are two numbers $x_1 \neq x_2$, with $0 \leq x_1, x_2 < p_i^{n_{L,p_i}}$, so that:

$$\pi(x_1) = \pi(x_2) = y\left(\text{mod } p_i^{n_{L,p_i}}\right) \tag{3.25}$$

Then, $\forall l \in \left\{0, 1, \ldots, \frac{L}{p_i^{n_{L,p_i}}} - 1\right\}$, we would have:

$$\pi\left(x_1 + l \cdot p_i^{n_{L,p_i}}\right)\left(\text{mod } p_i^{n_{L,p_i}}\right) =$$

$$= \pi\left(x_2 + l \cdot p_i^{n_{L,p_i}}\right)\left(\text{mod } p_i^{n_{L,p_i}}\right) = y\left(\text{mod } p_i^{n_{L,p_i}}\right) \tag{3.26}$$

Thereby, it would follow that a total of $\dfrac{2 \cdot L}{p_i^{n_{L,p_i}}}$ different numbers from the set $\mathbb{Z}_L$ are equal modulo $p_i^{n_{L,p_i}}$. However, this cannot be true, because $\pi(x)$ is PP modulo L and, consequently, just $\dfrac{L}{p_i^{n_{L,p_i}}}$ different numbers from the set $\mathbb{Z}_L$ are equal modulo $p_i^{n_{L,p_i}}$. Therefore, the initial assumption is false and $\pi(x)$ is PP modulo $p_i^{n_{L,p_i}}$.

"$\Leftarrow$" Let L_1 and L_2 be two relatively prime positive integers. Assume that $\pi(x)$ is PP both modulo L_1 and modulo L_2. Assume that $\pi(x)$ is no PP modulo $L_1 \cdot L_2$. Then, there are two positive integers x and x', in the set $\mathbb{Z}_{L_1 \cdot L_2}$, so that $x \neq x'$ and

$$\pi(x) \ (\text{mod } (L_1 \cdot L_2)) = \pi(x') \ (\text{mod } (L_1 \cdot L_2)) \tag{3.27}$$

Let there be $x_1 = x \ (\text{mod } L_1)$, $x_1' = x' \ (\text{mod } L_1)$, $x_2 = x \ (\text{mod } L_2)$, $x_2' = x' \ (\text{mod } L_2)$. Evaluating (3.27) modulo L_1 and modulo L_2, respectively, we get:

$$\pi(x_1) \ (\text{mod } L_1) = \pi(x_1') \ (\text{mod } L_1) \tag{3.28}$$

and

$$\pi(x_2) \ (\text{mod } L_2) = \pi(x_2') \ (\text{mod } L_2) \tag{3.29}$$

respectively.

Because $\pi(x)$ is PP both modulo L_1 and modulo L_2, we should have:

$$x_1 \ (\text{mod } L_1) = x_1' \ (\text{mod } L_1) \tag{3.30}$$

and

$$x_2 \ (\text{mod } L_2) = x_2' \ (\text{mod } L_2) \tag{3.31}$$

respectively.

Because L_1 and L_2 are relatively prime, from the Chinese remainder theorem (see Theorem 3.49 at the end of this Chapter), we have:

$$(x_1 \cdot x_2) \ (\text{mod } (L_1 \cdot L_2)) = (x_1' \cdot x_2') \ (\text{mod } (L_1 \cdot L_2)) \tag{3.32}$$

$$\Leftrightarrow x \pmod{(L_1 \cdot L_2)} = x' \pmod{(L_1 \cdot L_2)} \tag{3.33}$$

But the equality (3.33) contradicts the initial assumption. Therefore, $\pi(x)$ is a PP modulo $L_1 \cdot L_2$. ■

Next, we give simple necessary and sufficient criteria on the coefficients of a polynomial of degrees 1, 2, 3, 4 and 5, so that it is PP.

3.5 Necessary and Sufficient Conditions on the Coefficients of a Polynomial of First Degree so that It Is PP Modulo Any Positive Integer

A PP of first degree is named LPP and it is of the form:

$$\pi(x) = (q_1 \cdot x) \pmod{L}, x = \overline{0, L-1} \tag{3.34}$$

Particularizing Theorem 3.7 for polynomials of degree one, we get the following theorem.

Theorem 3.9 *For any* $L = \prod_{i=1}^{n_L} p_i^{n_{L,p_i}}$, *where* $n_L \in \mathbb{N}^*$, p_i, *with* $i = \overline{1, n_L}$, *are distinct prime numbers and* $n_{L,p_i} \in \mathbb{N}^*$, $\forall i = \overline{1, n_L}$, *a polynomial of degree one is LPP modulo L if and only if* $q_1 \neq 0 \pmod{p_i}$, $\forall i = \overline{1, n_L}$.

Proof According to Theorem 3.8 we must show only that $\pi(x)$ in (3.34) is LPP modulo $p_i^{n_{L,p_i}}$, $\forall i = \overline{1, n_L}$. Further, we consider that the index i is any of the set $\{1, 2, \ldots, n_L\}$.

"$\Rightarrow$" Assume that $\pi(x)$ in (3.34) is LPP modulo $p_i^{n_{p_i}}$. Considering Theorem 3.7, the formal derivative of the polynomial in (3.34) is

$$\pi'(x) = q_1 \neq 0 \pmod{p_i} \tag{3.35}$$

"$\Leftarrow$" Assume that $q_1 \neq 0 \pmod{p_i}$. This means that q_1 and $p_i^{n_{L,p_i}}$ are relatively prime. Therefore, from Theorem 57 in Hardy and Wright (1975), the congruence equation

$$q_1 \cdot x = y \left(\bmod\ p_i^{n_{L,p_i}}\right) \tag{3.36}$$

has only one solution modulo $p_i^{n_{L,p_i}}$, in variable x. Because the solutions are distinct for each $y \in \mathbb{Z}_{p_i^{n_{L,p_i}}}$ it means that $\pi(x)$ from (3.34) is LPP modulo $p_i^{n_{L,p_i}}$. ■

3.6 Necessary and Sufficient Conditions on the Coefficients of a Polynomial of Second Degree so that It Is PP Modulo Any Positive Integer

A PP of degree two is named QPP and it is of the form:

$$\pi(x) = (q_1 \cdot x + q_2 \cdot x^2) \pmod{L}, x = \overline{0, L-1} \tag{3.37}$$

Particularizing Theorem 3.8 for polynomials of degree two, the necessary and sufficient conditions for a polynomial of degree two to be QPP can be tested using the following algorithm in three steps which was given in Chen et al. (2006) for the coefficients of a CPP.

(1) Decompose the interleaver length into prime factors of the form $L = \prod_{i=1}^{n_L} p_i^{n_{L,p_i}}$, where $n_L \in \mathbb{N}^*$, p_i, with $i = \overline{1, n_L}$, are different prime numbers and $n_{L,p_i} \in \mathbb{N}^*$, $\forall i = \overline{1, n_L}$.
(2) For any p_i, with $i = \overline{1, n_L}$, and the corresponding exponent n_{L,p_i} from the previous step, check if the conditions from Table 3.1 are accomplished.
(3) $\pi(x)$ from (3.37) is QPP if and only if the conditions in step (2) are fulfilled for all p_i, with $i = \overline{1, n_L}$.

The proofs for the conditions on the coefficients given in Table 3.1 are presented below.

3.6.1 *Case $p = 2$ and $n_{L,2} = 1$*

This case results by particularizing Lemma 3.1 for polynomials of second degree, requiring that $q_1 + q_2$ has to be odd.

3.6.2 *Case $p = 2$ and $n_{L,2} > 1$, and $p > 2$ and $n_{L,p} \geq 1$*

Particularizing Theorem 3.7 for polynomials of second degree, we get the following Corrolary (Sun and Takeshita 2005).

Table 3.1 Conditions for coefficients q_1, q_2 so that $\pi(x)$ in (3.37) is a QPP

1(a)	$p = 2$	$n_{L,2} = 1$	$(q_1 + q_2) \neq 0 \pmod 2$
1(b)		$n_{L,2} > 1$	$q_1 \neq 0, q_2 = 0 \pmod 2$
(2)	$p > 2$	$n_{L,p} \geq 1$	$q_1 \neq 0, q_2 = 0 \pmod p$

Corollary 3.10 *Let the interleaver length be of the form $L = p^n$, with p a prime number and $n \in \mathbb{N}^*$, $n \geq 1$ when $p > 2$, and $n > 1$ when $p = 2$. A polynomial of second degree is a QPP modulo $L = p^n$ if and only if $q_1 \neq 0 \pmod{p}$ and $q_2 = 0 \pmod{p}$.*

Proof For $p = 2$ and $n > 1$, this corollary is a particular case of Theorem 3.6, from where we have that q_1 is odd and q_2 is even, which is equivalent to $q_1 \neq 0 \pmod{2}$ and $q_2 = 0 \pmod{2}$. For this reason we will consider only the case when $p > 2$. We check the conditions given in Theorem 3.7 both for direct and reverse proof. The formal derivative of the polynomial of second degree is:

$$\pi'(x) = q_1 + 2 \cdot q_2 \cdot x \tag{3.38}$$

" $\Rightarrow$" Consider that $\pi(x)$ in (3.37) is QPP modulo $L = p^n$. From Theorem 3.7 we have

$$\pi'(x) = q_1 + 2 \cdot q_2 \cdot x \neq 0 \pmod{p} \tag{3.39}$$

Replacing $x = 0$ in (3.39), it follows that $q_1 \neq 0 \pmod{p}$.

Further, assume that $q_2 \neq 0 \pmod{p}$. Then, it follows that $2 \cdot q_2$ and p are relatively prime. Therefore, $2 \cdot q_2 \cdot x$ is PP modulo p. This means there is a number x in the set $\mathbb{Z}_p$ so that $q_1 + 2 \cdot q_2 \cdot x = 0 \pmod{p}$, which contradicts relation (3.39). Thus, it follows that $q_2 = 0 \pmod{p}$.

" $\Leftarrow$" We consider that the coefficients of the polynomial of second degree satisfy the conditions $q_1 \neq 0 \pmod{p}$ and $q_2 = 0 \pmod{p}$. Then

$$\pi(x) \pmod{p} = (q_1 \cdot x) \pmod{p} \tag{3.40}$$

As q_1 and p are relatively prime, it follows that $\pi(x)$ is PP modulo p. The formal derivative of the polynomial in (3.40) is

$$\pi'(x) = q_1 \neq 0 \pmod{p} \tag{3.41}$$

Therefore, from Theorem 3.7, it results that $\pi(x)$ is a QPP modulo p^n. ∎

The conditions in Table 3.1 are the same as in Proposition 1 in Takeshita (2006) or Corrolary 2.7 in Ryu and Takeshita (2006).

For example, if $L = 256$, then from the case Sect. 3.6.2, for $p = 2$ and $n_{L,2} = 8$, it follows that $q_1 \in \{1, 2, 5, \ldots, 255\}$ (the set of numbers relatively prime with 256) and $q_2 \in \{2, 4, 6, \ldots, 254\}$ (the set of numbers containing 2 as a factor). This leads to $128 \times 127 = 16256$ possible pairs of coefficients (q_1, q_2) which determine that $\pi(x)$ is QPP.

Further, we give the definition of normalized permutation polynomial and two propositions that follow from Dickson (1896, 1901) needed to establish the conditions on the coefficients of a polynomial of degree greater than two, so that it is PP.

Definition 3.11 The polynomial $\overline{\pi}(x) = \sum_{k=1}^{d} q_k x^k (\mod p^n)$ is a normalized PP if $q_d = 1, \overline{\pi}(0) = 0$, and, if $p \nmid d$ then $q_{d-1} = 0$.

Proposition 3.12 *A polynomial $\pi(x)$ of degree d is a PP $\pmod p$, with $p \nmid d$, if and only if $a \cdot \pi(x+b) + c$ is a PP for all $a \neq 0, b, c \in \mathbb{Z}_p$.*

Proposition 3.13 *A polynomial $\pi(x)$ of degree d is a PP $\pmod p$, with $p \mid d$, if and only if $a \cdot \pi(x) + c$ is a PP for all $a \neq 0, c \in \mathbb{Z}_p$.*

3.7 Necessary and Sufficient Conditions on the Coefficients of a Polynomial of Third Degree so that It Is PP Modulo Any Positive Integer

A PP of third degree is named CPP and it is of the form:

$$\pi(x) = (q_1 \cdot x + q_2 \cdot x^2 + q_3 \cdot x^3) \pmod L, x = \overline{0, L-1} \tag{3.42}$$

Particularizing Theorem 3.8 for polynomials of degree three, we can test whether a polynomial of degree three is CPP using the same algorithm in three steps from Chen et al. (2006) given in Sect. 3.6. The conditions on the coefficients used in step 2 are given in Table 3.2.

Table 3.2 can be considered a simpler equivalent test of Theorem 3.8. The proof is given below.

Table 3.2 Conditions for coefficients q_1, q_2, q_3 so that $\pi(x)$ in (3.42) is a CPP

1(a)	$p = 2$	$n_{L,2} = 1$	$(q_1 + q_2 + q_3) \neq 0 \pmod 2$
1(b)		$n_{L,2} > 1$	$q_1 \neq 0, q_2 = 0, q_3 = 0 \pmod 2$
2(a)	$p = 3$	$n_{L,3} = 1$	$(q_1 + q_3) \neq 0, q_2 = 0 \pmod 3$
2(b)		$n_{L,3} > 1$	$q_1 \neq 0, (q_1 + q_3) \neq 0, q_2 = 0 \pmod 3$
3(a)	$3 \mid (p-1)$	$n_{L,p} = 1$	$q_1 \neq 0, q_2 = 0, q_3 = 0 \pmod p$
3(b)		$n_{L,p} > 1$	$q_1 \neq 0, q_2 = 0, q_3 = 0 \pmod p$
4(a)	$3 \nmid (p-1)$, $p > 3$	$n_{L,p} = 1$	(1) $q_2^2 = 3q_1q_3 \pmod p$ if $q_3 \neq 0 \pmod p$ and (2) $q_1 \neq 0 \pmod p$, $q_2 = 0 \pmod p$ if $q_3 = 0 \pmod p$
4(b)		$n_{L,p} > 1$	$q_1 \neq 0, q_2 = 0, q_3 = 0 \pmod p$

3.7.1 Case $p = 2$

This case follows immediately from Lemma 3.1 for $n_{L,2} = 1$ and Theorem 3.6 for $n_{L,2} > 1$.

3.7.2 Case $p = 3$

3.7.2.1 Subcase $p = 3$ and $n_{L,3} = 1$

Theorem 3.14 *The polynomial* $\pi(x) = (q_1 \cdot x + q_2 \cdot x^2 + q_3 \cdot x^3) \pmod 3$ *is a CPP if and only if* $(q_1 + q_3) \neq 0 \pmod 3$ *and* $q_2 = 0 \pmod 3$.

Proof It is obvious that $\pi(0) = 0$. Then $\pi(x)$ is CPP if $\pi(1) \neq 0$, $\pi(2) \neq 0$ and $\pi(1) \neq \pi(2)$. Replacing $x = 1$ and $x = 2$ in (3.42), respectively, and evaluating the results modulo 3, we have

$$(q_1 + q_2 + q_3) \neq 0 \pmod 3 \tag{3.43}$$

and

$$(2q_1 + q_2 + 2q_3) \neq 0 \pmod 3 \tag{3.44}$$

Then, condition $\pi(1) \neq \pi(2)$ is equivalent to

$$(q_1 + q_3) \neq 0 \pmod 3 \tag{3.45}$$

From (3.43)–(3.44) and (3.45) it follows

$$q_2 = 0 \pmod 3 \tag{3.46}$$

The converse implication follows the steps from the direct implication in reverse order. ∎

3.7.2.2 Subcase $p = 3$ and $n_{L,3} > 1$

Theorem 3.15 *The polynomial* $\pi(x) = (q_1 \cdot x + q_2 \cdot x^2 + q_3 \cdot x^3) \pmod{3^{n_{L,3}}}$ *with* $n_{L,3} \in \mathbb{N}^*$, $n_{L,3} > 1$ *is a CPP if and only if* $q_1 \neq 0 \pmod 3$, $(q_1 + q_3) \neq 0 \pmod 3$ *and* $q_2 = 0 \pmod 3$.

Proof " $\Rightarrow$" Let $\pi(x)$ be a CPP modulo $3^{n_{L,3}}$. Then, from Theorem 3.7, it follows that $\pi(x)$ is CPP modulo 3 and

$$\pi'(x) = (q_1 + 2q_2 \cdot x + 3q_3 \cdot x^2) \neq 0 \pmod 3, \forall x \in \mathbb{Z}_3 \tag{3.47}$$

As $\pi(x)$ is a CPP modulo 3, from Theorem 3.14 we have $(q_1 + q_3) \neq 0 \pmod 3$ and $q_2 = 0 \pmod 3$. Replacing $x = 0$ in (3.47) we get $q_1 \neq 0 \pmod 3$.

"$\Leftarrow$" Let there be $q_1 \neq 0 \pmod 3$, $(q_1 + q_3) \neq 0 \pmod 3$ and $q_2 = 0 \pmod 3$. As $(q_1 + q_3) \neq 0 \pmod 3$ and $q_2 = 0 \pmod 3$, from Theorem 3.14 we have that $\pi(x)$ is CPP modulo 3. The formal derivative of $\pi(x)$ in (3.47), evaluated modulo 3, is for $q_2 = 0 \pmod 3$, $\pi'(x) = q_1 \neq 0 \pmod 3$. Therefore, from Theorem 3.7 it follows that $\pi(x)$ is CPP modulo $3^{n_{L,3}}$. ∎

3.7.3 *Case* 3 | (*p* − 1), *with p* > 3

3.7.3.1 Subcase 3 | (*p* − 1), with *p* > 3, and $n_{L,p} = 1$

Theorem 3.16 *Let p be a prime number so that* $3 \mid (p-1)$. *The polynomial* $\pi(x) = (q_1 \cdot x + q_2 \cdot x^2 + q_3 \cdot x^3) \pmod p$ *is a CPP if and only if* $q_1 \neq 0 \pmod p$, $q_2 = 0 \pmod p$ *and* $q_3 = 0 \pmod p$.

Proof From the Corrolary given in Lidl and GL Mullen (1988) we have that if a polynomial of degree $d > 1$ is PP modulo p then $d \nmid (p-1)$. Therefore, if the degree $d \mid (p-1)$, then the degree of the PP is smaller then d, or, in other words, $q_d = 0 \pmod p$. For polynomial degree equal to 3, if $3 \mid (p-1)$ is equivalent to $q_3 = 0 \pmod p$. This means that the polynomial of degree 3 modulo p is equivalent to a polynomial of degree 2. As p is odd, it follows that $2 \mid (p-1)$. From the same Corrolary from Lidl and GL Mullen (1988) we have that $q_2 = 0 \pmod p$. Then, the polynomial of degree 3 modulo p is equivalent to the polynomial of degree 1, $(q_1 \cdot x) \pmod p$, which, according to Theorem 3.9 is PP if and only if $q_1 \neq 0 \pmod p$. ∎

3.7.3.2 Subcase 3 | (*p* − 1) and $n_{L,p} > 1$

Theorem 3.17 *Let p be a prime number so that* $3 \mid (p-1)$. *The polynomial* $\pi(x) = (q_1 \cdot x + q_2 \cdot x^2 + q_3 \cdot x^3) \pmod{p^{n_{L,p}}}$, *with* $n_{L,p} \in \mathbb{N}^*$, $n_{L,p} > 1$, *is a CPP if and only if* $q_1 \neq 0 \pmod p$, $q_2 = 0 \pmod p$ *and* $q_3 = 0 \pmod p$.

Proof "$\Rightarrow$" Let $\pi(x)$ be a CPP modulo $p^{n_{L,p}}$. Then, from Theorem 3.8, it follows that $\pi(x)$ is CPP modulo p. From Theorem 3.16 we have that $q_1 \neq 0 \pmod p$, $q_2 = 0 \pmod p$ and $q_3 = 0 \pmod p$.

"$\Leftarrow$" Let there be $q_1 \neq 0 \pmod p$, $q_2 = 0 \pmod p$ and $q_3 = 0 \pmod p$. Then, from the Theorem 3.16 we have that $\pi(x)$ is CPP modulo p. The formal derivative of $\pi(x)$, evaluated modulo p, for $q_2 = 0 \pmod p$ and $q_3 = 0 \pmod p$, is $\pi'(x) = q_1 \neq 0 \pmod p$. Therefore, from Theorem 3.7 we have that $\pi(x)$ is CPP modulo $p^{n_{L,p}}$. ∎

3.7.4 *Case* $3 \nmid (p - 1)$ *with* $p > 3$

3.7.4.1 Subcase $3 \nmid (p - 1)$ with $p > 3$ and $n_{L,p} = 1$

In 1897 Dickson gave all normalized polynomials of degree at most six, except those of degree 6 modulo a power of 2 (Dickson 1896). From this list we can see that the only normalized CPP modulo $p > 3$, with $3 \nmid (p - 1)$, is $\overline{\pi}(x) = x^3 \pmod p$ (see Table 3.7). This case is given in the next proposition.

Proposition 3.18 *The only normalized CPP modulo* $p > 3$*, with* $3 \nmid (p - 1)$*, is* $\overline{\pi}(x) = x^3 \pmod p$.

The following lemma is required to get the general result from Theorem 3.20 for this subcase.

Lemma 3.19 *The polynomial* $\pi(x) = (q_1 \cdot x + q_2 \cdot x^2 + q_3 \cdot x^3) \pmod p$*, where* $q_3 \neq 0 \pmod p$*, can be factorized in the form* $\pi(x) = a \cdot (x + b)^3 + c \pmod p$ *if and only if* $(q_2)^2 = 3q_1q_3 \pmod p$.

Proof " ⇒" Assume that:

$$\pi(x) = a \cdot (x + b)^3 + c \pmod p \tag{3.48}$$

We have to prove that $(q_2)^2 = 3q_1q_3 \pmod p$. Equation (3.48) can be written as:

$$\begin{aligned} \pi(x) &= a \cdot (x^3 + 3x^2b + 3xb^2 + b^3) + c \pmod p = \\ &= ax^3 + 3abx^2 + 3ab^2x + ab^3 + c \pmod p \end{aligned} \tag{3.49}$$

Identifying the coefficients of terms of degrees 3 and 2, we have:

$$q_3 = a \pmod p \tag{3.50}$$

$$q_2 = 3ab \pmod p \tag{3.51}$$

Considering (3.50) and that $q_3 \neq 0 \pmod p$, (3.51) can be written as:

$$\frac{q_2}{3q_3} = b \pmod p \tag{3.52}$$

Identifying the coefficients for degree 1, we have:

$$q_1 = 3ab^2 \pmod p \tag{3.53}$$

or, considering (3.50) and (3.52)

$$q_1 = 3q_3\left(\frac{q_2}{3q_3}\right)^2 \pmod p \Leftrightarrow (q_2)^2 = 3q_1q_3 \pmod p \tag{3.54}$$

Identifying the coefficient of the free terms, we have:

$$ab^3 + c = 0 \pmod p \Leftrightarrow c = -ab^3 \pmod p \tag{3.55}$$

or, considering (3.50) and (3.52)

$$c = -\frac{(q_2)^3}{27(q_3)^2} \pmod p \tag{3.56}$$

Therefore, we have

$$\begin{aligned}\pi(x) &= (q_1 \cdot x + q_2 \cdot x^2 + q_3 \cdot x^3) \pmod p = \\ &= q_3 \cdot \left(x + \frac{q_2}{3q_3}\right)^3 - \frac{(q_2)^3}{27(q_3)^2} \pmod p\end{aligned} \tag{3.57}$$

whence, the obvious equality

$$q_1 = \frac{(q_2)^2}{3q_3} \pmod p \Leftrightarrow (q_2)^2 = 3q_1q_3 \pmod p \tag{3.58}$$

" $\Leftarrow$" The reciprocal is shown in reverse way. Thus, assuming that equality (3.58) is true, $\pi(x)$ can be written as in (3.57), as follows

$$\begin{aligned}\pi(x) &= (q_1 \cdot x + q_2 \cdot x^2 + q_3 \cdot x^3) \pmod p = \\ &= \left(\frac{(q_2)^2}{3q_3} \cdot x + q_2 \cdot x^2 + q_3 \cdot x^3\right) \pmod p = \\ &= q_3 \cdot \left(x^3 + 3 \cdot \frac{q_2}{3q_3} \cdot x^2 + 3 \cdot \left(\frac{q_2}{3q_3}\right)^2 \cdot x + \left(\frac{q_2}{3q_3}\right)^3\right) - \\ &- \frac{(q_2)^3}{27(q_3)^2} \pmod p = q_3 \cdot \left(x + \frac{q_2}{3q_3}\right)^3 - \frac{(q_2)^3}{27(q_3)^2} \pmod p\end{aligned} \tag{3.59}$$

Therefore $\pi(x)$ can be written as in (3.48), where a, b, and c, are those from (3.50), (3.52), and (3.56), respectively. ■

Theorem 3.20 *The polynomial* $\pi(x) = (q_1 \cdot x + q_2 \cdot x^2 + q_3 \cdot x^3) \pmod p$, *with* $3 \nmid (p-1)$, *is PP if and only if one of the following conditions is fulfilled:*

(1) For $q_3 \neq 0 \pmod p$, *then* $(q_2)^2 = 3q_1q_3 \pmod p$.
(2) For $q_3 = 0 \pmod p$, *then* $q_2 = 0 \pmod p$ *and* $q_1 \neq 0 \pmod p$.

Proof From Propositions 3.12 and 3.18 we have that all CPPs modulo $p > 3$, with $3 \nmid (p-1)$, when $q_3 \neq 0 \pmod p$, can be obtained using the formula $\pi(x) = a \cdot (x+b)^3 + c \pmod p$, for all $a \neq 0, b, c \in \mathbb{Z}_p$. From Lemma 3.19 it follows that for $q_3 \neq 0 \pmod p$, $\pi(x)$ is CPP if and only if $(q_2)^2 = 3q_1q_3 \pmod p$.

If $q_3 = 0 \pmod{p}$, then we can apply the conditions for QPPs from Corrolary 3.10, getting the conditions $q_2 = 0 \pmod{p}$ and $q_1 \neq 0 \pmod{p}$. ■

3.7.4.2 Subcase $3 \nmid (p-1)$ with $p > 3$ and $n_{L,p} > 1$

The next proposition and lemma are needed to get the general result from Theorem 3.23 for this subcase.

Proposition 3.21 *The equation $y^2 = 0 \pmod{p}$ has $y = 0 \pmod{p}$ as single solution.*

Lemma 3.22 *Let $p > 3$ be a prime number so that $3 \nmid (p-1)$. If $(q_2)^2 = 3q_1q_3 \pmod{p}$, with $q_3 \neq 0 \pmod{p}$, then the equation $\pi'(x) = (q_1 + 2q_2 \cdot x + 3q_3 \cdot x^2) = 0 \pmod{p}$ has always the solution $x = -\dfrac{q_2}{3q_3} \pmod{p}$.*

Proof We rearrange $\pi'(x)$ in terms of a perfect square:

$$\pi'(x) = q_1 + 2q_2 \cdot x + 3q_3 \cdot x^2 =$$
$$= 3q_3 \cdot \left(x^2 + \tfrac{2q_2}{3q_3} \cdot x\right) + q_1 =$$
$$= 3q_3 \cdot \left(x^2 + 2 \cdot \tfrac{q_2}{3q_3} \cdot x + \left(\tfrac{q_2}{3q_3}\right)^2\right) + q_1 - \tfrac{(q_2)^2}{3q_3} =$$
$$= 3q_3 \cdot \left(x + \frac{q_2}{3q_3}\right)^2 + q_1 - \frac{(q_2)^2}{3q_3} \tag{3.60}$$

Considering that $(q_2)^2 = 3q_1q_3 \pmod{p}$, (3.60) becomes

$$\pi'(x) = 3q_3 \cdot \left(x + \frac{q_2}{3q_3}\right)^2 + q_1 - q_1 = 3q_3 \cdot \left(x + \frac{q_2}{3q_3}\right)^2 \tag{3.61}$$

We make the change of variable $y = x + \dfrac{q_2}{3q_3}$ in (3.61), and the equation $\pi'(x) = 0 \pmod{p}$ becomes $y^2 = 0 \pmod{p}$. From Proposition 3.21 we have that equation $y^2 = 0 \pmod{p}$ has always the solution $y = 0 \pmod{p}$. Therefore, the equation $\pi'(x) = 0 \pmod{p}$ has always the solution $x = -\dfrac{q_2}{3q_3}$. ■

Theorem 3.23 *The polynomial $\pi(x) = (q_1 \cdot x + q_2 \cdot x^2 + q_3 \cdot x^3) \pmod{p^{n_{L,p}}}$, with $3 \nmid (p-1)$ and $n_{L,p} > 1$, is PP if and only if $q_3 = q_2 = 0 \pmod{p}$ and $q_1 \neq 0 \pmod{p}$.*

Proof "$\Rightarrow$" Let $\pi(x)$ be a PP $\pmod{p^{n_{L,p}}}$, with $3 \nmid (p-1)$ and $n_{L,p} > 1$. From Theorem 3.7 we have that $\pi(x)$ is PP $\pmod{p}$ and $\pi'(x) = q_1 + 2q_2 \cdot x + 3q_3 \cdot x^2 \neq 0 \pmod{p}$, $\forall x \in \mathbb{Z}_p$.

As $\pi(x)$ is PP (mod p), from Theorem 3.20 we have that when $q_3 \neq 0 \pmod p$, the equality $(q_2)^2 = 3q_1q_3 \pmod p$ is true. But from Lemma 3.22, equation $\pi'(x) = q_1 + 2q_2 \cdot x + 3q_3 \cdot x^2 = 0 \pmod p$ has always one solution in this case. Therefore, there is no CPP (mod $p^{n_{L,p}}$), with $3 \nmid (p-1)$ and $n_{L,p} > 1$, when $q_3 \neq 0 \pmod p$, because, otherwise Theorem 3.7 would be contradicted.

When $q_3 = 0 \pmod p$, from Theorem 3.20 we have that $q_2 = 0 \pmod p$ and $q_1 \neq 0 \pmod p$. In this case $\pi'(x) = q_1 \neq 0 \pmod p$, $\forall x \in \mathbb{Z}_p$. Thus, if $\pi(x)$ is PP (mod $p^{n_{L,p}}$), with $3 \nmid (p-1)$ and $n_{L,p} > 1$, then $q_3 = q_2 = 0 \pmod p$ and $q_1 \neq 0 \pmod p$.

"$\Leftarrow$" Converse proof is obvious, considering Theorem 3.7. Indeed, for $q_3 = q_2 = 0 \pmod p$ and $q_1 \neq 0 \pmod p$ we have that $\pi(x) = q_1 x \pmod p$, which, for $q_1 \neq 0 \pmod p$, is a LPP, according to Theorem 3.9, and $\pi'(x) = q_1 \neq 0 \pmod p$, $\forall x \in \mathbb{Z}_p$. ■

3.8 Necessary and Sufficient Conditions on the Coefficients of a Polynomial of Fourth Degree so that It Is PP Modulo Any Positive Integer

A PP of fourth degree is denoted by 4-PP and it is of the form:

$$\pi(x) = (q_1 \cdot x + q_2 \cdot x^2 + q_3 \cdot x^3 + q_4 \cdot x^4) \pmod L,$$

$$x = \overline{0, L-1} \tag{3.62}$$

Particularizing Theorem 3.8 for polynomials of degree four, the necessary and sufficient conditions for a polynomial of degree four to be a 4-PP can be tested using the same algorithm in three steps from Chen et al. (2006) given in Sect. 3.6, where the conditions on the coefficients are given in Table 3.3 (Trifina and Tarniceriu 2016).

In the following, we prove the conditions on the coefficients given in Table 3.3. Because of the similarities of the cases $p = 7$ and $n_{L,7} > 1$, $3 \mid (p-1)$ with $p > 7$ and $n_{L,p} \geq 1$, $3 \nmid (p-1)$ with $p = 5$ or $p > 7$ and $n_{L,p} > 1$, they are addressed together in Sect. 3.8.4.

3.8.1 *Case* $p = 2$

This case follows immediately from Lemma 3.1, for $n_{L,2} = 1$ and from Theorem 3.6 for $n_{L,2} > 1$.

Table 3.3 Conditions for coefficients q_1, q_2, q_3, q_4 so that $\pi(x)$ in (3.62) is a 4-PP

1(a)	$p = 2$	$n_{L,2} = 1$	$(q_1 + q_2 + q_3 + q_4) \neq 0 \pmod 2$
1(b)		$n_{L,2} > 1$	$q_1 \neq 0 \pmod 2$, $(q_2 + q_4) = 0 \pmod 2$, $q_3 = 0 \pmod 2$
2(a)	$p = 3$	$n_{L,3} = 1$	$(q_1 + q_3) \neq 0 \pmod 3$, $(q_2 + q_4) = 0 \pmod 3$
2(b)		$n_{L,3} > 1$	$q_1 \neq 0 \pmod 3$, $(q_1 + q_3) \neq 0 \pmod 3$, $q_2 = q_4 = 0 \pmod 3$
3(a)	$p = 7$	$n_{L,7} = 1$	(1) If $q_4 \neq 0 \pmod 7$, then (1.1) $3(q_3)^2 = q_2q_4 \pmod 7$, and (1.2) $2q_1(q_4)^2 =$ $= (q_3)^3 + (q_4)^3 \pmod 7$ or $2q_1(q_4)^2 =$ $= (q_3)^3 + 6(q_4)^3 \pmod 7$ (2) If $q_4 = 0 \pmod 7$ then $q_1 \neq 0 \pmod 7$ and $q_2 = q_3 = 0 \pmod 7$
3(b)		$n_{L,7} > 1$	$q_1 \neq 0 \pmod 7$, $q_2 = q_3 = q_4 = 0 \pmod 7$
4(a)	$3 \nmid (p-1)$ ($p = 5$ or $p > 7$)	$n_{L,p} = 1$	$q_4 = 0 \pmod p$ and (1) If $q_3 = 0 \pmod p$ then $q_1 \neq 0 \pmod p$ and $q_2 = 0 \pmod p$ (2) If $q_3 \neq 0 \pmod p$ then $(q_2)^2 = 3q_1q_3 \pmod p$
4(b)		$n_{L,p} > 1$	$q_1 \neq 0 \pmod p$, $q_2 = q_3 = q_4 = 0 \pmod p$
5	$3 \mid (p-1)$ ($p > 7$)	$n_{L,p} \geq 1$	$q_1 \neq 0 \pmod p$, $q_2 = q_3 = q_4 = 0 \pmod p$

3.8.2 *Case $p = 3$*

3.8.2.1 Subcase $p = 3$ and $n_{L,3} = 1$

Theorem 3.24 $\pi(x) = (q_1x + q_2x^2 + q_3x^3 + q_4x^4) \pmod 3$ *is a 4-PP if and only if* $(q_1 + q_3) \neq 0 \pmod 3$ *and* $(q_2 + q_4) = 0 \pmod 3$.

Proof As $\pi(0) = 0$, it is required that

$$\pi(1) = q_1 + q_2 + q_3 + q_4 \neq 0 \pmod 3, \tag{3.63}$$

$$\pi(2) = 2q_1 + q_2 + 2q_3 + q_4 \neq 0 \pmod 3, \tag{3.64}$$

and

$$\pi(1) \neq \pi(2) \pmod 3. \tag{3.65}$$

Replacing (3.63) and (3.64) in (3.65), we have

$$(q_1 + q_3) \neq 0 \pmod 3. \tag{3.66}$$

If $q_1 + q_3 = 1 \pmod 3$, then, from (3.63) it follows that $q_2 + q_4 = 0 \pmod 3$ or $q_2 + q_4 = 1 \pmod 3$, and from (3.64) it follows that $q_2 + q_4 = 0 \pmod 3$ or $q_2 + q_4 = 2 \pmod 3$. Therefore, $q_2 + q_4 = 0 \pmod 3$. For case $q_1 + q_3 = 2 \pmod 3$ we can obtain the same result in a similar way. ■

3.8.2.2 Subcase $p = 3$ and $n_{L,3} > 1$

Theorem 3.25 *$\pi(x) = (q_1x + q_2x^2 + q_3x^3 + q_4x^4) \pmod{3^{n_{L,3}}}$, with $n_{L,3} \in \mathbb{N}$, $n_{L,3} > 1$, is 4-PP if and only if $q_1 \neq 0 \pmod 3$, $(q_1 + q_3) \neq 0 \pmod 3$ and $q_2 = q_4 = 0 \pmod 3$.*

Proof "$\Rightarrow$" For the direct proof, we consider that $\pi(x)$ is a PP $\pmod{3^{n_{L,3}}}$, with $n_{L,3} > 1$. Then, according to Theorem 3.7, $\pi(x)$ is a PP $\pmod 3$ and

$$\pi'(x) = q_1 + 2q_2x + 3q_3x^2 + 4q_4x^3 \pmod 3 =$$

$$= q_1 + 2q_2x + q_4x^3 \neq 0 \pmod 3. \tag{3.67}$$

As $\pi(x)$ is a PP $\pmod 3$, from Theorem 3.24, we have $(q_1+q_3) \neq 0 \pmod 3$ and $(q_2+q_4) = 0 \pmod 3$. Replacing $x = 0$ in (3.67), we have $\pi'(0) = q_1 \neq 0 \pmod 3$. Replacing $x = 1$ in (3.67), we have $\pi'(1) = q_1 + 2q_2 + q_4 \neq 0 \pmod 3$. As $(q_2 + q_4) = 0 \pmod 3$, it follows that

$$\pi'(1) = q_1 + q_2 \neq 0 \pmod 3. \tag{3.68}$$

Replacing $x = 2$ in (3.67), we have $\pi'(2) = q_1 + q_2 + 2q_4 \neq 0 \pmod 3$ and, because $(q_2 + q_4) = 0 \pmod 3$, it follows that

$$\pi'(2) = q_1 + q_4 \neq 0 \pmod 3. \tag{3.69}$$

Relations (3.68) and (3.69) must hold for any $q_1 \neq 0 \pmod 3$. For $q_1 = 1 \pmod 3$, from (3.68) it follows that $q_2 = 0 \pmod 3$ or $q_2 = 1 \pmod 3$, and from (3.69) that $q_4 = 0 \pmod 3$ or $q_4 = 1 \pmod 3$. For $q_1 = 2 \pmod 3$, from (3.68), it follows that $q_2 = 0 \pmod 3$ or $q_2 = 2 \pmod 3$, and from (3.69) that $q_4 = 0 \pmod 3$ or $q_4 = 2 \pmod 3$. Therefore, only the values $q_2 = 0 \pmod 3$ and $q_4 = 0 \pmod 3$ meet (3.68) and (3.69) for any $q_1 \neq 0 \pmod 3$.

"$\Leftarrow$" For the converse proof, because $(q_1 + q_3) \neq 0 \pmod 3$ and $q_2 = q_4 = 0 \pmod 3$, from Theorem 3.24 it follows that $\pi(x)$ is a PP $\pmod 3$.

For $q_2 = q_4 = 0 \pmod 3$, from (3.67) we have that $\pi'(x) = q_1 \neq 0 \pmod 3$. Then, according to Theorem 3.7, it results that $\pi(x)$ is a PP $\pmod{3^{n_{L,3}}}$, with $n_{L,3} > 1$. ■

3.8.3 *Case $p = 7$ and $n_{L,7} = 1$*

From the list with normalized PPs from Dickson (1896) we see that the only normalized 4-PPs modulo $p = 7$ are $\bar{\pi}(x) = x^4 \pm 3x \pmod 7$ (see Table 3.7). This aspect is presented in the following proposition.

Proposition 3.26 *The only normalized 4-PPs* $\pmod 7$ *are* $\bar{\pi}(x) = x^4 \pm 3x \pmod 7$.

The next lemma is needed to get the general result from Theorem 3.28 for this subcase.

Lemma 3.27 *Let there be* $\pi(x) = q_1 x + q_2 x^2 + q_3 x^3 + q_4 x^4 \pmod 7$, *where* $q_4 \neq 0 \pmod 7$. *Then,* $\pi(x)$ *can be factorized as* $\pi(x) = a\big((x+b)^4 \pm 3(x+b)\big) + c \pmod 7$, *if and only if the following two conditions are fulfilled:*

(1) $3(q_3)^2 = q_2 q_4 \pmod 7$ *and*
(2) $2q_1(q_4)^2 = (q_3)^3 + (q_4)^3 \pmod 7$ *or*
$2q_1(q_4)^2 = (q_3)^3 + 6(q_4)^3 \pmod 7$.

Proof We consider that $\pi(x) = a\big((x+b)^4 \pm 3(x+b)\big) + c \pmod 7$. Then, we can write

$$\pi(x) = ax^4 + 4abx^3 + 6ab^2x^2 +$$

$$+ a(4b^3 \pm 3)x + ab^4 \pm 3ab + c \pmod 7 \tag{3.70}$$

By identifying the coefficients of terms of degree 4, 3, and 0 and using $q_4 \neq 0 \pmod 7$, we have

$$a = q_4, \tag{3.71}$$

$$b = \frac{q_3}{4q_4} \pmod 7, \tag{3.72}$$

$$c = -q_4\left(\frac{q_3}{4q_4}\right)^4 \mp 3q_4\left(\frac{q_3}{4q_4}\right) \pmod 7. \tag{3.73}$$

Then, we have

$$\pi(x) = q_4\left(\left(x + \frac{q_3}{4q_4}\right)^4 \pm 3\left(x + \frac{q_3}{4q_4}\right)\right) -$$

$$-q_4\left(\frac{q_3}{4q_4}\right)^4 \mp 3q_4\left(\frac{q_3}{4q_4}\right) \pmod 7 =$$

$$= q_4x^4 + q_3x^3 + \frac{3(q_3)^2}{q_4}x^2 + \left(4\frac{(q_3)^3}{(q_4)^2} \pm 3q_4\right)x \pmod 7. \quad (3.74)$$

Therefore, it is required that

$$q_2 = \frac{3(q_3)^2}{q_4} \pmod 7, \quad (3.75)$$

$$q_1 = 4\frac{(q_3)^3}{(q_4)^2} \pm 3q_4 \pmod 7. \quad (3.76)$$

Equation (3.75) is equivalent to

$$3(q_3)^2 = q_2q_4 \pmod 7, \quad (3.77)$$

and (3.76) is equivalent to

$$2q_1(q_4)^2 = (q_3)^3 + (q_4)^3 \pmod 7, \quad (3.78)$$

or

$$2q_1(q_4)^2 = (q_3)^3 + 6(q_4)^3 \pmod 7. \quad (3.79)$$

The reciprocal is proved in the reverse way. ■

Theorem 3.28 $\pi(x) = q_1x + q_2x^2 + q_3x^3 + q_4x^4 \pmod 7$ *is a PP if and only if:*

(1) If $q_4 \neq 0 \pmod 7$, *then*

(1.1) $3(q_3)^2 = q_2q_4 \pmod 7$ *and*
(1.2) $2q_1(q_4)^2 = (q_3)^3 + (q_4)^3 \pmod 7$ *or* $2q_1(q_4)^2 = (q_3)^3 + 6(q_4)^3 \pmod 7$.

(2) If $q_4 = 0 \pmod 7$, *then* $q_1 \neq 0 \pmod 7$ *and* $q_2 = q_3 = 0 \pmod 7$.

Proof From Propositions 3.12 and 3.26, we have that all 4-PPs, when $q_4 \neq 0 \pmod 7$, can be obtained with the formula $a\bar{\pi}(x + b) + c$, where $\bar{\pi}(x) = x^4 \pm 3x \pmod 7$ and $a \neq 0, b, c \in \mathbb{Z}_7$. Then, according to Lemma 3.27, $\pi(x)$ is a $\pmod 7$ PP if and only if the equalities (1.1) and (1.2) from the theorem statement are true. When $q_4 = 0 \pmod 7$, we can use the test for CPP coefficients when $3 \mid (p-1)$ and $n_{L,p} = 1$, given in Sect. 3.7.3.1. This is $q_1 \neq 0 \pmod 7$, $q_2 = q_3 = 0 \pmod 7$. ■

3.8.4 Cases $p = 7$ and $n_{L,7} > 1$, $3 \mid (p-1)$ with $p > 7$ and $n_{L,p} \geq 1$, $3 \nmid (p-1)$ with $p = 5$, or $p > 7$ and $n_{L,p} > 1$

Theorem 3.29 *Let p be a prime number and $n_{L,p}$ a positive integer so that: a) $3 \mid (p-1)$ and $n_{L,p} > 1$ if $p = 7$ and $n_{L,p} \geq 1$ if $p > 7$ or b) $3 \nmid (p-1)$ and $n_{L,p} > 1$, with $p = 5$ or $p > 7$. Then, $\pi(x) = q_1x + q_2x^2 + q_3x^3 + q_4x^4 \pmod{p^{n_{L,p}}}$ is a PP if and only if $q_1 \neq 0 \pmod p$ and $q_2 = q_3 = q_4 = 0 \pmod p$.*

Proof Because all 4-PPs $\pmod 7$, when $q_4 \neq 0 \pmod 7$, can be obtained with the formula $\pi(x) = a\bar{\pi}(x+b)+c$, where $\bar{\pi}(x) = x^4 \pm 3x \pmod 7$ and $a \neq 0, b, c \in \mathbb{Z}_7$, when $p = 7$ and $n_{L,7} > 1$, we have to show that equation $\bar{\pi}'(x) = 4x^3 \pm 3 = 0$ always has solutions $\pmod 7$. Obviously, if x is a solution of equation $\bar{\pi}'(x) = 0 \pmod 7$, then $x - b$ is a solution of equation $\pi'(x) = 0 \pmod 7$. It is easy to check that $x = 1$, $x = 2$, and $x = 4$ are solutions for the equation $4x^3 + 3 = 0 \pmod 7$ and that $x = 3$, $x = 5$, and $x = 6$ are solutions for the equation $4x^3 - 3 = 0 \pmod 7$. Because in the others cases there are no fourth degree normalized PPs, it follows that $q_4 = 0 \pmod p$ and, therefore, we can use the test for CPP coefficients when $3 \mid (p-1)$, $p > 3$, and $n_{L,p} \geq 1$, given in Sect. 3.7.3.1, or when $3 \nmid (p-1)$ and $n_{L,p} > 1$, given in Sect. 3.7.4.2. This is $q_1 \neq 0 \pmod p$, $q_2 = q_3 = 0 \pmod p$. ■

3.8.5 Cases $3 \nmid (p-1)$ with $p = 5$ or $p > 7$ and $n_{L,p} = 1$

Theorem 3.30 *Let p be a prime number so that $3 \nmid (p-1)$ ($p = 5$ or $p > 7$). Then, $\pi(x) = q_1x + q_2x^2 + q_3x^3 + q_4x^4 \pmod p$ is a PP, if and only if $q_4 = 0 \pmod p$ and*

(1) If $q_3 = 0 \pmod p$, then $q_1 \neq 0 \pmod p$ and $q_2 = 0 \pmod p$.
(2) If $q_3 \neq 0 \pmod p$, then $(q_2)^2 = 3q_1q_3 \pmod p$.

Proof Since in this case there are also no fourth degree normalized PPs, we have that $q_4 = 0 \pmod p$ and, therefore, we can use the test for the CPP coefficients when $3 \nmid (p-1)$ and $n_{L,p} = 1$, given in Sect. 3.7.4.1, which represents the statement of the theorem. ■

3.9 Necessary and Sufficient Conditions on the Coefficients of a Polynomial of Fifth Degree so that It Is PP Modulo Any Positive Integer

A PP of fifth degree is denoted by 5-PP (named quintic PP) and it is of the form:

$$\pi(x) = (q_1 \cdot x + q_2 \cdot x^2 + q_3 \cdot x^3 + q_4 \cdot x^4 + q_5 \cdot x^5) \pmod{L},$$
$$x = \overline{0, L-1} \tag{3.80}$$

Particularizing Theorem 3.8 for polynomials of degree five, the necessary and sufficient conditions for a polynomial of degree five to be 5-PP can be tested using the same algorithm in three steps from Chen et al. (2006) given in Sect. 3.6, where the conditions on the coefficients are given in Table 3.4 (Trifina and Tarniceriu 2018).

The proofs for the conditions on the coefficients are given in the following. The cases $p = 2$, $p = 3$, $p = 5$, $p = 7$, and $p = 13$ are addressed in Sects. 3.9.1–3.9.5, respectively, and the cases $p = 1 \pmod{5}$, $p = 2, 3 \pmod{5}$, with $p > 13$, and $p = 4 \pmod{5}$, in Sects. 3.9.6–3.9.8, respectively.

3.9.1 *Case $p = 2$*

This case is identical to that from Lemma 3.1 for $n_{L,2} = 1$ and to that from Theorem 3.6 for $n_{L,2} > 1$.

3.9.2 *Case $p = 3$*

3.9.2.1 Subcase $p = 3$ and $n_{L,3} = 1$

Theorem 3.31 $\pi(x) = q_1x + q_2x^2 + q_3x^3 + q_4x^4 + q_5x^5 \pmod{3}$ *is a PP if and only if* $(q_1 + q_3 + q_5) \neq 0 \pmod{3}$ *and* $(q_2 + q_4) = 0 \pmod{3}$.

Proof As $\pi(0) = 0$, it is required that

$$\pi(1) = q_1 + q_2 + q_3 + q_4 + q_5 \neq 0 \pmod{3}, \tag{3.81}$$

$$\pi(2) = 2q_1 + q_2 + 2q_3 + q_4 + 2q_5 \neq 0 \pmod{3}, \tag{3.82}$$

and

$$\pi(1) \neq \pi(2) \pmod{3}. \tag{3.83}$$

Table 3.4 Conditions for coefficients q_1, q_2, q_3, q_4, q_5 so that $\pi(x)$ in (3.80) is a 5-PP

1(a)	$p = 2$	$n_{L,2} = 1$	$(q_1 + q_2 + q_3 + q_4 + q_5) = 1 \pmod 2$
1(b)		$n_{L,2} > 1$	$q_1 = 1 \pmod 2$, $(q_2 + q_4) = 0 \pmod 2$ and $(q_3 + q_5) = 0 \pmod 2$
3(a)	$p = 3$	$n_{L,3} = 1$	$(q_1 + q_3 + q_5) \neq 0 \pmod 3$ and $(q_2 + q_4) = 0 \pmod 3$
3(b)		$n_{L,3} > 1$	$q_1 \neq 0 \pmod 3$,$(q_1 + q_3 + q_5) \neq 0 \pmod 3$, $(q_2 + q_4) = 0 \pmod 3$, $(q_1 + q_2 + 2q_5) \neq 0 \pmod 3$ and $(q_1 + q_4 + 2q_5) \neq 0 \pmod 3$
4(a)	$p = 5$	$n_{L,5} = 1$	(1) $q_4 = q_2 = 0 \pmod 5$ and $(q_1 + q_5) \neq 0 \pmod 5$, when $q_3 = 0 \pmod 5$, (2) $q_4 = 0 \pmod 5$ and $(q_2)^2 = 3(q_1 + q_5)q_3 \pmod 5$, when $q_3 \neq 0 \pmod 5$
4(b)		$n_{L,5} > 1$	(1) $q_4 = q_3 = q_2 = 0 \pmod 5$, $q_1 \neq 0 \pmod 5$ and $(q_1 + q_5) \neq 0 \pmod 5$, or (2) $q_4 = 0 \pmod 5$, $(q_2)^2 = 3(q_1 + q_5)q_3 \pmod 5$ and (2.1) $(q_3 + q_5) = 0 \pmod 5$, or (2.2) $(q_3 - q_5) = 0 \pmod 5$, when $q_5 \neq 0 \pmod 5$, or (3) $q_4 = q_3 = q_2 = 0 \pmod 5$ and $q_1 \neq 0 \pmod 5$, when $q_5 = 0 \pmod 5$
5(a)	$p = 7$	$n_{L,7} = 1$	(1) $4q_2(q_5)^2 = 2(q_4)^3 \pmod 7$ and $6q_1(q_5)^3 = (q_4)^4 \pmod 7$, or (2) $4q_2(q_5)^2 = 2(q_4)^3 \pm (q_5)^3 \pmod 7$ and $6q_1(q_5)^3 = (q_4)^4 \pm 2q_4(q_5)^3 \pmod 7$, when $q_5 \neq 0 \pmod 7$ and $5q_3q_5 = 2(q_4)^2 \pmod 7$, or (3) $4q_2(q_5)^2 = 2(q_4)^3 + \alpha q_4(q_5)^2 \pmod 7$ and $6q_1(q_5)^3 = (q_4)^4 + \alpha(q_4)^2(q_5)^2 + +4\alpha^2(q_5)^4 \pmod 7$, where $\alpha = (q_3q_5 + (q_4)^2) \cdot ((q_5)^2)^{-1} \pmod 7$, when $q_5 \neq 0 \pmod 7$ and $5q_3q_5 \neq 2(q_4)^2 \pmod 7$, or (4) $4q_2(q_5)^2 = 2(q_4)^3 + \alpha q_4(q_5)^2 \pm \pm 4(q_5)^3 \pmod 7$ and $6q_1(q_5)^3 = (q_4)^4 + \alpha(q_4)^2(q_5)^2 \pm$ $\pm q_4(q_5)^3 + 4\alpha^2(q_5)^4 \pmod 7$, where $\alpha = (q_3q_5 + (q_4)^2) \cdot ((q_5)^2)^{-1} \pmod 7$, when $\alpha \in \{3, 5, 6\}$, $q_5 \neq 0 \pmod 7$ and $5q_3q_5 \neq 2(q_4)^2 \pmod 7$, or (5) $3(q_3)^2 = q_2q_4 \pmod 7$ and $2q_1(q_4)^2 = (q_3)^3 + (q_4)^3$, or (6) $3(q_3)^2 = q_2q_4 \pmod 7$ and $2q_1(q_4)^2 = (q_3)^3 + 6(q_4)^3 \pmod 7$, when $q_5 = 0 \pmod 7$ and $q_4 \neq 0 \pmod 7$, or (7) $q_3 = q_2 = 0 \pmod 7$ and $q_1 \neq 0 \pmod 7$, when $q_5 = q_4 = 0 \pmod 7$

(continued)

Table 3.4 (continued)

5(b)		$n_{L,7} > 1$	(1) $5q_3q_5 \neq 2(q_4)^2$ (mod 7), $4q_2(q_5)^2 = 2(q_4)^3 + \alpha q_4(q_5)^2$ (mod 7) and $6q_1(q_5)^3 = (q_4)^4 +$ $+\alpha(q_4)^2(q_5)^2 + 4\alpha^2(q_5)^4$ (mod 7), where $\alpha = (q_3q_5 + (q_4)^2) \cdot ((q_5)^2)^{-1}$ (mod 7), when $q_5 \neq 0$ (mod 7), or (2) $q_4 = q_3 = q_2 = 0$ (mod 7) and $q_1 \neq 0$ (mod 7), when $q_5 = 0$ (mod 7)
6(a)	$p = 13$	$n_{L,13} = 1$	(1) $12q_2(q_5)^2 = 2(q_4)^3$ (mod 13) and $8q_1(q_5)^3 = (q_4)^4$ (mod 13), when $q_5 \neq 0$ (mod 13) and $5q_3q_5 = 2(q_4)^2$ (mod 13), or (2) $12q_2(q_5)^2 = 2(q_4)^3 + +2\alpha q_4(q_5)^2$ (mod 13) and $8q_1(q_5)^3 = (q_4)^4 + 2\alpha(q_4)^2(q_5)^2 +$ $+12\alpha^2(q_5)^4$ (mod 13), where $\alpha == (q_3q_5 + 10(q_4)^2) \cdot ((q_5)^2)^{-1}$ (mod 13), when $q_5 \neq 0$ (mod 13) and $5q_3q_5 \neq 2(q_4)^2$ (mod 13), or (3) $12q_2(q_5)^2 = 2(q_4)^3 + +2\alpha q_4(q_5)^2$ (mod 13) and $8q_1(q_5)^3 = (q_4)^4 + 2\alpha(q_4)^2(q_5)^2 +$ $+11\alpha^2(q_5)^4$ (mod 13), where $\alpha == (q_3q_5 + 10(q_4)^2) \cdot ((q_5)^2)^{-1}$ (mod 13), when $\alpha \in \{2, 5, 6, 7, 8, 11\}$, $q_5 \neq 0$ (mod 13) and $5q_3q_5 \neq 2(q_4)^2$ (mod 13), or (4) $q_4 = q_3 = q_2 = 0$ (mod 13) and $q_1 \neq 0$ (mod 13), when $q_5 = 0$ (mod 13)
6(b)		$n_{L,13} > 1$	(1) $5q_3q_5 \neq 2(q_4)^2$ (mod 13), $12q_2(q_5)^2 = 2(q_4)^3 + 2\alpha q_4(q_5)^2$ (mod 13) and $8q_1(q_5)^3 = (q_4)^4 + 2\alpha(q_4)^2(q_5)^2 + +12\alpha^2(q_5)^4$ (mod 13), where $\alpha =$ $= (q_3q_5 + 10(q_4)^2) \cdot ((q_5)^2)^{-1}$ (mod 13), when $q_5 \neq 0$ (mod 13), or (2) $q_4 = q_3 = q_2 = 0$ (mod 13) and $q_1 \neq 0$ (mod 13), when $q_5 = 0$ (mod 13)
7(a)	$p = 1$ (mod p)	$n_{L,p} = 1$	(1) $q_5 = q_4 = q_2 = 0$ (mod p) and $q_1 \neq 0$ (mod p), when $3 \nmid (p-1)$ and $q_3 = 0$ (mod p), or (2) $q_5 = q_4 = 0$ (mod p) and $(q_2)^2 = 3q_1q_3$ (mod p), when $3 \nmid (p-1)$ and $q_3 \neq 0$ (mod p), or (3) $q_5 = q_4 = q_3 = q_2 = 0$ (mod p) and $q_1 \neq 0$ (mod p), when $3 \mid (p-1)$

(continued)

Table 3.4 (continued)

7(b)		$n_{L,p} > 1$	$q_5 = q_4 = q_3 = q_2 = 0 \pmod p$ and $q_1 \neq 0 \pmod p$
8(a)	$p = 2, 3 \pmod p$	$n_{L,p} = 1$	(1) $25q_2(q_5)^2 = 2(q_4)^3 \pmod p$ and $125q_1(q_5)^3 = (q_4)^4 \pmod p$, when $q_5 \neq 0 \pmod p$ and $5q_3q_5 = 2(q_4)^2 \pmod p$, or (2) $25q_2(q_5)^2 = 2(q_4)^3 + +15\alpha q_4(q_5)^2 \pmod p$ and $125q_1(q_5)^3 = (q_4)^4 + 15\alpha(q_4)^2(q_5)^2 + +25\alpha^2(q_5)^4 \pmod p$, where $\alpha = (5q_3q_5 + (p-2)\cdot(q_4)^2)\cdot \cdot(5(q_5)^2)^{-1} \pmod p$, when $q_5 \neq 0 \pmod p$ and $5q_3q_5 \neq 2(q_4)^2 \pmod p$, or (3) $q_4 = q_2 = 0 \pmod p$ and $q_1 \neq 0 \pmod p$, when $3 \nmid (p-1)$ and $q_5 = q_3 = 0 \pmod p$, or (4) $q_4 = 0 \pmod p$ and $(q_2)^2 = 3q_1q_3 \pmod p$, when $3 \nmid (p-1)$, $q_5 = 0 \pmod p$ and $q_3 \neq 0 \pmod p$, or (5) $q_4 = q_3 = q_2 = 0 \pmod p$ and $q_1 \neq 0 \pmod p$, when $3 \mid (p-1)$ and $q_5 = 0 \pmod p$
8(b)		$n_{L,p} > 1$	(1) $5q_3q_5 \neq 2(q_4)^2 \pmod p$, $25q_2(q_5)^2 = 2(q_4)^3 + 15\alpha q_4(q_5)^2 \pmod p$ and $125q_1(q_5)^3 = (q_4)^4 + +15\alpha(q_4)^2(q_5)^2 + 25\alpha^2(q_5)^4 \pmod p$, where $\alpha = (5q_3q_5 + +(p-2)\cdot(q_4)^2)\cdot\left(5(q_5)^2\right)^{-1} \pmod p$, when $q_5 \neq 0 \pmod p$, or (2) $q_4 = q_3 = q_2 = 0 \pmod p$ and $q_1 \neq 0 \pmod p$, when $q_5 = 0 \pmod p$
9(a)	$p = 4 \pmod p$	$n_{L,p} = 1$	(1) $5q_3q_5 = 2(q_4)^2 \pmod p$, $25q_2(q_5)^2 = 2(q_4)^3 \pmod p$ and $125q_1(q_5)^3 = (q_4)^4 \pmod p$, when $q_5 \neq 0 \pmod p$, or (2) $q_4 = q_2 = 0 \pmod p$ and $q_1 \neq 0 \pmod p$, when $3 \nmid (p-1)$ and $q_5 = q_3 = 0 \pmod p$, or (3) $q_4 = 0 \pmod p$ and $(q_2)^2 = 3q_1q_3 \pmod p$, when $3 \nmid (p-1)$, $q_5 = 0 \pmod p$ and $q_3 \neq 0 \pmod p$, or (4) $q_4 = q_3 = q_2 = 0 \pmod p$ and $q_1 \neq 0 \pmod p$, when $3 \mid (p-1)$ and $q_5 = 0 \pmod p$
9(b)		$n_{L,p} > 1$	$q_5 = q_4 = q_3 = q_2 = 0 \pmod p$ nd $q_1 \neq 0 \pmod p$

Replacing (3.81) and (3.82) in (3.83), we have

$$(q_1 + q_3 + q_5) \neq 0 \pmod 3. \tag{3.84}$$

If $(q_1+q_3+q_5) = 1 \pmod 3$, then, from (3.81) it follows that $(q_2+q_4) = 0 \pmod 3$ or $(q_2 + q_4) = 1 \pmod 3$, and from (3.82) it follows that $(q_2 + q_4) = 0 \pmod 3$ or $(q_2+q_4) = 2 \pmod 3$. Therefore, $(q_2+q_4) = 0 \pmod 3$. The case $(q_1+q_3+q_5) = 2 \pmod 3$ is approached similarly and leads to the same result. ■

3.9.2.2 Subcase $p = 3$ and $n_{L,3} > 1$

Theorem 3.32 $\pi(x) = q_1x+q_2x^2+q_3x^3+q_4x^4+q_5x^5 \pmod{3^{n_{L,3}}}$, *with* $n_{L,3} > 1$, *is a PP if and only if* $q_1 \neq 0 \pmod 3$, $(q_1+q_3+q_5) \neq 0 \pmod 3$, $(q_2+q_4) = 0 \pmod 3$, $(q_1 + q_2 + 2 \cdot q_5) \neq 0 \pmod 3$ *and* $(q_1 + q_4 + 2 \cdot q_5) \neq 0 \pmod 3$.

Proof " ⇒" To prove the necessity, we assume that $\pi(x)$ is a PP $\pmod{3^{n_{L,3}}}$, with $n_{L,3} > 1$. Then, according to Theorem 3.7, $\pi(x)$ is a PP $\pmod 3$ and

$$\begin{aligned}\pi'(x) &= q_1 + 2q_2x + 3q_3x^2 + 4q_4x^3 + 5q_5x^4 \pmod 3 = \\ &= q_1 + 2q_2x + q_4x^3 + 2q_5x^4 \neq 0 \pmod 3\end{aligned} \tag{3.85}$$

As $\pi(x)$ is a PP $\pmod 3$, from Theorem 3.31, we have that $(q_1 + q_3 + q_5) \neq 0 \pmod 3$ and $(q_2 + q_4) = 0 \pmod 3$. Replacing $x = 0$ in (3.85), we have that $\pi'(0) = q_1 \neq 0 \pmod 3$. Replacing $x = 1$ in (3.85), we have that $\pi'(1) = q_1 + 2q_2 + q_4 + 2q_5 \neq 0 \pmod 3$. As $(q_2 + q_4) = 0 \pmod 3$, it follows that

$$\pi'(1) = q_1 + q_2 + 2 \cdot q_5 \neq 0 \pmod 3 \tag{3.86}$$

Replacing $x = 2$ in (3.85), we have that $\pi'(2) = q_1 + q_2 + 2q_4 + 2q_5 \neq 0 \pmod 3$ and, as $(q_2 + q_4) = 0 \pmod 3$, it follows that

$$\pi'(2) = q_1 + q_4 + 2 \cdot q_5 \neq 0 \pmod 3 \tag{3.87}$$

" ⇐" To prove the sufficiency, we assume that $q_1 \neq 0 \pmod 3$, $(q_1 + q_3 + q_5) \neq 0 \pmod 3$, $(q_2 + q_4) = 0 \pmod 3$, $(q_1 + q_2 + 2 \cdot q_5) \neq 0 \pmod 3$ and $(q_1+q_4+2\cdot q_5) \neq 0 \pmod 3$. Then, from Theorem 3.31, it follows that $\pi(x)$ is a PP $\pmod 3$. For $x = 0$, from (3.85), we have that $\pi'(0) = q_1 \neq 0 \pmod 3$. For $x = 1$ and $x = 2$, and taking into account the equality $(q_2 + q_4) = 0 \pmod 3$, from (3.85), we have that $\pi'(1) = q_1+q_2+2\cdot q_5 \neq 0 \pmod 3$ and $\pi'(2) = q_1+q_4+2\cdot q_5 \neq 0 \pmod 3$, respectively. Then, according to Theorem 3.7, it results that $\pi(x)$ is a PP $\pmod{3^{n_{L,3}}}$. ■

3.9.3 Case $p = 5$

3.9.3.1 Subcase $p = 5$ and $n_{L,5} = 1$

Proposition 3.33 (Dickson 1896) *The only normalized quintic PPs* (mod 5) *are* $\bar{\pi}(x) = x^5 \pmod 5$, $\bar{\pi}(x) = x^5 - \alpha x \pmod 5$ *(*α *not a fourth power) and* $\bar{\pi}(x) = x^5 - 2\alpha x^3 + \alpha^2 x \pmod 5$ *(*α *not a square).*

Theorem 3.34 $\pi(x) = q_1 x + q_2 x^2 + q_3 x^3 + q_4 x^4 + q_5 x^5 \pmod 5$ *is PP if and only if:*

(1) $q_4 = q_2 = 0 \pmod 5$ *and* $(q_1 + q_5) \neq 0 \pmod 5$, *when* $q_3 = 0 \pmod 5$,
or if and only if:
(2) $q_4 = 0 \pmod 5$ *and* $(q_2)^2 = 3(q_1 + q_5)q_3 \pmod 5$, *when* $q_3 \neq 0 \pmod 5$.

Proof From Propositions 3.13 and 3.33, we have that, when $q_5 \neq 0 \pmod 5$, all 5-PPs can be obtained with the formula $a\bar{\pi}(x) + c$, where $\bar{\pi}(x) = x^5 \pmod 5$ or $\bar{\pi}(x) = x^5 - \alpha x \pmod 5$ (α not a fourth power) or $\bar{\pi}(x) = x^5 - 2\alpha x^3 + \alpha^2 x \pmod 5$ (α not a square). We note that if α is not a fourth power modulo 5, then $\alpha \in \{2, 3, 4\}$ and, if α is not a square modulo 5, then $\alpha \in \{2, 3\}$.

When $\bar{\pi}(x) = x^5 \pmod 5$, it follows that $q_5 \neq 0 \pmod 5$, $q_4 = 0 \pmod 5$, $q_3 = 0 \pmod 5$, $q_2 = 0 \pmod 5$ and $q_1 = 0 \pmod 5$.

When $\bar{\pi}(x) = x^5 - \alpha x \pmod 5$ ($\alpha \in \{2, 3, 4\}$), it follows that $q_5 \neq 0 \pmod 5$, $q_4 = 0 \pmod 5$, $q_3 = 0 \pmod 5$, $q_2 = 0 \pmod 5$ and $q_1 + \alpha q_5 = 0 \pmod 5$ for only one $\alpha \in \{2, 3, 4\}$.

When $\bar{\pi}(x) = x^5 - 2\alpha x^3 + \alpha^2 x \pmod 5$ ($\alpha \in \{2, 3\}$), we have that $q_5 \neq 0 \pmod 5$, $q_4 = 0 \pmod 5$, $q_3 = -2\alpha q_5 \pmod 5$, $q_2 = 0 \pmod 5$ and $q_1 = \alpha^2 q_5 = 4q_5 \pmod 5$. For $\alpha \in \{2, 3\}$, the equality $q_3 = -2\alpha q_5 \pmod 5$ is equivalent to $q_3 + q_5 = 0 \pmod 5$ or $q_3 - q_5 = 0 \pmod 5$, and the equality $q_1 = 4q_5 \pmod 5$ is equivalent to $(q_1 + q_5) = 0 \pmod 5$.

There are four null quintic polynomials modulo 5 ($\pi(x)$ is a null polynomial modulo L if $\pi(x) = 0 \pmod L$, $\forall x \in \mathbb{Z}_L$). These null polynomials modulo 5 are $x^5 + 4x \pmod 5$, $2x^5 + 3x \pmod 5$, $3x^5 + 2x \pmod 5$ and $4x^5 + x \pmod 5$, and therefore we have that a quintic polynomial $\pi(x) = q_1 x + q_2 x^2 + q_3 x^3 + q_4 x^4 + q_5 x^5 \pmod 5$ is equivalent to the polynomial $\pi(x) + q_5 x + (5 - q_5)x^5 \pmod 5$, $\forall q_5 \in \mathbb{Z}_5^*$. There are two normalized PPs modulo 5 of degree less than five (Dickson 1896), i.e. $\bar{\pi}(x) = x \pmod 5$ and $\bar{\pi}(x) = x^3 \pmod 5$. Therefore, a 5-PP can also result when $\pi(x) + q_5 x + (5 - q_5)x^5 \pmod 5 = a(x + b) + c$, or $\pi(x) + q_5 x + (5 - q_5)x^5 \pmod 5 = a(x + b)^3 + c$, with $a \neq 0, b, c \in \mathbb{Z}_5$.

Considering the analysis for CPPs from Sect. 3.7 or for 4-PPs from Sect. 3.8, for the normalized PP $\bar{\pi}(x) = x \pmod 5$, it follows that $q_5 \neq 0 \pmod 5$, $q_4 = 0 \pmod 5$, $q_3 = 0 \pmod 5$, $q_2 = 0 \pmod 5$ and $q_1 + q_5 \neq 0 \pmod 5$.

We note that for the normalized PPs $\bar{\pi}(x) = x^5 \pmod 5$, $\bar{\pi}(x) = x^5 - \alpha x \pmod 5$ ($\alpha \in \{2, 3, 4\}$) and $\bar{\pi}(x) = x \pmod 5$, the common conditions on the coefficients q_5, q_4, q_3, q_2 are $q_5 \neq 0 \pmod 5$, $q_4 = 0 \pmod 5$, $q_3 = 0 \pmod 5$ and $q_2 = 0 \pmod 5$. The condition for the coefficient q_1 is just $q_1 + q_5 \neq 0 \pmod 5$, because it includes the conditions for all the normalized PPs above. Indeed, for $q_1 = 0 \pmod 5$, we have $q_1 + q_5 \neq 0 \pmod 5$, $\forall q_5 \neq 0 \pmod 5$, and the condition $q_1 + \alpha q_5 = 0 \pmod 5$ for any $\alpha \in \{2, 3, 4\}$ and for only one $q_5 \neq 0 \pmod 5$ is equivalent to $q_1 \in \{\mathbb{Z}_5^* - \{-q_5\}\}$ and, therefore, $q_1 + q_5 \neq 0 \pmod 5$. Thus, for the three normalized PPs, the conditions in (1) result for $q_5 \neq 0 \pmod 5$.

For the normalized PP $\bar{\pi}(x) = x^3 \pmod 5$, we have $q_5 \neq 0 \pmod 5$, $q_4 = 0 \pmod 5$, $q_3 \neq 0 \pmod 5$ and $(q_2)^2 = 3(q_1 + q_5)q_3 \pmod 5$. We note that for the normalized PP $\bar{\pi}(x) = x^5 - 2\alpha x^3 + \alpha^2 x \pmod 5$ ($\alpha \in \{2, 3\}$), the conditions on the coefficients are included in those for the normalized PP $\bar{\pi}(x) = x^3 \pmod 5$, because for $q_2 = 0 \pmod 5$ and $q_1 + q_5 = 0 \pmod 5$, we have that $(q_2)^2 = 3(q_1 + q_5)q_3 \pmod 5 = 0 \pmod 5$. Thus, for the two normalized PPs, conditions (2) result for $q_5 \neq 0 \pmod 5$.

When $q_5 = 0 \pmod 5$, we can use the test on the coefficients of a 4-PP from Sect. 3.8.5, for the case $3 \nmid (p-1)$ and $n_{L,p} = 1$. This is given by conditions (1) or (2) for $q_5 = 0 \pmod 5$. ■

3.9.3.2 Subcase $p = 5$ and $n_{L,5} > 1$

Theorem 3.35 $\pi(x) = q_1 x + q_2 x^2 + q_3 x^3 + q_4 x^4 + q_5 x^5 \pmod{5^{n_{L,5}}}$, *with* $n_{L,5} > 1$, *is PP if and only if:*

(1) $q_4 = q_3 = q_2 = 0 \pmod 5$, $q_1 \neq 0 \pmod 5$ *and* $(q_1 + q_5) \neq 0 \pmod 5$,
or if and only if:
(2) $q_4 = 0 \pmod 5$, $(q_2)^2 = 3(q_1 + q_5)q_3 \pmod 5$ *and*

(2.1) $q_3 + q_5 = 0 \pmod 5$,
or
(2.2) $q_3 - q_5 = 0 \pmod 5$,

when $q_5 \neq 0 \pmod 5$,
or if and only if:
(3) $q_4 = q_3 = q_2 = 0 \pmod 5$ *and* $q_1 \neq 0 \pmod 5$, *when* $q_5 = 0 \pmod 5$.

Proof "⇒" To prove the necessity, we assume that $\pi(x)$ is a PP $\pmod{5^{n_{L,5}}}$, with $n_{L,5} > 1$. Then, from Theorem 3.7, it follows that $\pi(x)$ is a PP $\pmod 5$ and

$$\begin{aligned}\pi'(x) &= q_1 + 2q_2 x + 3q_3 x^2 + 4q_4 x^3 + 5q_5 x^4 \pmod 5 = \\ &= q_1 + 2q_2 x + 3q_3 x^2 + 4q_4 x^3 \neq 0 \pmod 5\end{aligned} \tag{3.88}$$

Because $\pi(x)$ is PP $\pmod 5$, $q_4 = 0 \pmod 5$ and, consequently

$$\pi'(x) = q_1 + 2q_2 x + 3q_3 x^2 \neq 0 \pmod 5. \tag{3.89}$$

As in the proof of Theorem 3.34, from Propositions 3.13 and 3.33, when $q_5 \neq 0 \pmod 5$, it follows that all 5-PPs can be obtained with the formula $a\bar{\pi}(x) + c$, with $a \neq 0, c \in \mathbb{Z}_5$, where $\bar{\pi}(x) = x^5 \pmod 5$ or $\bar{\pi}(x) = x^5 - \alpha x \pmod 5$ ($\alpha \in \{2, 3, 4\}$) or $\bar{\pi}(x) = x^5 - 2\alpha x^3 + \alpha^2 x \pmod 5$ ($\alpha \in \{2, 3\}$). Thus, we have that $\pi'(x) = a\bar{\pi}'(x) = q_5\bar{\pi}'(x)$. In the following we will consider each normalized PP.

When $\bar{\pi}(x) = x^5 \pmod 5$, it follows that $\pi'(x) = q_5\bar{\pi}'(x) = 5q_5x^4 \pmod 5 = 0 \pmod 5$. Therefore, the value $q_1 = 0 \pmod 5$, under conditions (1) from Theorem 3.34, is invalid for this case.

When $\bar{\pi}(x) = x^5 - \alpha x \pmod 5$ ($\alpha \in \{2, 3, 4\}$), we have $\pi'(x) = q_5\bar{\pi}'(x) = q_5(5x^4 - \alpha) \pmod 5 = -\alpha q_5 \pmod 5 \neq 0 \pmod 5$, $\forall q_5 \neq 0 \pmod 5$ for $\alpha \in \{2, 3, 4\}$.

When $\bar{\pi}(x) = x^5 - 2\alpha x^3 + \alpha^2 x \pmod 5$ ($\alpha \in \{2, 3\}$), we have $\pi'(x) = q_5\bar{\pi}'(x) = q_5(5x^4 - 2\alpha \cdot 3x^2 + \alpha^2) \pmod 5 = 4q_5(\alpha x^2 + 1) \pmod 5$. It is easy to verify that the equation $(\alpha x^2 + 1) = 0 \pmod 5$ has no solutions for $\alpha = 2$ or $\alpha = 3$. Because in this case $\pi'(x) \neq 0, \forall x \in \mathbb{Z}_5$, conditions (2) from Theorem 3.34, for $q_5 \neq 0 \pmod 5$, are still valid when $q_3 + q_5 = 0 \pmod 5$ or $q_3 - q_5 = 0 \pmod 5$, and result conditions (2) from Theorem 3.35.

For the two normalized PPs modulo 5 of degree less than 5, when $q_5 \neq 0 \pmod 5$, all 5-PPs can be obtained with the formula $\pi(x) + q_5 x + (5 - q_5)x^5 \pmod 5 = a\bar{\pi}(x + b) + c$, with $a \neq 0, b, c \in \mathbb{Z}_5$. Thus, it follows that $\pi'(x) = (a\bar{\pi}'(x+b) - q_5) \pmod 5$.

When $\bar{\pi}(x) = x \pmod 5$, $a = q_1 + q_5$ and we have that $\pi'(x) = q_1 \pmod 5$. Therefore, beside conditions (1) from Theorem 3.34, when $q_5 \neq 0 \pmod 5$ we have additionally to impose $q_1 \neq 0 \pmod 5$, resulting in conditions (1) from Theorem 3.35.

When $\bar{\pi}(x) = x^3 \pmod 5$, then $a = q_3$ and we have $\pi'(x) = q_3\bar{\pi}'(x + b) - q_5 = 3q_3(x + b)^2 - q_5$. As $q_3 \neq 0 \pmod 5$ then from the proof of Lemma 3.19, it follows that $b = \dfrac{q_2}{3q_3}$. The equation $3q_3(x + b)^2 - q_5 = 0 \pmod 5$ is equivalent to $(x + b)^2 = \dfrac{q_5}{3q_3} \pmod 5$. For this equation to have no solution, $\dfrac{q_5}{3q_3}$ cannot be a square modulo 5, that is $\dfrac{q_5}{3q_3} = 2 \pmod 5$ or $\dfrac{q_5}{3q_3} = 3 \pmod 5$. These equalities are equivalent to $q_3 - q_5 = 0 \pmod 5$ or $q_3 + q_5 = 0 \pmod 5$, respectively. Therefore, besides equalities (2) from Theorem 3.34, when $q_5 \neq 0 \pmod 5$, we have to impose the equalities $q_3 + q_5 = 0 \pmod 5$ or $q_3 - q_5 = 0 \pmod 5$, resulting in conditions (2) of Theorem 3.35.

When $q_5 = 0 \pmod 5$, we can apply the coefficient test on a 4-PP when $3 \nmid (p - 1)$ and $n_{L,p} > 1$ from Sect. 3.8.4. This is given by conditions (3) from Theorem 3.35.

"$\Leftarrow$" To prove the sufficiency, we assume that the conditions on the coefficients from the theorem statement are fulfilled. From these conditions, according to Theorem 3.34, we have that $\pi(x)$ is PP $\pmod 5$. According to Theorem 3.7, we still need to show that $\pi'(x) \neq 0 \pmod 5$, $\forall x \in \mathbb{Z}_5$, where $\pi'(x)$ is that from (3.89).

In cases (1) and (3), as $q_3 = q_2 = 0 \pmod 5$, we have that $\pi'(x) = q_1 \neq 0 \pmod 5$, $\forall x \in \mathbb{Z}_5$.

Case (2) follows from the equivalence of equations $\pi'(x) = 0 \pmod 5$ and $(x + b)^2 = \dfrac{q_5}{3q_3} \pmod 5$, with $b = \dfrac{q_2}{3q_3}$. Therefore, for conditions (2.1) or (2.2) the

Table 3.5 Quintic Normalized PPs modulo p for $p > 5$

Normalized PP	p
$\bar{\pi}(x) = x^5$	$p \neq 1 \pmod 5$
$\bar{\pi}(x) = x^5 \pm 2x^2$	$p = 7$
$\bar{\pi}(x) = x^5 + \alpha x^3 \pm x^2 + 3\alpha^2 x$, α not a square	$p = 7$
$\bar{\pi}(x) = x^5 + \alpha x^3 + 5^{-1}\alpha^2 x$, α arbitrary	$p = 2, 3 \pmod 5$
$\bar{\pi}(x) = x^5 + \alpha x^3 + 3\alpha^2 x$, α not a square	$p = 13$

equation has no solution modulo 5. Thus, $\pi'(x) \neq 0 \pmod 5$, $\forall x \in \mathbb{Z}_5$, also in this case. ■

Because for $p > 5$ there are seven normalized quintic PPs (Dickson 1896) (given in Table 3.5), we give a unified approach for the conditions on the coefficients of a 5-PP, when the normalized PP has the form:

$$\bar{\pi}(x) = x^5 + a_3 x^3 + a_2 x^2 + a_1 x. \tag{3.90}$$

Lemma 3.36 *Let there be* $\pi(x) = q_1 x + q_2 x^2 + q_3 x^3 + q_4 x^4 + q_5 x^5 \pmod p$, *where* $q_5 \neq 0 \pmod p$. *Then,* $\pi(x)$ *can be factorized as* $\pi(x) = a\big((x+b)^5 + a_3(x+b)^3 + a_2(x+b)^2 + a_1(x+b)\big) + c \pmod p$ *if and only if the following three conditions are fulfilled:*

(1) $5q_3q_5 = 2(q_4)^2 + 5a_3(q_5)^2 \pmod p$,
(2) $25q_2(q_5)^2 = 2(q_4)^3 + 15a_3q_4(q_5)^2 + 25a_2(q_5)^3 \pmod p$,
(3) $125q_1(q_5)^3 = (q_4)^4 + 15a_3(q_4)^2(q_5)^2 + 50a_2q_4(q_5)^3 +$
$+125a_1(q_5)^4 \pmod p$.

Proof We consider that $\pi(x) = a\big((x+b)^5 + a_3(x+b)^3 + a_2(x+b)^2 + a_1(x+b)\big) + c \pmod p$. Then, we can write

$$\begin{aligned}\pi(x) = ax^5 + 5abx^4 + a(10b^2 + a_3)x^3 +\\ +a(10b^3 + 3a_3b + a_2)x^2 +\\ +a(5b^4 + 3a_3b^2 + 2a_2b + a_1)x +\\ +a(b^5 + a_3b^3 + a_2b^2 + a_1b)x + c \pmod p\end{aligned} \tag{3.91}$$

By identifying the coefficients of terms of degree 5, 4, 3, 2, 1 and 0, we have

$$a = q_5, \tag{3.92}$$

$$b = \frac{q_4}{5q_5} \pmod p, \tag{3.93}$$

$$q_3 = 10q_5\left(\frac{q_4}{5q_5}\right)^2 + a_3q_5 \pmod{p}, \tag{3.94}$$

$$q_2 = 10q_5\left(\frac{q_4}{5q_5}\right)^3 + 3a_3q_5\frac{q_4}{5q_5} + a_2q_5 \pmod{p}, \tag{3.95}$$

$$q_1 = 5q_5\left(\frac{q_4}{5q_5}\right)^4 + 3a_3q_5\left(\frac{q_4}{5q_5}\right)^2 + 2a_2q_5\frac{q_4}{5q_5} + a_1q_5 \pmod{p}, \tag{3.96}$$

$$c = -q_5\left(\left(\frac{q_4}{5q_5}\right)^5 + a_3\left(\frac{q_4}{5q_5}\right)^3 + a_2\left(\frac{q_4}{5q_5}\right)^2 + a_1\frac{q_4}{5q_5}\right) \pmod{p}, \tag{3.97}$$

Equations (3.94)–(3.96) are equivalent to:

$$5q_3q_5 = 2(q_4)^2 + 5a_3(q_5)^2 \pmod{p}, \tag{3.98}$$

$$5q_2(q_5)^2 = 2(q_4)^3 + 15a_3q_4(q_5)^2 + 25a_2(q_5)^3 \pmod{p}, \tag{3.99}$$

and

$$125q_1(q_5)^3 = (q_4)^4 + 15a_3(q_4)^2(q_5)^2 + 50a_2q_4(q_5)^3 + 125a_1(q_5)^4 \pmod{p}, \tag{3.100}$$

respectively.

Then, we have

$$\begin{aligned}\pi(x) = q_5&\left(\left(x+\frac{q_4}{5q_5}\right)^5 + a_3\left(x+\frac{q_4}{5q_5}\right)^3 + a_2\left(x+\frac{q_4}{5q_5}\right)^2 +\right.\\ &\left.+a_1\left(x+\tfrac{q_4}{5q_5}\right)\right) - q_5\left(\left(\tfrac{q_4}{5q_5}\right)^5 + a_3\left(\tfrac{q_4}{5q_5}\right)^3 + a_2\left(\tfrac{q_4}{5q_5}\right)^2 + a_1\tfrac{q_4}{5q_5}\right)\\ &\pmod{p} = q_5x^5 + q_4x^4 + \left(10q_5\left(\tfrac{q_4}{5q_5}\right)^2 + a_3q_5\right)x^3 +\\ &+\left(10q_5\left(\tfrac{q_4}{5q_5}\right)^3 + 3a_3q_5\tfrac{q_4}{5q_5} + a_2q_5\right)x^2 +\\ &+\left(5q_5\left(\frac{q_4}{5q_5}\right)^4 + 3a_3q_5\left(\frac{q_4}{5q_5}\right)^2 + 2a_2q_5\frac{q_4}{5q_5} + a_1q_5\right)x \pmod{p}\end{aligned} \tag{3.101}$$

Therefore, conditions (3.98), (3.99) and (3.100) have to be met, that is, the three conditions of the lemma statement.

The reciprocal is proved in the reverse way. ■

To facilitate the handling of cases for PPs modulo $p^{n_{L,p}}$, with $p > 5$ and $n_{L,p} > 1$, we remark that when a quintic PP has a corresponding normalized quintic PP, $\bar{\pi}(x)$, according to Proposition 3.12, it is of the form $\pi(x) = a\bar{\pi}(x + b) + c$, with $a \neq 0, b, c \in \mathbb{Z}_p$. According to Theorem 3.7, it is required that $\pi'(x) = a\bar{\pi}'(x + b) \neq 0 \pmod{p}, \forall x \in \mathbb{Z}_p$. Therefore, if x is a solution for the equation $\bar{\pi}'(x) = 0 \pmod{p}$, then $x - b$ is a solution for the equation $\pi'(x) = 0 \pmod{p}$. The next lemma addresses the solving of equation $\bar{\pi}'(x) = 0 \pmod{p}$, for each normalized PP from Table 3.5.

We note that if α is not a square modulo 7, then $\alpha \in \{3, 5, 6\}$, and if α is not a square modulo 13, then $\alpha \in \{2, 5, 6, 7, 8, 11\}$.

Lemma 3.37 *Let $\bar{\pi}(x)$ be a normalized PP from Table 3.5. The equation $\bar{\pi}'(x) = 0 \pmod{p}$ has always solutions modulo p, except for $\bar{\pi}(x) = x^5 + \alpha x^3 + 5^{-1}\alpha^2 x$, with $\alpha \in \mathbb{Z}_p^*$ and $p = 2, 3 \pmod{5}$.*

Proof For $\bar{\pi}(x) = x^5 \pmod{p}$, with $p \neq 1 \pmod{5}$, we have $\bar{\pi}'(x) = 5x^4 = 0 \pmod{p}$, with solution $x = 0$.

For $\bar{\pi}(x) = x^5 \pm 2x^2 \pmod{7}$, we have $\bar{\pi}'(x) = 5x^4 \pm 4x = 0 \pmod{7}$, with solution $x = 0$.

For $\bar{\pi}(x) = x^5 + \alpha x^3 \pm x^2 + 3\alpha^2 x \pmod{7}$, with $\alpha \in \{3, 5, 6\}$, we have $\bar{\pi}'(x) = 5x^4 + 3\alpha x^2 \pm 2x + 3\alpha^2 = 0 \pmod{7}$. It can be easily verified that for $\alpha = 3$, the solutions are $x = 2$ and $x = 5$, for $\alpha = 5$, the solutions are $x = 3$ and $x = 4$, and for $\alpha = 6$, the solutions are $x = 1$ and $x = 6$.

For $\bar{\pi}(x) = x^5 + \alpha x^3 + 3\alpha^2 x \pmod{13}$, with $\alpha \in \{2, 5, 6, 7, 8, 11\}$, we have $\bar{\pi}'(x) = 5x^4 + 3\alpha x^2 + 3\alpha^2 = 0 \pmod{13}$. It can be easily verified that for $\alpha = 2$, the solutions are $x = 6$ and $x = 7$, for $\alpha = 5$, the solutions are $x = 5$ and $x = 8$, for $\alpha = 6$, the solutions are $x = 2$ and $x = 11$, for $\alpha = 7$, the solutions are $x = 3$ and $x = 10$, for $\alpha = 8$, the solutions are $x = 1$ and $x = 12$, and for $\alpha = 11$, the solutions are $x = 4$ and $x = 9$.

For $\bar{\pi}(x) = x^5 + \alpha x^3 + 5^{-1}\alpha^2 x \pmod{p}$, with $p = 2, 3 \pmod{5}$ and arbitrary α, we have $\bar{\pi}'(x) = 5x^4 + 3\alpha x^2 + 5^{-1}\alpha^2 = 0 \pmod{p}$. We use the substitution $x^2 = y$ and one of the following equivalent equations result: $5y^2 + 3\alpha y + 5^{-1}\alpha^2 = 0 \pmod{p}$ or $25y^2 + 15\alpha y + \alpha^2 = 0 \pmod{p}$ or $(5y)^2 + 2 \cdot 5y \cdot 2^{-1} \cdot 3\alpha + (2^{-1} \cdot 3\alpha)^2 + \alpha^2 - (2^{-1} \cdot 3\alpha)^2 = 0 \pmod{p}$ or $(5y + 2^{-1} \cdot 3\alpha)^2 + \alpha^2 - (2^{-1} \cdot 3\alpha)^2 = 0 \pmod{p}$ or $(10y + 3\alpha)^2 + 4\alpha^2 - (3\alpha)^2 = 0 \pmod{p}$ or $(10y + 3\alpha)^2 = 5\alpha^2 \pmod{p}$. The last equation has solutions modulo p for $\alpha \neq 0 \pmod{p}$ if $5\alpha^2$ is a quadratic residue modulo p. As α^2 is a quadratic residue, according to Theorem 85 from Hardy and Wright (1975), $5\alpha^2$ is a quadratic residue, only if 5 is a quadratic residue. But, according to Theorem 97 in Hardy and Wright (1975), 5 is a quadratic non-residue for $p = 2, 3 \pmod{5}$. Therefore, the equation $(10y + 3\alpha)^2 = 5\alpha^2 \pmod{p}$ has no solution for $p = 2, 3 \pmod{5}$ and $\alpha \neq 0 \pmod{p}$. ■

3.9.4 Case $p = 7$

3.9.4.1 Subcase $p = 7$ and $n_{L,7} = 1$

Theorem 3.38 $\pi(x) = q_1x + q_2x^2 + q_3x^3 + q_4x^4 + q_5x^5 \pmod 7$ *is PP if and only if:*

(1) $4q_2(q_5)^2 = 2(q_4)^3 \pmod 7$ *and* $6q_1(q_5)^3 = (q_4)^4 \pmod 7$,
or if and only if:
(2) $4q_2(q_5)^2 = 2(q_4)^3 \pm (q_5)^3 \pmod 7$ *and*
$6q_1(q_5)^3 = (q_4)^4 \pm 2q_4(q_5)^3 \pmod 7$,
when $q_5 \neq 0 \pmod 7$ *and* $5q_3q_5 = 2(q_4)^2 \pmod 7$,
or if and only if:
(3) $4q_2(q_5)^2 = 2(q_4)^3 + \alpha q_4(q_5)^2 \pmod 7$ *and*
$6q_1(q_5)^3 = (q_4)^4 + \alpha(q_4)^2(q_5)^2 + 4\alpha^2(q_5)^4$
$\pmod 7$, *where* $\alpha = (q_3q_5 + (q_4)^2) \cdot \left((q_5)^2\right)^{-1} \pmod 7$,
when $q_5 \neq 0 \pmod 7$ *and* $5q_3q_5 \neq 2(q_4)^2 \pmod 7$,
or if and only if:
(4) $4q_2(q_5)^2 = 2(q_4)^3 + \alpha q_4(q_5)^2 \pm 4(q_5)^3 \pmod 7$ *and*
$6q_1(q_5)^3 = (q_4)^4 + \alpha(q_4)^2(q_5)^2 \pm q_4(q_5)^3 + 4\alpha^2(q_5)^4 \pmod 7$,
where $\alpha = (q_3q_5 + (q_4)^2) \cdot \left((q_5)^2\right)^{-1} \pmod 7$,
when $\alpha \in \{3, 5, 6\}$, $q_5 \neq 0 \pmod 7$ *and* $5q_3q_5 \neq 2(q_4)^2 \pmod 7$,
(5) $3(q_3)^2 = q_2q_4 \pmod 7$ *and* $2q_1(q_4)^2 = (q_3)^3 + (q_4)^3$,
or if and only if:
(6) $3(q_3)^2 = q_2q_4 \pmod 7$ *and* $2q_1(q_4)^2 = (q_3)^3 + 6(q_4)^3 \pmod 7$,
when $q_5 = 0 \pmod 7$ *and* $q_4 \neq 0 \pmod 7$,
or if and only if:
(7) $q_3 = q_2 = 0 \pmod 7$ *and* $q_1 \neq 0 \pmod 7$,
when $q_5 = q_4 = 0 \pmod 7$.

Proof If $q_5 \neq 0 \pmod 7$, considering Proposition 3.12, Lemma 3.36 and the normalized PPs modulo 7 from Table 3.5, the next conditions result:

(1) $5q_3q_5 = 2(q_4)^2 \pmod 7$, $4q_2(q_5)^2 = 2(q_4)^3 \pm (q_5)^3 \pmod 7$ and
$6q_1(q_5)^3 = (q_4)^4 \pm 2q_4(q_5)^3 \pmod 7$,
or
(2) $5q_3q_5 = 2(q_4)^2 + 5\alpha(q_5)^2 \pmod 7$,
$4q_2(q_5)^2 = 2(q_4)^3 + \alpha q_4(q_5)^2 \pm 4(q_5)^3 \pmod 7$ and
$6q_1(q_5)^3 = (q_4)^4 + \alpha(q_4)^2(q_5)^2 \pm q_4(q_5)^3 + 4\alpha^2(q_5)^4 \pmod 7$, for only one $\alpha \in \{3, 5, 6\}$,
or
(3) $5q_3q_5 = 2(q_4)^2 + 5\alpha(q_5)^2 \pmod 7$,
$4q_2(q_5)^2 = 2(q_4)^3 + \alpha q_4(q_5)^2 \pmod 7$ and
$6q_1(q_5)^3 = (q_4)^4 + \alpha(q_4)^2(q_5)^2 + 4\alpha^2(q_5)^4 \pmod 7$, for only one $\alpha \in \mathbb{Z}_7$.

Because conditions (2) and (3) above need up to three and seven sets of checking conditions, respectively, it is more efficient to compute the value of α from the first congruence equation in these sets of conditions. This congruence equation is equivalent to:

$$\alpha(q_5)^2 = q_3q_5 + (q_4)^2 \pmod 7 \tag{3.102}$$

Because $q_5 \neq 0 \pmod 7$, we have $(q_5)^2 \neq 0 \pmod 7$. Then, the congruence equation (3.102) has only one solution (Hardy and Wright 1975), which is $\alpha = 0 \pmod 7$ if $5q_3q_5 = 2(q_4)^2 \pmod 7$, and $\alpha \neq 0 \pmod 7$ if $5q_3q_5 \neq 2(q_4)^2 \pmod 7$. To find out the solution, we need to compute the inverse modulo 7 of $(q_5)^2$. An algorithm for finding the arithmetic inverse of an integer modulo another integer is given in Table II from Ryu and Takeshita (2006) (see Algorithm 2 from Chap. 5). The six values of the inverses modulo 7 for $\{1, 2, 3, 4, 5, 6\}$ are $\{1, 4, 5, 2, 3, 6\}$, respectively, in this order. These values can be stored in an array before proceeding to find 5-PPs modulo a number which contains 7 as a prime factor. Thus, if $q_5 \neq 0 \pmod 7$, the conditions (1) or (2) and (3) or (4) from the theorem result.

If $q_5 = 0 \pmod 7$, we can apply the test coefficient for 4-PPs from Sect. 3.8.3, resulting the conditions (5) or (6) or (7) from Theorem 3.38. ■

3.9.4.2 Subcase $p = 7$ and $n_{L,7} > 1$

Theorem 3.39 $\pi(x) = q_1x+q_2x^2+q_3x^3+q_4x^4+q_5x^5 \pmod{7^{n_{L,7}}}$, *with* $n_{L,7} > 1$, *is PP if and only if:*

(1) $5q_3q_5 \neq 2(q_4)^2 \pmod 7$, $4q_2(q_5)^2 = 2(q_4)^3 + \alpha q_4(q_5)^2 \pmod 7$ *and* $6q_1(q_5)^3 = (q_4)^4 + \alpha(q_4)^2(q_5)^2 + 4\alpha^2(q_5)^4 \pmod 7$, *where* $\alpha = (q_3q_5 + (q_4)^2) \cdot ((q_5)^2)^{-1} \pmod 7$, *when* $q_5 \neq 0 \pmod 7$,
or if and only if:

(2) $q_4 = q_3 = q_2 = 0 \pmod 7$ *and* $q_1 \neq 0 \pmod 7$, *when* $q_5 = 0 \pmod 7$.

Proof If $q_5 \neq 0 \pmod 7$, according to Theorem 3.7 and Lemma 3.37, $\pi(x)$ is PP if and only if the normalized PP leading to $\pi(x) \pmod 7$ is $\bar{\pi}(x) = x^5 + \alpha x^3 + 5^{-1}\alpha^2 x \pmod 7$, with $\alpha \in \mathbb{Z}_7^*$. Because $\alpha \neq 0 \pmod 7$, it is required that $5q_3q_5 \neq 2(q_4)^2 \pmod 7$. Then, the conditions on the coefficients are those from (3) in Theorem 3.38.

If $q_5 = 0 \pmod 7$, we can apply the conditions on the coefficients for 4-PPs $\pmod{7^{n_{L,7}}}$ with $n_{L,7} > 1$ from Sect. 3.8.4. They are those in (2) from Theorem 3.39. ■

3.9.5 Case $p = 13$

3.9.5.1 Subcase $p = 13$ and $n_{L,13} = 1$

Theorem 3.40 $\pi(x) = q_1x + q_2x^2 + q_3x^3 + q_4x^4 + q_5x^5 \pmod{13}$ *is PP if and only if:*

(1) $12q_2(q_5)^2 = 2(q_4)^3 \pmod{13}$ *and* $8q_1(q_5)^3 = (q_4)^4 \pmod{13}$,
when $q_5 \neq 0 \pmod{13}$ *and* $5q_3q_5 = 2(q_4)^2 \pmod{13}$,
or if and only if:
(2) $12q_2(q_5)^2 = 2(q_4)^3 + 2\alpha q_4(q_5)^2 \pmod{13}$ *and*
$8q_1(q_5)^3 = (q_4)^4 + 2\alpha(q_4)^2(q_5)^2 + 12\alpha^2(q_5)^4 \pmod{13}$,
where $\alpha = (q_3q_5 + 10(q_4)^2) \cdot \left((q_5)^2\right)^{-1} \pmod{13}$,
when $q_5 \neq 0 \pmod{13}$ *and* $5q_3q_5 \neq 2(q_4)^2 \pmod{13}$,
or if and only if:
(3) $12q_2(q_5)^2 = 2(q_4)^3 + 2\alpha q_4(q_5)^2 \pmod{13}$ *and*
$8q_1(q_5)^3 = (q_4)^4 + 2\alpha(q_4)^2(q_5)^2 + 11\alpha^2(q_5)^4 \pmod{13}$,
where $\alpha = (q_3q_5 + 10(q_4)^2) \cdot \left((q_5)^2\right)^{-1} \pmod{13}$,
when $\alpha \in \{2, 5, 6, 7, 8, 11\}$, $q_5 \neq 0 \pmod{13}$ *and*
$5q_3q_5 \neq 2(q_4)^2 \pmod{13}$,
or if and only if:
(4) $q_4 = q_3 = q_2 = 0 \pmod{13}$ *and* $q_1 \neq 0 \pmod{13}$,
when $q_5 = 0 \pmod{13}$.

Proof If $q_5 \neq 0 \pmod{13}$, considering Proposition 3.12, Lemma 3.36 and the normalized PPs modulo 13 from Table 3.5, the next conditions result:

(1) $5q_3q_5 = 2(q_4)^2 + 5\alpha(q_5)^2 \pmod{13}$,
$12q_2(q_5)^2 = 2(q_4)^3 + 2\alpha q_4(q_5)^2 \pmod{13}$ and
$8q_1(q_5)^3 = (q_4)^4 + 2\alpha(q_4)^2(q_5)^2 + 12\alpha^2(q_5)^4 \pmod{13}$, for only one $\alpha \in \mathbb{Z}_{13}$,
or
(2) $5q_3q_5 = 2(q_4)^2 + 5\alpha(q_5)^2 \pmod{13}$,
$12q_2(q_5)^2 = 2(q_4)^3 + 2\alpha q_4(q_5)^2 \pmod{13}$ and
$8q_1(q_5)^3 = (q_4)^4 + 2\alpha(q_4)^2(q_5)^2 + 11\alpha^2(q_5)^4 \pmod{13}$, for only one $\alpha \in \{2, 5, 6, 7, 8, 11\}$.

Because the conditions above need up to 13 and six sets of checking conditions, respectively, it is more efficient to compute the value of α from the first congruence equation in these sets of conditions. This congruence equation is equivalent to:

$$\alpha(q_5)^2 = q_3q_5 + 10(q_4)^2 \pmod{13} \tag{3.103}$$

Because $q_5 \neq 0 \pmod{13}$, we have $(q_5)^2 \neq 0 \pmod{13}$. Then, the congruence equation (3.103) has only one solution (Hardy and Wright 1975), which is $\alpha = 0 \pmod{13}$ if $5q_3q_5 = 2(q_4)^2 \pmod{13}$, and $\alpha \neq 0 \pmod{13}$ if $5q_3q_5 \neq 2(q_4)^2 \pmod{13}$. The 12 values of the inverses modulo 13 for $\{1, 2, 3, 4, 5, 6, 7, 8, 9, 10, 11, 12\}$ are

$\{1, 7, 9, 10, 8, 11, 2, 5, 3, 4, 6, 12\}$, in this order, and they can be stored in an array before to proceed for finding 5-PPs modulo a number which contains 13 as prime factor. Thus, if $q_5 \neq 0 \pmod{13}$, the conditions (1) or (2) or (3) in the theorem result.

If $q_5 = 0 \pmod{13}$, we can apply the test for 4-PPs from Sect. 3.8.4. This is given by conditions (4) from Theorem 3.40. ■

3.9.5.2 Subcase $p = 13$ and $n_{L,13} > 1$

Theorem 3.41 $\pi(x) = q_1x + q_2x^2 + q_3x^3 + q_4x^4 + q_5x^5 \pmod{13^{n_{L,13}}}$, *with* $n_{L,13} > 1$, *is PP if and only if:*

(1) $5q_3q_5 \neq 2(q_4)^2 \pmod{13}$,
$12q_2(q_5)^2 = 2(q_4)^3 + 2\alpha q_4(q_5)^2 \pmod{13}$ *and*
$8q_1(q_5)^3 = (q_4)^4 + 2\alpha(q_4)^2(q_5)^2 + 12\alpha^2(q_5)^4 \pmod{13}$,
where $\alpha = (q_3q_5 + 10(q_4)^2) \cdot \left((q_5)^2\right)^{-1} \pmod{13}$,
when $q_5 \neq 0 \pmod{13}$,
or if and only if:
(2) $q_4 = q_3 = q_2 = 0 \pmod{13}$ *and* $q_1 \neq 0 \pmod{13}$,
when $q_5 = 0 \pmod{13}$.

Proof If $q_5 \neq 0 \pmod{13}$, according to Theorem 3.7 and Lemma 3.37, $\pi(x)$ is PP if and only if the normalized PP leading to $\pi(x) \pmod{13}$ is $\bar{\pi}(x) = x^5 + \alpha x^3 + 5^{-1}\alpha^2 x \pmod{13}$, with $\alpha \in \mathbb{Z}_{13}^*$. Because $\alpha \neq 0 \pmod{13}$, it is required that $5q_3q_5 \neq 2(q_4)^2 \pmod{13}$. Then, the conditions on the coefficients are those from (2) in Theorem 3.40.

If $q_5 = 0 \pmod{13}$, we can apply the conditions on the coefficients for 4-PPs $\pmod{13^{n_{L,13}}}$ with $n_{L,13} > 1$ from Sect. 3.8.4. They are those in (2) from Theorem 3.41. ■

3.9.6 *Case $p = 1 \pmod 5$*

3.9.6.1 Subcase $p = 1 \pmod 5$ and $n_{L,p} = 1$

Theorem 3.42 $\pi(x) = q_1x + q_2x^2 + q_3x^3 + q_4x^4 + q_5x^5 \pmod{p}$, *with* $p = 1 \pmod 5$, *is PP if and only if:*

(1) $q_5 = q_4 = q_2 = 0 \pmod{p}$ *and* $q_1 \neq 0 \pmod{p}$,
when $3 \nmid (p-1)$ *and* $q_3 = 0 \pmod{p}$,
or if and only if:
(2) $q_5 = q_4 = 0 \pmod{p}$ *and* $(q_2)^2 = 3q_1q_3 \pmod{p}$,
when $3 \nmid (p-1)$ *and* $q_3 \neq 0 \pmod{p}$,
or if and only if:

(3) $q_5 = q_4 = q_3 = q_2 = 0 \pmod{p}$ *and* $q_1 \neq 0 \pmod{p}$, *when* $3 \mid (p-1)$.

Proof Because in this case there are no normalized PPs of fifth or fourth degree, we can apply the coefficient test for CPPs from Sect. 3.7.4.1. ■

3.9.6.2 Subcase $p = 1 \pmod{5}$ and $n_{L,p} > 1$

Theorem 3.43 $\pi(x) = q_1x + q_2x^2 + q_3x^3 + q_4x^4 + q_5x^5 \pmod{p^{n_{L,p}}}$, *with* $p = 1 \pmod{5}$ *and* $n_{L,p} > 1$, *is PP if and only if* $q_5 = q_4 = q_3 = q_2 = 0 \pmod{p}$ *and* $q_1 \neq 0 \pmod{p}$.

Proof Because in this case there are no normalized PPs of fifth or fourth degree, we can apply the coefficient test for CPPs from Sect. 3.7.4.2. ■

3.9.7 *Case* $p = 2, 3 \pmod{5}$ *with* $p > 13$

3.9.7.1 Subcase $p = 2, 3 \pmod{5}$ with $p > 13$ and $n_{L,p} = 1$

Theorem 3.44 $\pi(x) = q_1x + q_2x^2 + q_3x^3 + q_4x^4 + q_5x^5 \pmod{p}$, *with* $p = 2, 3 \pmod{5}$ *and* $p > 13$, *is PP if and only if:*

(1) $25q_2(q_5)^2 = 2(q_4)^3 \pmod{p}$ *and* $125q_1(q_5)^3 = (q_4)^4 \pmod{p}$, *when* $q_5 \neq 0 \pmod{p}$ *and* $5q_3q_5 = 2(q_4)^2 \pmod{p}$, *or if and only if:*

(2) $25q_2(q_5)^2 = 2(q_4)^3 + 15\alpha q_4(q_5)^2 \pmod{p}$ *and* $125q_1(q_5)^3 = (q_4)^4 + 15\alpha(q_4)^2(q_5)^2 + 25\alpha^2(q_5)^4 \pmod{p}$, *where* $\alpha = (5q_3q_5 + (p-2)\cdot(q_4)^2)\cdot\left(5(q_5)^2\right)^{-1} \pmod{p}$, *when* $q_5 \neq 0 \pmod{p}$ *and* $5q_3q_5 \neq 2(q_4)^2 \pmod{p}$, *or if and only if:*

(3) $q_4 = q_2 = 0 \pmod{p}$ *and* $q_1 \neq 0 \pmod{p}$, *when* $3 \nmid (p-1)$ *and* $q_5 = q_3 = 0 \pmod{p}$, *or if and only if:*

(4) $q_4 = 0 \pmod{p}$ *and* $(q_2)^2 = 3q_1q_3 \pmod{p}$, *when* $3 \nmid (p-1)$, $q_5 = 0 \pmod{p}$ *and* $q_3 \neq 0 \pmod{p}$, *or if and only if:*

(5) $q_4 = q_3 = q_2 = 0 \pmod{p}$ *and* $q_1 \neq 0 \pmod{p}$, *when* $3 \mid (p-1)$ *and* $q_5 = 0 \pmod{p}$.

Proof If $q_5 \neq 0 \pmod{p}$, by considering Proposition 3.12, Lemma 3.36 and the normalized PPs modulo p from Table 3.5, when $p = 2, 3 \pmod{5}$ and $p > 13$, that is $\bar{\pi}(x) = x^5 \pmod{p}$ and $\bar{\pi}(x) = x^5 + \alpha x^3 + 5^{-1}\alpha^2 x \pmod{p}$, with $\alpha \in \mathbb{Z}_p^*$, the next conditions result:

$5q_3q_5 = 2(q_4)^2 + 5\alpha(q_5)^2 \pmod{p}$,
$25q_2(q_5)^2 = 2(q_4)^3 + 15\alpha q_4(q_5)^2 \pmod{p}$ and
$125q_1(q_5)^3 = (q_4)^4 + 15\alpha(q_4)^2(q_5)^2 + 25\alpha^2(q_5)^4 \pmod{p}$, for only one $\alpha \in \mathbb{Z}_p$.

Because the conditions above need up to p sets of checking conditions, it is more efficient to compute the value of α from the first congruence equation from this set of conditions. This congruence equation is equivalent to:

$$\alpha \cdot 5(q_5)^2 = 5q_3q_5 + (p-2) \cdot (q_4)^2 \pmod{p} \tag{3.104}$$

Because $q_5 \neq 0 \pmod{p}$, we have $(q_5)^2 \neq 0 \pmod{p}$. Then, the congruence equation (3.104) has only one solution (Hardy and Wright 1975), which is $\alpha = 0 \pmod{p}$ if $5q_3q_5 = 2(q_4)^2 \pmod{p}$, and $\alpha \neq 0 \pmod{p}$ if $5q_3q_5 \neq 2(q_4)^2 \pmod{p}$. Thus, if $q_5 \neq 0 \pmod{p}$, conditions (1) or (2) in the theorem result.

If $q_5 = 0 \pmod{p}$, we can apply the test for 4-PPs from Sect. 3.8.4. This is given by conditions (3) or (4) or (5) from Theorem 3.44. ■

3.9.7.2 Subcase $p = 2, 3 \pmod{5}$ with $p > 13$ and $n_{L,p} > 1$

Theorem 3.45 $\pi(x) = q_1x + q_2x^2 + q_3x^3 + q_4x^4 + q_5x^5 \pmod{p^{n_{L,p}}}$, *with* $p = 2, 3 \pmod{5}$, $p > 13$ *and* $n_{L,p} > 1$, *is PP if and only if:*

(1) $5q_3q_5 \neq 2(q_4)^2 \pmod{p}$,
$25q_2(q_5)^2 = 2(q_4)^3 + 15\alpha q_4(q_5)^2 \pmod{p}$ *and*
$125q_1(q_5)^3 = (q_4)^4 + 15\alpha(q_4)^2(q_5)^2 + 25\alpha^2(q_5)^4 \pmod{p}$,
where $\alpha = (5q_3q_5 + (p-2) \cdot (q_4)^2) \cdot \left(5(q_5)^2\right)^{-1} \pmod{p}$,
when $q_5 \neq 0 \pmod{p}$,
or if and only if:
(2) $q_4 = q_3 = q_2 = 0 \pmod{p}$ *and* $q_1 \neq 0 \pmod{p}$,
when $q_5 = 0 \pmod{p}$.

Proof If $q_5 \neq 0 \pmod{p}$, according to Theorem 3.7 and Lemma 3.37, $\pi(x)$ is PP if and only if the normalized PP leading to $\pi(x) \pmod{p}$ is $\bar{\pi}(x) = x^5 + \alpha x^3 + 5^{-1}\alpha^2 x \pmod{p}$, with $\alpha \in \mathbb{Z}_p^*$. Because $\alpha \neq 0 \pmod{p}$, it is required that $5q_3q_5 \neq 2(q_4)^2 \pmod{p}$. Then, the conditions for the coefficients are those in (2) from Theorem 3.44.

If $q_5 = 0 \pmod{p}$, we can apply the conditions on coefficients for 4-PPs $\pmod{p^{n_{L,p}}}$ with $n_{L,p} > 1$ from Sect. 3.8.4. They are those in (2) from Theorem 3.45. ■

3.9.8 Case $p = 4 \pmod 5$

3.9.8.1 Subcase $p = 4 \pmod 5$ and $n_{L,p} = 1$

Theorem 3.46 $\pi(x) = q_1x + q_2x^2 + q_3x^3 + q_4x^4 + q_5x^5 \pmod p$, *with* $p = 4 \pmod 5$, *is PP if and only if:*

(1) $5q_3q_5 = 2(q_4)^2 \pmod p$, $25q_2(q_5)^2 = 2(q_4)^3 \pmod p$ *and* $125q_1(q_5)^3 = (q_4)^4 \pmod p$, *when* $q_5 \neq 0 \pmod p$, *or if and only if:*
(2) $q_4 = q_2 = 0 \pmod p$ *and* $q_1 \neq 0 \pmod p$, *when* $3 \nmid (p-1)$ *and* $q_5 = q_3 = 0 \pmod p$, *or if and only if:*
(3) $q_4 = 0 \pmod p$ *and* $(q_2)^2 = 3q_1q_3 \pmod p$, *when* $3 \nmid (p-1)$, $q_5 = 0 \pmod p$ *and* $q_3 \neq 0 \pmod p$, *or if and only if:*
(4) $q_4 = q_3 = q_2 = 0 \pmod p$ *and* $q_1 \neq 0 \pmod p$, *when* $3 \mid (p-1)$ *and* $q_5 = 0 \pmod p$.

Proof If $q_5 \neq 0 \pmod p$, the conditions in the theorem result by considering Proposition 3.12, Lemma 3.36 and the normalized PP modulo p from Table 3.5, when $p = 4 \pmod 5$, i.e. $\bar{\pi}(x) = x^5 \pmod p$.

If $q_5 = 0 \pmod p$, we can apply the test for 4-PPs from Sect. 3.8.4. ■

3.9.8.2 Subcase $p = 4 \pmod 5$ and $n_{L,p} > 1$

Theorem 3.47 $\pi(x) = q_1x + q_2x^2 + q_3x^3 + q_4x^4 + q_5x^5 \pmod{p^{n_{L,p}}}$, *with* $p = 4 \pmod 5$ *and* $n_{L,p} > 1$, *is PP if and only if* $q_5 = q_4 = q_3 = q_2 = 0 \pmod p$ *and* $q_1 \neq 0 \pmod p$.

Proof Because in this case the only normalized quintic PP is $\bar{\pi}(x) = x^5 \pmod p$ and the equation $\bar{\pi}'(x) = 0 \pmod p$ has always solutions, we have that $q_5 = 0 \pmod p$ and thus, we can apply the coefficient test for 4-PPs from Sect. 3.8.4. ■

3.10 Sufficient Conditions on the Coefficients of a Polynomial of Any Degree so that It Is PP Modulo Any Positive Integer

Hongyu Zhao and Pingzhi Fan gave in (2007) sufficient conditions on the coefficients of a polynomial of any degree so that it is PP modulo any positive integer. These are given in Table 3.6, which is similar to Theorem 1 in Zhao and Fan (2007).

Table 3.6 Sufficient conditions for coefficients $q_1, q_2, \ldots, q_d$ so that $\pi(x)$ in (3.1) is a PP

1(a)	$p = 2$	$n_{L,2} = 1$	$(q_1 + q_2 + \cdots + q_d) \neq 0 \pmod 2$
1(b)		$n_{L,2} > 1$	$q_1 \neq 0, q_2 + q_4 + q_6 + \cdots = 0 \pmod 2$, $q_3 + q_5 + q_7 + \cdots = 0 \pmod 2$
(2)	$p > 2$	$n_{L,p} \geq 1$	$q_1 \neq 0, q_2 = q_3 = \cdots = q_d = 0 \pmod p$

In the following we give the proof of the sufficiency for conditions on the coefficients given in Table 3.6.

3.10.1 *Case* $p = 2$

This case is identical to that from Lemma 3.1 for $n_{L,2} = 1$ and to that from Theorem 3.6 for $n_{L,2} > 1$.

3.10.2 *Case* $p > 2$

Theorem 3.48 *$\pi(x)$ from (3.1) is PP modulo $p^{n_{L,p}}$, with $p > 2$ and $n_{L,p} \geq 1$, if $q_1 \neq 0 \pmod p$ and $q_2 = q_3 = \cdots = q_d = 0 \pmod p$.*

Proof In this case, as $q_2 = q_3 = \cdots = q_d = 0 \pmod p$, for $n_{L,p} = 1$ we have

$$\pi(x) = q_1 \cdot x \pmod p \tag{3.105}$$

As $q_1 \neq 0 \pmod p$, from Theorem 3.9 it results that $\pi(x)$ is PP $\pmod p$.

For $n_{L,p} > 1$, because the conditions are the same as for $n_{L,p} = 1$, it follows that $\pi(x)$ is PP $\pmod p$. According to Theorem 3.7, we have to show additionally that $\pi'(x) \neq 0, \forall x \in \mathbb{Z}_p$. Considering $\pi(x)$ from (3.105), it follows that

$$\pi'(x) = q_1 \pmod p \tag{3.106}$$

But $q_1 \neq 0 \pmod p$, thus also $\pi'(x) \neq 0, \forall x \in \mathbb{Z}_p$. Therefore, $\pi(x)$ is PP $\pmod{p^{n_{L,p}}}$. ∎

Considering the conditions on the coefficients from Theorem 3.6 for PPs of any degree modulo 2^n, from Theorem 3.9 for LPPs and from Table 3.1 for QPPs, it follows that for these particular cases, the sufficient conditions of Zhao and Fan, from Table 3.6, also become necessary.

We mention that Zhao and Fan sufficient conditions were also given, in a different form, by Jing Sun and Oscar Y. Takeshita in Corollarry 2.5 from Sun and Takeshita (2005).

3.11 Getting all Permutation Polynomials of Degrees up to Five by Weng and Dong Algorithm

In Weng and Dong (2008) an algorithm for getting all PPs of degree no more than six over $\mathbb{Z}_L$ is derived on the base of normalized PPs, Theorems 3.7 and 3.8, Propositions 3.12 and 3.13 and on the Chinese Remainder Theorem (Theorem 3.49 below).

In Dickson (1896), all normalized PPs of degree no more than six are given, except for odd powers of two. The PPs modulo p, for $p > 5$, of degrees up to five, which are of interest in this chapter, are given in Table 3.7. We mention that the normalized PP of degree three from Table 3.7 was used in Sect. 3.7.4.1 to obtain the coefficient conditions for a CPP modulo $p > 3$, with $3 \nmid (p-1)$. The two normalized PPs of degree four from Table 3.7 were used in Sect. 3.8.3 to obtain the coefficient conditions for a 4-PP modulo 7. The normalized PPs of degree five from the last five rows of Table 3.7 were used in Sect. 3.9 to obtain the coefficient conditions for a 5-PP modulo p with $p > 5$, or for a 5-PP modulo $p^{n_{L,p}}$ with $p > 5$ and $n_{L,p} > 1$.

For the primes $p = 2$, $p = 3$ and $p = 5$, the coefficient conditions for PPs of degrees no more than five were given in Table 3.4. We mention that, unlike Table II from Weng and Dong (2008), the coefficient conditions from Table 3.4 provide all PPs modulo 2, 3 or 5.

Considering Theorem 3.7, the permutation polynomials from Table 3.7 can be used to get the PPs over a prime at a power greater than one, by imposing the condition $\pi'(x) \neq 0 \pmod{p}$ for every integer x. The PPs modulo p, for $p > 5$, of degrees up to five, that fulfill this constraint are given in Table 3.8. The proof that the PP from

Table 3.7 Normalized PPs modulo p of degree up to five, for $p > 5$

Normalized PP	p
$\bar{\pi}(x) = x$	any p
$\bar{\pi}(x) = x^3$	$p \neq 1 \pmod{3}$
$\bar{\pi}(x) = x^4 \pm 3x^2$	$p = 7$
$\bar{\pi}(x) = x^5$	$p \neq 1 \pmod{5}$
$\bar{\pi}(x) = x^5 \pm 2x^2$	$p = 7$
$\bar{\pi}(x) = x^5 + \alpha x^3 \pm x^2 + 3\alpha^2 x$, α not a square	$p = 7$
$\bar{\pi}(x) = x^5 + \alpha x^3 + 5^{-1}\alpha^2 x$, $\alpha \in \mathbb{Z}_p$	$p = 2, 3 \pmod{5}$
$\bar{\pi}(x) = x^5 + \alpha x^3 + 3\alpha^2 x$, α not a square	$p = 13$

Table 3.8 Normalized PPs modulo p of degree up to five, for $p > 5$, that permute $\mathbb{Z}_{p^{n_{N,p}}}$, with $n_{N,p} > 1$

Normalized PP	p
$\bar{\pi}(x) = x$	any p
$\bar{\pi}(x) = x^5 + \alpha x^3 + 5^{-1}\alpha^2 x$, $\alpha \in \mathbb{Z}_p^*$	$p = 2, 3 \pmod{5}$

the third line in Table 3.8 fulfills the constraint $\pi'(x) \neq 0 \pmod{p}, \forall x \in \mathbb{Z}_p$, was given in Lemma 3.37. The proof for the PP $\bar{\pi}(x) = x \pmod{p}$ is trivial. For powers greater than one of the primes 2, 3 or 5, the coefficient conditions are also given in Table 3.4. Unlike Table IV from Weng and Dong (2008), these conditions provide all PPs modulo a power greater than one of the primes 2, 3 or 5.

If we consider $c = 0$ and $\bar{\pi}(x)$ any normalized PP from Table 3.7, from Proposition 3.12 all PPs $\pmod{p}$ of degree d $(d \leq 5)$ are obtained with formula $a\bar{\pi}(x+b)$ for all $a \neq 0, b \in \mathbb{Z}_p$, when $p \nmid d$. When $p \mid d$, from Proposition 3.13 all PPs $\pmod{p}$ of degree d $(d \leq 5)$ are obtained with formula $a\bar{\pi}(x)$ for all $a \in \mathbb{Z}_p^*$. We recall that for primes 2, 3 or 5, Table 3.4 provides all PPs modulo 2, 3 or 5. Further we consider the PPs obtained above where $\bar{\pi}(x)$ is any normalized PP given in Table 3.8. To obtain all PPs $\pmod{p^{n_{L,p}}}$ with $n_{L,p} > 1$, we have to add to the previously obtained PPs a polynomial $p \cdot \rho(x)$, where $\rho(x)$ is any polynomial over $\mathbb{Z}_{p^{n_{L,p}-1}}$. In this way we have $(\pi(x)+p\cdot\rho(x))' \pmod{p} = \pi'(x) \pmod{p} + p\cdot\rho'(x) \pmod{p} = \pi'(x) \pmod{p} \neq 0 \pmod{p}$.

In the following, we recall the result of the Chinese remainder theorem.

Theorem 3.49 (Chinese remainder theorem) *Suppose $n_1, n_2, \ldots, n_k$ are k positive integers ($k \in \mathbb{N}^*$) that are pairwise coprime. Then, for any given sequence of integers $(a_1, a_2, \ldots, a_k)$, there is an integer x solving the following system of simultaneous congruencies:*

$$\begin{cases} x = a_1 \pmod{n_1} \\ x = a_2 \pmod{n_2} \\ \ldots\ldots\ldots\ldots\ldots\ldots \\ x = a_k \pmod{n_k} \end{cases} \tag{3.107}$$

Furthermore, all solutions x of this system are congruent modulo the product $N = n_1 \cdot n_2 \cdot \ \ldots \ \cdot n_k$. Hence

$$x = y \pmod{n_i}, \textit{for an } i, \textit{with } 1 \leq i \leq k \Leftrightarrow x = y \pmod{N} \tag{3.108}$$

Theorem 3.49 ensures that for two different sequences $a_1, a_2, \ldots, a_k$, two distinct modulo N solutions exist.

Below, we give the algorithm for getting all PPs of degrees up to five. This is a detailed version of the algorithm from Weng and Dong (2008), where we have considered $c = 0$.

1. Factor the interleaver length as

$$L = \prod_{k=1}^{n_{L1}} p_k \cdot \prod_{k=n_{L1}+1}^{n_{L1}+n_{L2}} p_k^{n_{L,p_k}}, \tag{3.109}$$

where $n_{L,p_k} > 1$, $\forall k = n_{L1}+1, \ldots, n_{L1}+n_{L2}$, $n_{L1} \geq 0$ is the number of primes at power of one from the decomposition of L, $n_{L2} \geq 0$ is the number of primes at powers greater than one from the decomposition of L, and $n_{L1} +$

$n_{L2} \geq 1$. Theorem 3.8 assures that if $\pi_1(x), \pi_2(x), \ldots, \pi_{n_{L1}}(x), \pi_{n_{L1}+1}(x), \ldots, \pi_{n_{L1}+n_{L2}}(x)$ is a set of PPs modulo $p_1, p_2, \cdots, p_{n_{L1}}, p_{n_{L1}+1}^{n_{L,p_{n_{L1}+1}}}, \cdots$, and $p_{n_{L1}+n_{L2}}^{n_{L,p_{n_{L1}+n_{L2}}}}$, respectively, then the corresponding PP $\pi(x) \pmod{L}$ is the polynomial which fulfills the following equalities:

$$\begin{cases} \pi(x) \pmod{p_1} = \pi_1(x) \pmod{p_1} \\ \pi(x) \pmod{p_2} = \pi_2(x) \pmod{p_2} \\ \cdots\cdots\cdots\cdots\cdots\cdots\cdots\cdots \\ \pi(x) \pmod{p_{n_{L1}}} = \pi_{n_{L1}}(x) \pmod{p_{n_{L1}}} \\ \pi(x) \left(\bmod p_{n_{L1}+1}^{n_{L,p_{n_{L1}+1}}}\right) = \pi_{n_{L1}+1}(x) \left(\bmod p_{n_{L1}+1}^{n_{L,p_{n_{L1}+1}}}\right) \\ \cdots\cdots\cdots\cdots\cdots\cdots\cdots\cdots\cdots\cdots \\ \pi(x) \left(\bmod p_{n_{L1}+n_{L2}}^{n_{L,p_{n_{L1}+n_{L2}}}}\right) = \pi_{n_{L1}+n_{L2}}(x) \left(\bmod p_{n_{L1}+n_{L2}}^{n_{L,p_{n_{L1}+n_{L2}}}}\right) \end{cases} \tag{3.110}$$

2. For $k = 1, \ldots, n_{L1}$, to get all PPs over $\mathbb{Z}_{p_k}$, if $p_k \geq 7$ one can use the following formula:
$$a\bar{\pi}(x+b), \quad \forall a \neq 0, b \in \mathbb{Z}_{p_k}, \tag{3.111}$$
where $\bar{\pi}(x)$ is any normalized PP $\pmod{p_k}$ from Table 3.7. If $p_k \in \{2, 3, 5\}$, Table 3.4 provides all PPs $\pmod{p_k}$.
3. For $k = n_{L1} + 1, \ldots, n_{L1} + n_{L2}$, to get all PPs over $\mathbb{Z}_{p_k^{n_{L,p_k}}}$, if $p_k \geq 7$ one can use the following formula:
$$a\bar{\pi}(x+b) + p \cdot \rho(x), \quad \forall a \neq 0, b \in \mathbb{Z}_{p_k}, \tag{3.112}$$
where $\bar{\pi}(x)$ is any normalized PP $\pmod{p_k}$ from Table 3.8 and $\rho(x)$ is any polynomial over $\mathbb{Z}_{p_k^{n_{L,p_k}-1}}$. If $p_k \in \{2, 3, 5\}$, Table 3.4 provides all PPs $\left(\bmod p_k^{n_{L,p_k}}\right)$.
4. Because the numbers $p_1, \ldots, p_{n_{L1}}, p_{n_{L1}+1}^{n_{L,p_{n_{L1}+1}}}, \ldots, p_{n_{L1}+n_{L2}}^{n_{L,p_{n_{L1}+n_{L2}}}}$, are pairwise coprime, after Steps 2 and 3, one can get all PPs over $\mathbb{Z}_L$ by the Chinese remainder theorem. Thus, if $q_{i,1}, q_{i,2}, \ldots, q_{i,n_{L1}}, q_{i,n_{L1}+1}, \ldots, q_{i,n_{L1}+n_{L2}}$, with $i \in \{1, 2, 3, 4, 5\}$, is a coefficients' set of PPs obtained in Steps 2 and 3, then the coefficient q_i of the resulting PP $\pmod{L}$ is given by the solution of the following system of simultaneous congruencies:
$$\begin{cases} q_i = q_{i,1} \pmod{p_1} \\ q_i = q_{i,2} \pmod{p_2} \\ \cdots\cdots\cdots\cdots\cdots\cdots \\ q_i = q_{i,n_{L1}} \pmod{p_{n_{L1}}} \\ q_i = q_{i,n_{L1}+1} \left(\bmod p_{n_{L1}+1}^{n_{L,p_{n_{L1}+1}}}\right) \\ \cdots\cdots\cdots\cdots\cdots\cdots\cdots\cdots \\ q_i = q_{i,n_{L1}+n_{L2}} \left(\bmod p_{n_{L1}+n_{L2}}^{n_{L,p_{n_{L1}+n_{L2}}}}\right) \end{cases} \tag{3.113}$$

At the end of this chapter we mention that the algorithm above gets all PPs of degree up to five without any coefficient ordering. If we want to see if a set of coefficients of a polynomial of degree no more than five determine a PP, then the coefficient conditions obtained in the previous sections of this chapter allow to directly decide in this matter. Using these coefficient conditions, the coefficients of PPs modulo any positive integer can be obtained in a desired order, which is tractable in computer processing.

References

Y.-L. Chen, J. Ryu, O.Y. Takeshita, A simple coefficient test for cubic permutation polynomials over integer rings. IEEE Commun. Lett. **10**(7), 549–551 (2006)

L.E. Dickson, The analytic representation of substitutions on a power of a prime number of letters with a discussion of the liner group. Ann. Math. **11**(1–6), 65–120 (1896)

L.E. Dickson, *Linear Groups: With an Exposition of the Galois Field Theory*, Dover Phoenix edn. (Dover, New York, 1901), https://ia801406.us.archive.org/22/items/lineargroupswith00dickuoft/lineargroupswith00dickuoft.pdf

G.H. Hardy, E.M. Wright, *An Introduction to the Theory of Numbers*, 4th edn. (Oxford University Press, Clarendon, 1975)

R. Lidl, G.L. Mullen, When does a polynomial over a finite field permute the elements of the field? Am. Math. Mon. **95**(3), 243–246 (1988)

G. Mullen, H. Stevens, Polynomial functions (mod m). Acta Math. Hung. **44**(3–4), 237–241 (1984)

W. Nöbauer, Über permutationspolynome und permutationsfunktionen für primzahlpotenzen. Mon. Math. **69**(3), 230–238 (1965)

R.L. Rivest, Permutation polynomials modulo 2w. Finite Fields Appl. **7**(2), 287–292 (2001)

J. Ryu, O.Y. Takeshita, On quadratic inverses for quadratic permutation polynomials over integers rings. IEEE Trans. Inf. Theory **52**(3), 1254–1260 (2006)

J. Sun, O.Y. Takeshita, Interleavers for turbo codes using permutation polynomial over integers rings. IEEE Trans. Inf. Theory **51**(1), 101–119 (2005)

J. Sun, O.Y. Takeshita, M.P. Fitz, Permutation polynomial based deterministic interleavers for turbo codes, in *IEEE International Symposium on Information Theory (ISIT)*, Yokohama, Japan, 29 June–4 July 2003, p. 319

O.Y. Takeshita, Maximum contention-free interleavers and permutation polynomials over integers rings. IEEE Trans. Inf. Theory **52**(3), 1249–1253 (2006)

O.Y. Takeshita, Permutation polynomial interleavers: an algebraic-geometric perspective. IEEE Trans. Inf. Theory **53**(6), 2116–2132 (2007)

L. Trifina, D. Tarniceriu, A coefficient test for fourth degree permutation polynomials over integer rings. AEÜ Int. J. Electron. Commun. **70**(11), 1565–1568 (2016)

L. Trifina, D. Tarniceriu, A coefficient test for quintic permutation polynomials over integer rings. IEEE Access **6**(2), 37893–37909 (2018). https://doi.org/10.1109/ACCESS.2018.2854373

G. Weng, C. Dong, A note on permutation polynomial over Zn. IEEE Trans. Inf. Theory **54**(9), 4388–4390 (2008)

H. Zhao, P. Fan, Simple method for generating mth-order permutation polynomials over integer rings. Electron. Lett. **43**(8), 449–451 (2007)

Chapter 4
Determining the Number of Permutation Polynomial-Based Interleavers in Terms of Their Length

4.1 Preliminaries

True PP interleavers are those that cannot be reduced to the same PP interleaver of smaller degree. This is of interest when we are looking for interleavers of a certain length.

These numbers may have a practical application in the search of PP interleavers for turbo codes, as follows. If we intend to optimize a PP interleaver of a certain length using a certain criterion, e.g., the distance spectrum of turbo codes, the optimization can be done among all possible interleavers, if their number is not too large. If the number of interleavers for a certain length is too large, a random search of a good interleaver or an optimization in a reduced search space, as in Tarniceriu et al. (2009, 2011), Trifina and Tarniceriu (2014, 2013), is the more practical solution.

The number of all true PP interleavers is given for at most degree 5. For degrees of PPs of 1, 2, and 3, we will firstly give explicit formulas for the number of all true different PPs, for each required prime factorization of interleaver length. For the true PP interleavers of degree from 3 to 5, we will also give and prove formulas for the number of such interleavers that fulfill only the sufficient conditions of Zhao and Fan from Sect. 3.10. In each case, according to the prime decomposition of the interleaver length, we identify the lengths for which the number of such true PPs is equal to zero. Some comments are made for the lengths used for QPP interleavers in the LTE standard (3GPP 2008).

The method used is based on the Chinese remainder theorem and is very simple to be applied.

Because we want to determine the true number of different PP interleavers of a certain degree we need equivalence conditions between PPs. They will be given in Sect. 4.2. In Sect. 4.3 we describe the method used, after which in the next subsections we apply this method and determine the number of all true different LPPs, QPPs and CPPs and then the number of true different CPPs, 4-PPs and 5-PPs under Zhao and Fan sufficient conditions.

L. Trifina and D. Tarniceriu, *Permutation Polynomial Interleavers for Turbo Codes*, Signals and Communication Technology,
https://doi.org/10.1007/978-981-13-2625-7_4

The number of PPs with degree at most six was addressed in Weng and Dong (2008), but this approach did not consider the equivalence conditions, nor the number of polynomials, separately, for each polynomial degree. At the end of this chapter, in Sect. 4.7, we will give an algorithm, based on the Weng and Dong algorithm given in Weng and Dong (2008), to compute the number of all true different PP interleavers of degree at most 5. The algorithm can be easily modified to compute the number of true different PPs under Zhao and Fan sufficient conditions. Reference Number of 1-5-PPs (2016) is a link to a file where we computed the number of true different LPPs, QPPs, CPPs, 4-PPs and 5-PPs, for any interleaver length $L \leq 100000$, using the mentioned algorithm. The number of true different LPPs, QPPs and CPPs determined with counting formulas given in Sects. 4.3–4.5 can be found in a file at the link in reference (Number of LPPs/QPPs/CPPs 2015) and they are the same as those from the file at the link in reference (Number of 1-5-PPs 2016).

4.2 The Equivalence Conditions Between Permutation Polynomials

The equivalence conditions between permutation polynomials modulo L involve the notion of NPs modulo L.

In the following we give some essential results regarding NPs from Li (2005), but with some different notations.

Definition 4.1 (*Definition 7 in* Li 2005) A polynomial $z(x) = \sum_{k=0}^{d} q_k x^k \pmod{L}$ of degree $d \geq 0$ is a NP of degree d modulo L if, $\forall x \in \mathbb{Z}, z(x) = 0 \pmod{L}$. Especially, $z(x) = 0$ is a trivial null polynomial of degree 0 modulo L.

Theorem 4.2 (Theorem 1 in Li 2005) *Assume $p_1, p_2, \ldots, p_{n_L}$ are n_L distinct prime numbers and $n_{L,p_1}, n_{L,p_2}, \ldots, n_{L,p_{n_L}} \geq 1$. A polynomial $z(x)$ is a NP modulo $L = \prod_{i=1}^{n_L} p_i^{n_{L,p_i}}$, if and only if $z(x)$ is a NP modulo $p_i^{n_{L,p_i}}$, $\forall i = \overline{1, n_L}$.*

Theorem 4.3 (Theorem 35 in Li 2005) *Let $\pi_1(x)$ and $\pi_2(x)$ denote two integer polynomials of any degrees modulo L, $\pi_1(x)$ is equivalent to $\pi_2(x)$, i.e. $\pi_1(x) \equiv \pi_2(x) \pmod{L}$ if and only if $\pi_1(x) - \pi_2(x) \pmod{L}$ is a NP modulo L.*

We mention that there are no non-trivial linear null polynomials (LNPs). This follows from the fact that the equation $q_1 \cdot x = 0 \pmod{L}$ has exactly $\gcd(q_1, L)$ solutions in variable x (Theorem 57 from Hardy and Wright 1975). For an LNP, the previous congruence equation must have L different solutions modulo L. Because $\gcd(q_1, L) = L$ only for $q_1 = 0, L, 2 \cdot L, \ldots$, i.e. for $q_1 \pmod{L} = 0$, only the trivial LNP results.

The necessary and sufficient conditions for quadratic null polynomials (QNPs) modulo L are given in the next theorem.

Theorem 4.4 (Theorem 5 in Zhao et al. 2010) $z(x) = q_0 + q_1 x + q_2 x^2 \pmod{L}$ *is a QNP different from the trivial* $z(x) = 0$ *if and only if* L *is an even integer,* $q_0 = 0$, *and* $q_1 = q_2 = L/2$.

Proof "$\Rightarrow$" To prove the sufficiency of the theorem we have to show that $\frac{L}{2} \cdot x + \frac{L}{2} \cdot x^2 = 0 \pmod{L}, \forall x \in \mathbb{Z}_L$. We can write $\frac{L}{2} \cdot x + \frac{L}{2} \cdot x^2 = \frac{L}{2} \cdot x \cdot (x+1)$. As $x \cdot (x+1)$ is the product of two successive numbers, it is always even and $\frac{x \cdot (x+1)}{2} \in \mathbb{N}$ and thus $\frac{L}{2} \cdot x \cdot (x+1) = 0 \pmod{L}, \forall x \in \mathbb{Z}_L$.

"$\Leftarrow$" To prove the necessity of the theorem, we assume that $z(x)$ is QNP. For $x = 0$, it follows that $z(0) = q_0 \pmod{L}$. From $z(0) = 0 \pmod{L}$, it follows that $q_0 = 0$. Therefore, from the condition $z(x) = 0 \pmod{L}, \forall x \in \mathbb{Z}_L^*$, it follows:

$$q_2 n^2 + q_1 n = 0 \pmod{L}, n = 1, 2, \ldots, L-1 \tag{4.1}$$

Summing up the equations in (4.1) for the first n natural numbers, we get the equivalent equations

$$q_2 \sum_{k=1}^{n} k^2 + q_1 \sum_{k=1}^{n} k = 0 \pmod{L}, n = 1, 2, \ldots, L-1 \tag{4.2}$$

For each $n = 1, 2, \ldots, L-1$, (4.2) is equivalent to

$$q_2 \cdot \frac{n(n+1)(2n+1)}{6} + q_1 \cdot \frac{n(n+1)}{2} = 0 \pmod{L} \tag{4.3}$$

or

$$\frac{n(n+1)}{2} \cdot \left(q_2 \cdot \frac{2n+1}{3} + q_1 \right) = 0 \pmod{L} \tag{4.4}$$

For $n = 1$ and $n = 2$, (4.4) becomes

$$q_2 + q_1 = 0 \pmod{L} \tag{4.5}$$

and

$$5q_2 + 3q_1 = 0 \pmod{L} \tag{4.6}$$

respectively.

Considering (4.5), (4.6) becomes

$$2q_2 = 0 \pmod{L} \tag{4.7}$$

According to Theorem 57 from Hardy and Wright (1975), Eq. (4.7) has solutions different from zero only if L is even. In this case $\gcd(2, L) = 2$ and the only solution different from zero of Eq. (4.7) is $q_2 = L/2$. Then, from (4.5), $q_1 = L/2$ follows naturally. Considering the sufficiency of the theorem, it is obvious that the solution $q_1 = q_2 = L/2$ satisfies all equations from (4.1) and thus also those from (4.4). In this way, the theorem is proved. ■

To prove the necessary and sufficient conditions for cubic null polynomials (CNPs) modulo L the next lemma is needed.

Lemma 4.5 *The sum of the first n natural numbers is:*

$$\frac{n(n+1)}{2} = 3m \text{ if } n = 3p \text{ or } n = 3p+2, \text{ with } m, p \in \mathbb{N} \tag{4.8}$$

or

$$\frac{n(n+1)}{2} = 3m+1 \text{ if } n = 3p+1, \text{ with } m, p \in \mathbb{N} \tag{4.9}$$

Proof If $n = 3p$, we have

$$\frac{n(n+1)}{2} = \frac{3p(3p+1)}{2} = 3m, m \in \mathbb{N} \tag{4.10}$$

because $p(3p+1)$ is an even number.

If $n = 3p+2$, we have

$$\frac{n(n+1)}{2} = \frac{(3p+2)(3p+3)}{2} = \frac{3(p+1)(3p+2)}{2} = 3m, m \in \mathbb{N} \tag{4.11}$$

because $(p+1)(3p+2)$ is an even number.

If $n = 3p+1$, we have

$$\frac{n(n+1)}{2} = \frac{(3p+1)(3p+2)}{2} = \frac{9p^2+9p+2}{2} = \frac{9p(p+1)}{2} + 1 =$$

$$= 3m+1, m \in \mathbb{N} \tag{4.12}$$

because $p(p+1)$ is an even number. ■

The necessary and sufficient conditions for CNPs modulo L are given in the next theorem from Trifina and Tarniceriu (2013) in a slightly changed form.

Theorem 4.6 *The third degree polynomial*

$$z(x) = q_0 + q_1 x + q_2 x^2 + q_3 x^3 \pmod{L}, x = 0, 1, \ldots, L-1 \tag{4.13}$$

is a CNP with $q_3 > 0$, if and only if $q_0 = 0$ and the coefficients q_1, q_2, q_3 are:

(a) if $2 \mid L$ *and* $3 \nmid L$ *those in cases (I) or (II)*

$$\text{(I)}\ q_1 = \frac{L}{2}, q_2 = 0, q_3 = \frac{L}{2}$$

$$\text{(II)}\ q_1 = 0, q_2 = \frac{L}{2}, q_3 = \frac{L}{2}$$

(b) if $2 \nmid L$ *and* $3 \mid L$ *those in cases (III) or (IV)*

$$\text{(III)}\ q_1 = \frac{2L}{3}, q_2 = 0, q_3 = \frac{L}{3}$$

$$\text{(IV)}\ q_1 = \frac{L}{3}, q_2 = 0, q_3 = \frac{2L}{3}$$

(c) if $6 \mid L$, *those in cases (I), (II), (III), (IV) or (V)–(X)*

$$\text{(V)}\ q_1 = \frac{5L}{6}, q_2 = 0, q_3 = \frac{L}{6}$$

$$\text{(VI)}\ q_1 = \frac{L}{3}, q_2 = \frac{L}{2}, q_3 = \frac{L}{6}$$

$$\text{(VII)}\ q_1 = \frac{L}{6}, q_2 = \frac{L}{2}, q_3 = \frac{L}{3}$$

$$\text{(VIII)}\ q_1 = \frac{5L}{6}, q_2 = \frac{L}{2}, q_3 = \frac{2L}{3}$$

$$\text{(IX)}\ q_1 = \frac{L}{6}, q_2 = 0, q_3 = \frac{5L}{6}$$

$$\text{(X)}\ q_1 = \frac{2L}{3}, q_2 = \frac{L}{2}, q_3 = \frac{5L}{6}$$

Proof The proof of the theorem is based on the idea in Zhao et al. (2010) used for QNP.

So, for a CNP with $q_0 = 0$, we must have

$$q_3 n^3 + q_2 n^2 + q_1 n = 0 \pmod{L}, n = 1, 2, \ldots, L-1 \tag{4.14}$$

Summing up the relations in (4.14) for the first n natural numbers, we have:

$$q_3 \sum_{k=1}^{n} k^3 + q_2 \sum_{k=1}^{n} k^2 + q_1 \sum_{k=1}^{n} k = 0 \pmod{L}, n = 1, 2, \ldots, L-1 \tag{4.15}$$

Equation (4.15) can be equivalently written as:

$$q_3 \cdot \left(\frac{n(n+1)}{2}\right)^2 + q_2 \cdot \frac{n(n+1)(2n+1)}{6} + q_1 \cdot \frac{n(n+1)}{2} =$$

$$= 0 \pmod{L}, n = 1, 2, \ldots, L-1 \tag{4.16}$$

or

$$\frac{n(n+1)}{2} \cdot \left(q_3 \cdot \frac{n(n+1)}{2} + q_2 \cdot \frac{2n+1}{3} + q_1 \right) = 0 \pmod{L},$$

$$n = 1, 2, \ldots, L-1 \tag{4.17}$$

" ⇒" In the following we prove the sufficiency for each of the cases I–X. Cases I and II follow directly because of the next relations:

$$\frac{L}{2}(x^3 + x) = \frac{L}{2}x(x^2+1) = 0 \pmod{L},\ x = 0, 1, \ldots, L-1 \tag{4.18}$$

$$\frac{L}{2}(x^3 + x^2) = \frac{L}{2}x^2(x+1) = 0 \pmod{L},\ x = 0, 1, \ldots, L-1 \tag{4.19}$$

They are true due to the fact that the numbers $x(x^2+1)$ and $x^2(x+1)$ are even, $\forall x \in \mathbb{N}$.

For cases III and IV, as $3 \mid L$, we consider $L = 3r, \forall r \in \mathbb{N}$. We have to check the condition in (4.17).

For case III:

- if condition (4.8) is fulfilled, the sum in (4.17) becomes

$$3m(r \cdot 3m + 2r) = 3rm(3m+2) \vdots (3r) \tag{4.20}$$

- if condition (4.9) is fulfilled, the sum in (4.17) becomes

$$(3m+1)(r \cdot (3m+1) + 2r) = 3r(m+1)(3m+2) \vdots (3r) \tag{4.21}$$

As $L = 3r$ the condition in (4.17) is fulfilled.

For case IV:

- if condition (4.8) is fulfilled, the sum in (4.17) becomes

$$3m(2r \cdot 3m + r) = 3rm(6m+1) \vdots (3r) \tag{4.22}$$

- if condition (4.9) is fulfilled, the sum in (4.17) becomes

$$(3m+1)(2r \cdot (3m+1) + r) = 3r(2m+1)(3m+1) \vdots (3r) \tag{4.23}$$

Thus, in this case the condition in (4.17) is fulfilled.

For cases V–X, as $6 \mid L$, we consider $L = 6r$, $r \in \mathbb{N}$. We have to check the condition in (4.17).

For case V:

- if condition (4.8) is fulfilled, the sum in (4.17) becomes

$$3m(r \cdot 3m + 5r) = 3rm(3m + 5) \vdots (6r) \tag{4.24}$$

because $m(3m + 5)$ is even.
- if condition (4.9) is fulfilled, the sum in (4.17) becomes

$$(3m + 1)(r \cdot (3m + 1) + 5r) = 3rm(3m + 1) + 6r(3m + 1) \vdots (6r) \tag{4.25}$$

because $m(3m + 1)$ is even.

For case VI:

- if condition (4.8) is fulfilled, the sum in (4.17) becomes

$$\begin{aligned} 3m(r \cdot 3m + r(2n + 1) + 2r) &= 9rm^2 + 6rmn + 9rm = \\ &= 9rm(m + 1) + 6rmn \vdots (6r) \end{aligned} \tag{4.26}$$

because $3 \mid 9$ and $m(m + 1)$ is even.
- if condition (4.9) is fulfilled, the sum in (4.17) becomes

$$\begin{aligned} &(3m + 1)(r \cdot (3m + 1) + r(2n + 1) + 2r) = \\ &= (3m + 1)(3rm + 2rn + 4r) = \\ &= 9rm^2 + 6rmn + 15rm + 2rn + 4r = \\ &= 6rmn + 3rm(3m + 5) + 2r(n + 2) = \\ &= 6rmn + 3rm(3m + 5) + 6r(p + 1) \vdots (6r) \end{aligned} \tag{4.27}$$

because $m(3m + 5)$ is even.

For case VII:

- if condition (4.8) is fulfilled, the sum in (4.17) becomes

$$\begin{aligned} 3m(2r \cdot 3m + r(2n + 1) + r) &= 18rm^2 + 6rmn + 6rm = \\ &= 6r(3m^2 + mn + m) \vdots (6r) \end{aligned} \tag{4.28}$$

- if condition (4.9) is fulfilled, the sum in (4.17) becomes

$$(3m+1)(2r\cdot(3m+1)+r(2n+1)+r)=$$
$$=(3m+1)(6rm+2rn+4r)=$$
$$=18rm^2+6rmn+18rm+2rn+4r=$$
$$=6rmn+18rm(m+1)+2r(n+2)=$$
$$=6rmn+18rm(m+1)+6r(p+1)=$$
$$=6r(mn+3rm(m+1)+p+1)\vdots(6r) \tag{4.29}$$

For case VIII:

- if condition (4.8) is fulfilled, the sum in (4.17) becomes

$$3m(4r\cdot 3m+r(2n+1)+5r)=36rm^2+6rmn+18rm=$$
$$=6r(6m^2+mn+3m)\vdots(6r) \tag{4.30}$$

- if condition (4.9) is fulfilled, the sum in (4.17) becomes

$$(3m+1)(4r\cdot(3m+1)+r(2n+1)+5r)=$$
$$=(3m+1)(12rm+2rn+10r)=$$
$$=36rm^2+6rmn+42rm+2rn+10r=$$
$$=6rmn+6rm(6m+7)+2r(n+5)=$$
$$=6rmn+6rm(6m+7)+6r(p+2)=$$
$$=6r(mn+m(6m+7)+p+2)\vdots(6r) \tag{4.31}$$

For case IX:

- if condition (4.8) is fulfilled, the sum in (4.17) becomes

$$3m(5r\cdot 3m+r)=3rm(15m+1)\vdots(6r) \tag{4.32}$$

because $m(15m+1)$ is even.

- if condition (4.9) is fulfilled, the sum in (4.17) becomes

$$(3m+1)(5r\cdot(3m+1)+r)=$$

$$=15rm(3m+1)+6r(3m+1)\vdots(6r) \tag{4.33}$$

because $3 \mid 15$ and $m(3m+1)$ is even.

For case X:

- if condition (4.8) is fulfilled, the sum in (4.17) becomes

$$3m(5r\cdot 3m+r(2n+1)+r)=45rm^2+6rmn+15rm=$$

$$=6rmn+15rm(3m+1)\vdots(6r) \tag{4.34}$$

because $3 \mid 15$ and $m(3m+1)$ is even.
- if condition (4.9) is fulfilled, the sum in (4.17) becomes

$$(3m+1)(5r\cdot(3m+1)+r(2n+1)+4r)=$$

$$=(3m+1)(15rm+2rn+10r)=$$

$$=45rm^2+6rmn+45rm+2rn+10r=$$

$$=6rmn+45rm(m+1)+2r(n+5)=$$

$$=6rmn+45rm(m+1)+6r(p+2)\vdots(6r) \tag{4.35}$$

because $3 \mid 45$ $m(m+1)$ is even (for this case $n=3p+1$).

In this way the sufficiency of the theorem is proved.

"$\Leftarrow$" In order to prove the necessity of the theorem, we write relation (4.17) for $n=1$, $n=2$ and $n=3$, obtaining

$$q_3+q_2+q_1=0 \pmod{L}, \tag{4.36}$$

$$9q_3+5q_2+3q_1=0 \pmod{L}, \tag{4.37}$$

$$36q_3+14q_2+6q_1=0 \pmod{L}. \tag{4.38}$$

Considering (4.36), relations (4.37) and (4.38) become:

$$6q_3+2q_2=0 \pmod{L}, \tag{4.39}$$

$$30q_3+8q_2=0 \pmod{L}. \tag{4.40}$$

Relations (4.39) and (4.40) are equivalent to

$$6q_3 + 2q_2 = k_1 L, k_1 \in \mathbb{N}^* \tag{4.41}$$

$$30q_3 + 8q_2 = k_2 L, k_2 \in \mathbb{N}^* \tag{4.42}$$

Multiplying (4.41) by 4 and subtracting it from (4.42), we get:

$$6q_3 = (k_2 - 4k_1)L \tag{4.43}$$

Obviously, from (4.41) and (4.42) we have $k_2 - 4k_1 > 0$, and, thus, (4.43) is equivalent to linear congruence equation:

$$6q_3 = 0 \ (\text{mod } L) \tag{4.44}$$

which has $\gcd(L, 6)$ distinct solutions modulo L (Theorem 57 in Hardy and Wright 1975). They are of the form:

$$q_3 = \frac{L \cdot i}{\gcd(L, 6)}, i = 0, 1, \ldots, \gcd(L, 6) - 1 \tag{4.45}$$

Considering (4.44), (4.39) becomes

$$2q_2 = 0 \ (\text{mod } L) \tag{4.46}$$

whose solutions are

$$q_2 = \frac{L \cdot i}{\gcd(L, 2)}, i = 0, 1, \ldots, \gcd(L, 2) - 1 \tag{4.47}$$

The solutions for q_1 are obtained from (4.36), taking into account (4.45) and (4.47). The fact that $q_0 = 0$ results from the condition $z(0) = 0 \ (\text{mod } L)$ and from (4.13) for $x = 0$. As $\gcd(L, 6)$ can take the values 1, 2, 3 or 6, and $\gcd(L, 2)$ can take the values 1 or 2, we see immediately that all solutions (different from zero) are those given in the theorem statement. Thus, the theorem is proven. ■

In Ryu (2007), Ryu and Takeshita (2011) Jonghoon Ryu gave the number and the general form of NPs of degree up to d. These are given in the next theorems.

Theorem 4.7 *The number of NPs modulo L of degree up to d is* $\prod_{k=1}^{d} \gcd(k!, L)$.

Theorem 4.8 *NPs of degree up to d are of the form*

$$z(x) = \sum_{k=1}^{d} \left\{ \frac{L}{\gcd(k!, L)} \cdot \tau_k \cdot \prod_{m=0}^{k-1} (x - m) \right\},$$

$$\text{where } 0 \le \tau_k \le \gcd(k!, L) - 1 \tag{4.48}$$

4.3 The Method Used for Determining the Number of True Different Permutation Polynomial-Based Interleavers Using the Chinese Remainder Theorem

The Chinese remainder theorem was given in Sect. 3.11 (Theorem 3.49). We recall that Theorem 3.49 ensures that if $n_1, n_2, \ldots, n_k$ are k positive integers ($k \in \mathbb{N}^*$) that are pairwise coprime and $(a_1, a_2, \ldots, a_k)$ is any given sequence of integers, then the following system of simultaneous congruencies

$$\begin{cases} x = a_1 \pmod{n_1} \\ x = a_2 \pmod{n_2} \\ \ldots\ldots\ldots\ldots\ldots\ldots \\ x = a_k \pmod{n_k} \end{cases} \tag{4.49}$$

has exactly one solution x modulo $N = n_1 \cdot n_2 \cdot \cdots \cdot n_k$. Theorem 3.49 also ensures that for two different sequences $a_1, a_2, \ldots, a_k$, exactly two distinct modulo $N = n_1 \cdot \cdots \cdot n_k$ solutions of system (4.49) exist.

Assume that $\pi(x)$ from (3.1) is a PP and the prime decomposition of L is $L = \prod_{j=1}^{n_L} p_j^{n_{L,p_j}}$. Let there be:

$$q_{i,j} = q_i \left(\text{mod } p_j^{n_{L,p_j}}\right), \forall j = \overline{1, n_L}, \forall i = \overline{1, d}. \tag{4.50}$$

Since the numbers $p_j^{n_{L,p_j}}$, $\forall j = \overline{1, n_L}$, are relatively prime to each other, from the Chinese remainder theorem it follows that for $\forall i = \overline{1, d}$, if we know the values $q_{i,j} \in \mathbb{Z}_{p_j^{n_{L,p_j}}}$, $\forall j = \overline{1, n_L}$, then there is a single number $q_i \in \mathbb{Z}_L$ that is precisely the coefficient q_i for the assumed PP.

Since $\left(\sum_{i=1}^{d} q_i \cdot x^i\right) \left(\text{mod } p_j^{n_{L,p_j}}\right) = \left(\sum_{i=1}^{d} q_{i,j} \cdot x^i\right) \left(\text{mod } p_j^{n_{L,p_j}}\right)$, from Theorem 3.8, it results that if the coefficients $q_{i,j} \in \mathbb{Z}_{p_j^{n_{L,p_j}}}$, $i = \overline{1, d}$, $j = \overline{1, n_L}$, are chosen so that the polynomials $\left(\sum_{i=1}^{d} q_{i,j} \cdot x^i\right) \left(\text{mod } p_j^{n_{L,p_j}}\right)$ are PPs modulo $p_j^{n_{L,p_j}}$, $\forall j = \overline{1, n_L}$, then the coefficients $q_i \in \mathbb{Z}_L$, $i = \overline{1, d}$, determine a modulo L PP. Therefore, we can determine the number of modulo L PPs in the following way:

(a) We decompose the interleaver length in prime factors
$L = \prod_{j=1}^{n_L} p_j^{n_{L,p_j}}$, so that $n_{L,p_j} \in \mathbb{N}^*$, $n_{L,p_j} \geq 1, \forall j = \overline{1, n_L}$.

(b) For any $j = \overline{1, n_L}$, we find all the coefficients $q_{i,j} \in \mathbb{Z}_{p_j^{n_{L,p_j}}}$, $i = \overline{1, d}$, so that the polynomial $\left(\sum_{i=1}^{d} q_{i,j} \cdot x^i\right) \left(\bmod\ p_j^{n_{L,p_j}}\right)$ is a PP modulo $p_j^{n_{L,p_j}}$. This can be done easily if we know the conditions the coefficients of the PP must meet depending on p_j and n_{L,p_j}. We calculate the number of such PPs modulo $p_j^{n_{L,p_j}}$ from the coefficient conditions.

(c) To determine the number of true different PPs, we must take into account the equivalence conditions imposed for PPs of degree d and considering $q_d \neq 0$. For QPPs and CPPs, these equivalence conditions are given in Theorems 4.3, 4.4 and 4.3, 4.6, respectively. The condition $q_d = 0$ is met only when $q_{d,j} = 0$, $\forall j = \overline{1, n_L}$. For the remaining number of coefficients $q_{i,j} \in \mathbb{Z}_{p_j^{n_{L,p_j}}}$, $j = \overline{1, n_L}$, we compute the total number of combinations that can lead to a PP modulo L. It will be the product of numbers of all coefficient combinations for $j = \overline{1, n_L}$.

In the next subsections, we apply the method described above for the case of QPPs and CPPs and determine the true number of different QPPs and CPPs, respectively, depending on the types of factors that appear in prime factor decomposition of L. As in Zhao et al. (2010), we denote by $\Phi(L)$ the Euler function, which is the number of numbers relatively prime with L, smaller than L. It is given by the following equation:

$$\Phi(L) = L \cdot \prod_{\substack{p \in \mathcal{P}, \\ p | L}} \left(1 - \frac{1}{p}\right), \tag{4.51}$$

where $\mathcal{P}$ is the set of prime numbers.

Considering Theorem 3.9, the number of LPPs is

$$C_{L,LPPs} = \Phi(L) \tag{4.52}$$

4.4 Determining the Number of All True Different Quadratic Permutation Polynomial-Based Interleavers

We mention that the equivalence condition of QPPs given in Theorems 4.3 and 4.4 requires that $q_2 < L/2$, when $2 \mid L$.

In the case of QPPs, we have three types of factors, as shown in Table 3.1. These are considered in the following and for each type of prime factor, the number of QPPs is determined. The prime factor 2 is considered the first one and the other prime factors are considered with arbitrary indices $j \geq 2$.

Case 1. (a) If $p = 2$ and $n_{L,2} = 1$, the coefficients $q_{i,1} \in \mathbb{Z}_2$, $i = \overline{1, 2}$. The condition $(q_{1,1} + q_{2,1}) \neq 0 \pmod 2$ is met for $q_{1,1} = 0$, $q_{2,1} = 1$ or $q_{1,1} = 1$, $q_{2,1} = 0$.

Since the two sets of coefficients lead to equivalent QPPs, from the two combinations only one must be kept, for example $q_{1,1} = 1$, $q_{2,1} = 0$, combined with other prime factors.

Case 1. (b) If $p = 2$ and $n_{L,2} > 1$, the coefficients $q_{i,1} \in \mathbb{Z}_{2^{n_{L,2}}}$, $i = \overline{1,2}$. The condition $q_{1,1} \neq 0 \pmod 2$ is met for $\Phi(2^{n_{L,2}}) = 2^{n_{L,2}-1}$ coefficients. The condition $q_{2,1} = 0 \pmod 2$ is met for $\dfrac{2^{n_{L,2}}}{2} = 2^{n_{L,2}-1}$ coefficients, out of which one is zero. In this case, from the equivalence conditions of QPPs it results that all values greater than or equal to $\dfrac{2^{n_{L,2}}}{2} = 2^{n_{L,2}-1}$ have to be removed, i.e. $\dfrac{2^{n_{L,2}-1}}{2} = 2^{n_{L,2}-2}$ values, leading to $\dfrac{2^{n_{L,2}-1}}{2} = 2^{n_{L,2}-2}$ values for $q_{2,1}$, out of which one is zero.

Case 2. If $p_j > 2$ and $n_{L,p_j} \geq 1$, the coefficients $q_{i,j} \in \mathbb{Z}_{p_j^{n_{L,p_j}}}$, $i = \overline{1,2}$. The condition $q_{1,j} \neq 0 \pmod{p_j}$ is met for $\Phi(p_j^{n_{L,p_j}}) = p_j^{n_{L,p_j}-1} \cdot (p_j - 1)$ coefficients. The condition $q_{2,j} = 0 \pmod{p_j}$ is met for $\dfrac{p_j^{n_{L,p_j}}}{p_j} = p_j^{n_{L,p_j}-1}$ coefficients, from which one is zero.

We apply the method described in Sect. 4.3 and distinguish three situations for the prime decomposition of L. The results regarding the number of all true different QPPs are given in Theorems 4.9–4.11 below and are summarized in Table 4.1 at the end of this section.

Table 4.1 Number of all true different QPP-based interleavers

Case	Decomposition of L	$C_{L,QPPs}$	Theorems
(1)	$L = \prod_{j=1}^{n_L} p_j^{n_{L,p_j}}$, $p_j > 2$ and $n_{L,p_j} \geq 1, \forall j = \overline{1, n_L}$	$C_{L,QPPs} = \prod_{j=1}^{n_L} \left(p_j^{n_{L,p_j}-1} \cdot (p_j - 1)\right) \cdot \left(\prod_{j=1}^{n_L} p_j^{n_{L,p_j}-1} - 1\right)$	4.9
(2)	$L = 2^{n_{L,2}} \cdot \prod_{j=2}^{n_L} p_j^{n_{L,p_j}}$, $n_{L,2} > 1$, $p_j > 2$ and $n_{L,p_j} \geq 1, \forall j = \overline{2, n_L}$	$C_{L,QPPs} = 2^{n_{L,2}-1} \cdot \prod_{j=2}^{n_L} \left(p_j^{n_{L,p_j}-1} \cdot (p_j - 1)\right) \cdot \left(2^{n_{L,2}-2} \cdot \prod_{j=2}^{n_L} p_j^{n_{L,p_j}-1} - 1\right)$	4.10
(3)	$L = 2 \cdot \prod_{j=2}^{n_L} p_j^{n_{L,p_j}}$, $p_j > 2$ and $n_{L,p_j} \geq 1, \forall j = \overline{2, n_L}$	$C_{L,QPPs} = \prod_{j=2}^{n_L} \left(p_j^{n_{L,p_j}-1} \cdot (p_j - 1)\right) \cdot \left(\prod_{j=2}^{n_L} p_j^{n_{L,p_j}-1} - 1\right)$	4.11

Theorem 4.9 *If* $2 \nmid L$, *i.e.* $L = \prod_{j=1}^{n_L} p_j^{n_{L,p_j}}$, *with* $p_j > 2$ *and* $n_{L,p_j} \geq 1, \forall j = \overline{1, n_L}$, *the number of QPPs will be equal to:*

$$C_{L,QPPs} = \prod_{j=1}^{n_L} \left(p_j^{n_{L,p_j}-1} \cdot (p_j - 1) \right) \cdot \left(\prod_{j=1}^{n_L} p_j^{n_{L,p_j}-1} - 1 \right) \tag{4.53}$$

Proof From case 2 above, the number of possible combinations for the coefficient q_1 results equal to $\prod_{j=1}^{n_L} \left(p_j^{n_{L,p_j}-1} \cdot (p_j - 1) \right)$ and the number of coefficients q_2 is equal to $\prod_{j=1}^{n_L} p_j^{n_{L,p_j}-1}$. The value $q_2 = 0$ results only when $q_{2,j} = 0, \forall j = \overline{1, n_L}$, that is for only one combination of the coefficients $q_{2,j}$, $j = \overline{1, n_L}$, which has to be removed. The number of QPPs will be that in (4.53). ■

Equation (4.53) is equivalent to Theorem 6, case (a), from Zhao et al. (2010).
From (4.53) we see that the number of QPPs is equal to 0, when the interleaver length is a product of prime numbers greater than 2, each of them to the power of 1.

Theorem 4.10 *If* $4 \mid L$, *i.e.* $L = 2^{n_{L,2}} \cdot \prod_{j=2}^{n_L} p_j^{n_{L,p_j}}$, *with* $n_{L,2} > 1$, $p_j > 2$ *and* $n_{L,p_j} \geq 1, \forall j = \overline{2, n_L}$, *the number of QPPs will be equal to:*

$$C_{L,QPPs} = 2^{n_{L,2}-1} \cdot \prod_{j=2}^{n_L} \left(p_j^{n_{L,p_j}-1} \cdot (p_j - 1) \right) \cdot \left(2^{n_{L,2}-2} \cdot \prod_{j=2}^{n_L} p_j^{n_{L,p_j}-1} - 1 \right) \tag{4.54}$$

Proof From cases 1.b and 2 above, the number of possible combinations for coefficient q_1 is equal to $2^{n_{L,2}-1} \cdot \prod_{j=2}^{n_L} \left(p_j^{n_{L,p_j}-1} \cdot (p_j - 1) \right)$ and the number of coefficients q_2 is equal to $2^{n_{L,2}-1} \cdot \prod_{j=2}^{n_L} p_j^{n_{L,p_j}-1}$. Since from the equivalence condition of QPPs we must have $q_2 < L/2$, a number of $\frac{1}{2} \cdot 2^{n_{L,2}-1} \cdot \prod_{j=2}^{n_L} p_j^{n_{L,p_j}-1} = 2^{n_{L,2}-2} \cdot \prod_{j=2}^{n_L} p_j^{n_{L,p_j}-1}$ coefficients remain, from which the value $q_2 = 0$ has to be removed, finally remaining $2^{n_{L,2}-2} \cdot \prod_{j=2}^{n_L} p_j^{n_{L,p_j}-1} - 1$ values for true different QPPs. The number of QPPs will be that in (4.54). ■

Equation (4.54) is equivalent to Theorem 6, case (b), from Zhao et al. (2010).

From (4.54), we see that the number of QPPs is equal to 0, when the interleaver length is a multiple of 4 of a product of prime numbers greater than 2, each of them to the power of 1.

Theorem 4.11 *If* $2 \mid L$ *and* $4 \nmid L$, *i.e.* $L = 2 \cdot \prod_{j=2}^{n_L} p_j^{n_{L,p_j}}$, *with* $p_j > 2$ *and* $n_{L,p_j} \geq 1$, $\forall j = \overline{2, n_L}$, *the number of QPPs will be equal to:*

$$C_{L,QPPs} = \prod_{j=2}^{n_L} \left(p_j^{n_{L,p_j}-1} \cdot (p_j - 1) \right) \cdot \left(\prod_{j=2}^{n_L} p_j^{n_{L,p_j}-1} - 1 \right) \tag{4.55}$$

Proof From cases 1. (a) and 2 above, the number of possible combinations for coefficient q_1 is equal to $\prod_{j=2}^{n_L} \left(p_j^{n_{L,p_j}-1} \cdot (p_j - 1) \right)$ and the number of coefficients q_2 is equal to $\prod_{j=2}^{n_L} p_j^{n_{L,p_j}-1}$. We mention that in this case, the $\prod_{j=2}^{n_L} p_j^{n_{L,p_j}-1}$ coefficients q_2 have $\prod_{j=2}^{n_L} p_j^{n_{L,p_j}-1}$ modulo L different values, all smaller than $L/2$. Therefore, we have only to remove the value $q_2 = 0$, finally leading to $\prod_{j=2}^{n_L} p_j^{n_{L,p_j}-1} - 1$ values for the coefficient q_2 of true different QPPs. The number of QPPs will be that in (4.55). ■

Equation (4.55) is equivalent to Theorem 6, case (c), from Zhao et al. (2010).

From (4.55) we see that the number of QPPs is equal to 0, when the interleaver length is a multiple of 2 of a product of prime numbers greater than 2, each of them to the power of 1.

From Theorems 4.9–4.11, we conclude that the number of true different QPPs is 0, when the interleaver length is

$$L = 2^{n_{L,2}} \cdot \prod_{j=2}^{n_L} p_j, \text{ with } n_{L,2} = \overline{0, 2},\ p_j > 2, \forall j = \overline{2, n_L} \tag{4.56}$$

Such lengths have to be avoided in designing QPP-based interleavers.

4.5 Determining the Number of All True Different Cubic Permutation Polynomial-Based Interleavers

We note that from the equivalence conditions for CPPs given in Theorems 4.3 and 4.6, we must have:

- $q_2 < L/2$ and $q_3 < L/2$, when $2 \mid L$ and $3 \nmid L$.
- $q_3 < L/3$, when $3 \mid L$ and $2 \nmid L$.
- $q_2 < L/2$ and $q_3 < L/6$, when $6 \mid L$.

In the case of CPPs, there are four types of prime factors, as shown in Table 3.2, each of them with two distinct sets of powers of the prime number. As the conditions for cases 3. (a), 3. (b) and 4. (b) are the same, they are analyzed together. We determine the number of CPPs for each type of prime factor. The prime factor 2 is considered the first one, the prime factor 3 is considered the second one and the other prime factors are considered with arbitrary indices $j \geq 3$.

Case 1. (a) If $p = 2$ and $n_{L,2} = 1$, the coefficients $q_{i,1} \in \mathbb{Z}_2$, $i = \overline{1,3}$. The condition $(q_{1,1} + q_{2,1} + q_{3,1}) \neq 0 \pmod 2$ is met for the following coefficient combinations $q_{i,1} \in \mathbb{Z}_2$, $i = \overline{1,3}$: $q_{1,1} = 0, q_{2,1} = 0, q_{3,1} = 1$ or $q_{1,1} = 0, q_{2,1} = 1, q_{3,1} = 0$ or $q_{1,1} = 1, q_{2,1} = 0, q_{3,1} = 0$ or $q_{1,1} = 1, q_{2,1} = 1, q_{3,1} = 1$. Since the four sets of coefficients lead to equivalent CPPs, only one must be kept in combination with other types of prime factors.

Case 1. (b) If $p = 2$ and $n_{L,2} > 1$, the coefficients $q_{i,1} \in \mathbb{Z}_{2^{n_{L,2}}}$, $i = \overline{1,3}$. The condition $q_{1,1} \neq 0 \pmod 2$ is met for $\Phi(2^{n_{L,2}}) = 2^{n_{L,2}-1}$ coefficients. The condition $q_{2,1} = 0 \pmod 2$ or $q_{3,1} = 0 \pmod 2$ is met for $\dfrac{2^{n_{L,2}}}{2} = 2^{n_{L,2}-1}$ coefficients. From the equivalence conditions of CPPs for $2 \mid L$ and $3 \nmid L$, it results that from the $2^{n_{L,2}-1}$ values of $q_{2,1}$ and $q_{3,1}$ only $\dfrac{2^{n_{L,2}-1}}{2} = 2^{n_{L,2}-2}$ lead to different permutations.

Case 2. (a) If $p = 3$ and $n_{L,3} = 1$, the coefficients $q_{i,2} \in \mathbb{Z}_3$, $i = \overline{1,3}$. The condition $(q_{1,2} + q_{3,2}) \neq 0 \pmod 3$ is met for the following coefficient combinations $(q_{1,2}, q_{3,2})$: $q_{1,2} = 0, q_{3,2} = 1$ or $q_{1,2} = 0, q_{3,2} = 2$ or $q_{1,2} = 1, q_{3,2} = 0$ or $q_{1,2} = 1$, $q_{3,2} = 1$ or $q_{1,2} = 2, q_{3,2} = 0$ or $q_{1,2} = 2, q_{3,2} = 2$. The condition $q_{2,2} = 0 \pmod 3$ is met only for $q_{2,2} = 0$. From the equivalence conditions of CPPs for $2 \nmid L$ and $3 \mid L$, it results that the six sets of coefficients $q_{i,2}$, $i = \overline{1,3}$, lead only to two distinct permutations. The corresponding two sets of coefficients can be considered $q_{1,2} = 1$, $q_{2,2} = 0$, $q_{3,2} = 0$, and $q_{1,2} = 2$, $q_{2,2} = 0$, $q_{3,2} = 0$. Because for these two sets we have $q_{3,2} = q_{2,2} = 0$, only $q_{1,2}$ being different, we must consider two coefficients for $q_{1,2}$ and only one for $q_{2,2}$ and $q_{3,2}$, respectively, in combination with other prime factors.

Case 2. (b) If $p = 3$ and $n_{L,3} > 1$, the coefficients $q_{i,2} \in \mathbb{Z}_{3^{n_{L,3}}}$, $i = \overline{1,3}$. The condition $q_{2,2} = 0 \pmod 3$ is met for $3^{n_{L,3}-1}$ values. The condition $q_{1,2} \neq 0 \pmod 3$ is met for $\Phi(3^{n_{L,3}}) = 2 \cdot 3^{n_{L,3}-1}$ values. The set of values for $q_{1,2}$ is $\{1, 2, 4, 5, 7, 8, \ldots, 3^{n_{L,3}} - 2, 3^{n_{L,3}} - 1\}$, out of which $3^{n_{L,3}-1}$ values are equal to 1 modulo 3 and also $3^{n_{L,3}-1}$ values are equal to 2 modulo 3. As $q_{3,2} \in \mathbb{Z}_{3^{n_{L,3}}}$, the condition $(q_{1,2} + q_{3,2}) \neq 0 \pmod 3$, for a fixed value of $q_{1,2}$, will be fulfilled for $3^{n_{L,3}-1} + 3^{n_{L,3}-1} = 2 \cdot 3^{n_{L,3}-1}$ coefficients $q_{3,2}$. However, when $3^{n_{L,3}-1}$ is multiple of 3, from the equivalence conditions of CPPs for $2 \nmid L$ and $3 \mid L$, it results that of the $2 \cdot 3^{n_{L,3}-1}$ coefficients $q_{3,2}$ only $\frac{1}{3} \cdot 2 \cdot 3^{n_{L,3}-1} = 2 \cdot 3^{n_{L,3}-2}$ lead to distinct permutations.

Cases 3. (a), 3. (b), 4. (b) If $p_j > 3$ and $n_{L,p_j} \geq 1$ when $p_j = 3k+1$, $k \in \mathbb{N}$ and $n_{L,p_j} > 1$ when $p_j = 3k+2$, $k \in \mathbb{N}$, the coefficients $q_{i,j} \in \mathbb{Z}_{p_j^{n_{L,p_j}}}$, $i = \overline{1,3}$. The condition $q_{1,j} \neq 0 \pmod{p_j}$ is met for $\Phi(p_j^{n_{L,p_j}}) = p_j^{n_{L,p_j}-1} \cdot (p_j - 1)$ coefficients. The condition $q_{2,j} = 0 \pmod{p_j}$ or $q_{3,j} = 0 \pmod{p_j}$ is met for $\dfrac{p_j^{n_{L,p_j}}}{p_j} = p_j^{n_{L,p_j}-1}$ coefficients, out of which one is zero.

Case 4. (a) If $3 \nmid (p_j - 1)$, $p_j > 3$ and $n_{L,p_j} = 1$, the coefficients $q_{i,j} \in \mathbb{Z}_{p_j}$, $i = \overline{1,3}$. The condition $q_{1,j} \neq 0 \pmod{p_j}$ is met for $\Phi(p_j) = p_j - 1$ coefficients. The condition $q_{2,j} = 0 \pmod{p_j}$ or $q_{3,j} = 0 \pmod{p_j}$ is obviously met only for the value zero. When $q_{3,j} \neq 0 \pmod{p_j}$ (for $\Phi(p_j) = p_j - 1$ values), the condition $q_{2,j}^2 = 3q_{1,j}q_{3,j} \pmod{p_j}$ has to be fulfilled. This congruence equation, for fixed $q_{2,j}$ and $q_{3,j}$, has only one solution modulo p_j in the variable $q_{1,j}$ (Hardy and Wright 1975). Therefore, by considering all the p_j possible values for $q_{2,j}$, a number of $p_j \cdot (p_j - 1)$ coefficient combinations $q_{i,j} \in \mathbb{Z}_{p_j}$, $i = \overline{1,3}$, results. The coefficients verify the condition $q_{2,j}^2 = 3q_{1,j}q_{3,j} \pmod{p_j}$.

For this case, it is useful to see how many coefficient combinations result when the factorization of L contains a product of factors of type 4. (a). The factors of type 4. (a) will be considered with the last $n_{4a} \in \mathbb{N}^*$ indices in writing of L as a product of prime factors. Thus, L will be of the form $L = \prod_{j=1}^{n_L - n_{4a}} p_j^{n_{L,p_j}} \cdot \prod_{j=n_L-n_{4a}+1}^{n_L} p_j$, with $n_L \in \mathbb{N}^*$, $n_L \geq n_{4a}$ (if $n_L = n_{4a}$ the first product in L is considered equal to 1), $p_j \geq 2$ and $n_{L,p_j} \geq 1$, $p_j \neq 3k+2$, $k \in \mathbb{N}^*$, if $n_{L,p_j} = 1$, $\forall j = \overline{1, n_L - n_{4a}}$, and $p_j = 3k+2$, $k \in \mathbb{N}^*$, $\forall j = \overline{n_L - n_{4a} + 1, n_L}$. We denote by $\prod_{j=n_L-n_{4a}+1}^{n_L} p_j = L_{(4a)}$ the product from the factorization of L consisting only of factors of type 4. (a). The product of the remaining factors from the decomposition of L will be $\prod_{j=1}^{n_L-n_{4a}} p_j^{n_{L,p_j}} = L/L_{(4a)}$.

The conditions $q_{1,j} \neq 0 \pmod{p_j}$, $q_{2,j} = 0 \pmod{p_j}$ and $q_{3,j} = 0 \pmod{p_j}$ have to be considered for each group of $n_{4a,0}$ prime factors of type 4. (a), with $n_{4a,0} = \overline{1, n_{4a}}$. We denote by $I_{n_{4a}} = \{n_L - n_{4a} + 1, n_L - n_{4a} + 2, \ldots, n_L\}$ the set of indices corresponding to those n_{4a} prime factors.

We firstly consider the case of groups consisting only of one prime factor, p_{j_1}, with $j_1 \in I_{n_{4a}}$. Thus, if $q_{3,j_1} = 0 \pmod{p_{j_1}}$, the following conditions must be met $q_{1,j_1} \neq 0 \pmod{p_{j_1}}$ and $q_{2,j_1} = 0 \pmod{p_{j_1}}$. For $\forall j \in I_{n_{4a}}$, $j \neq j_1$ we have $q_{3,j_1} \neq 0 \pmod{p_{j_1}}$, and the condition $q_{2,j}^2 = 3q_{1,j}q_{3,j} \pmod{p_j}$, $\forall j \in I_{n_{4a}}$, $j \neq j_1$

must be met. The first set of conditions is met for $p_{j_1} - 1$ coefficients q_1 and a single value for q_2 and q_3, respectively, which is zero. The second set of conditions is met for $\prod_{\substack{j=n_L-n_{4a}+1,\\ j\neq j_1}}^{n_L} (p_j - 1)$ coefficients q_3, and the congruence equation has one solution in the variable $q_{1,j}$, for each of the $\prod_{\substack{j=n_L-n_{4a}+1,\\ j\neq j_1}}^{n_L} p_j$ coefficients q_2, and the $\prod_{\substack{j=n_L-n_{4a}+1,\\ j\neq j_1}}^{n_L} (p_j - 1)$ coefficients q_3. Therefore, in total, for the groups consisting of one factor p_{j_1}, for which $q_{3,j_1} = 0 \pmod{p_{j_1}}$, we have $(p_{j_1} - 1) \cdot \prod_{\substack{j=n_L-n_{4a}+1,\\ j\neq j_1}}^{n_L} p_j \cdot \prod_{\substack{j=n_L-n_{4a}+1,\\ j\neq j_1}}^{n_L} (p_j - 1) = \prod_{\substack{j=n_L-n_{4a}+1,\\ j\neq j_1}}^{n_L} p_j \cdot \prod_{j=n_L-n_{4a}+1}^{n_L} (p_j - 1)$ combinations of coefficients q_i, $i = \overline{1, 3}$.

In the following, we consider the case of groups consisting of two prime factors, p_{j_1} and p_{j_2}, with $j_1, j_2 \in I_{n_{4a}}$ and $j_1 \neq j_2$. Thus, if $q_{3,j} = 0 \pmod{p_j}$, $\forall j \in \{j_1, j_2\}$ the conditions $q_{1,j} \neq 0 \pmod{p_j}$ and $q_{2,j} = 0 \pmod{p_j}$ must be met for $j \in \{j_1, j_2\}$. For $\forall j \in I_{n_{4a}}$, $j \neq j_1$ and $j \neq j_2$, we have $q_{3,j} \neq 0 \pmod{p_j}$, and the condition $q_{2,j}^2 = 3q_{1,j}q_{3,j} \pmod{p_j}$, $\forall j \in I_{n_{4a}}$, $j \neq j_1$ and $j \neq j_2$, must be met. The first set of conditions is met for $(p_{j_1} - 1) \cdot (p_{j_2} - 1)$ coefficients q_1 and a single value for q_2 and q_3, respectively, which is zero. The second set of conditions is met for $\prod_{\substack{j=n_L-n_{4a}+1,\\ j\neq j_1, j\neq j_2}}^{n_L} (p_j - 1)$ coefficients q_3, and the congruence equation has one solution in the variable $q_{1,j}$, for each of the $\prod_{\substack{j=n_L-n_{4a}+1,\\ j\neq j_1, j\neq j_2}}^{n_L} p_j$ coefficients q_2 and the $\prod_{\substack{j=n_L-n_{4a}+1,\\ j\neq j_1, j\neq j_2}}^{n_L} (p_j - 1)$ coefficients q_3. Thus, in total, for the groups consisting of two factors, p_{j_1} and p_{j_2}, for which $q_{3,j} = 0 \pmod{p_j}$, $\forall j \in \{j_1, j_2\}$, we have $(p_{j_1} - 1) \cdot (p_{j_2} - 1) \cdot \prod_{\substack{j=n_L-n_{4a}+1,\\ j\neq j_1, j\neq j_2}}^{n_L} p_j \cdot \prod_{\substack{j=n_L-n_{4a}+1,\\ j\neq j_1, j\neq j_2}}^{n_L} (p_j - 1) = \prod_{\substack{j=n_L-n_{4a}+1,\\ j\neq j_1, j\neq j_2}}^{n_L} p_j \cdot \prod_{j=n_L-n_{4a}+1}^{n_L} (p_j - 1)$ combinations of coefficients q_i, $i = \overline{1, 3}$.

Now, let there be the set $I_{n_{4a,0}} \subseteq I_{n_{4a}}$, with $1 \leq n_{4a,0} \leq n_{4a}$, consisting of $n_{4a,0}$ different indices (the notation 0 derives from the fact that $q_{3,j} = 0 \pmod{p_j}$, for $j \in I_{n_{4a,0}}$). Then, if there are groups of $n_{4a,0}$ prime factors of type 4. (a), it means that the following conditions have to be met: $q_{1,j} \neq 0 \pmod{p_j}$, $q_{2,j} = 0 \pmod{p_j}$, $q_{3,j} = 0 \pmod{p_j}$, $\forall j \in I_{n_{4a,0}}$, and $q_{2,j}^2 = 3q_{1,j}q_{3,j} \pmod{p_j}$, $q_{3,j} \neq 0 \pmod{p_j}$, $\forall j \in \left\{I_{n_{4a}} - I_{n_{4a,0}}\right\}$. The first set of conditions is met for $\prod_{j \in I_{n_{4a,0}}} (p_j - 1)$ coefficients q_1 and one value for q_2 and q_3, respectively, which is zero. The second set of conditions is met for $\prod_{j \in \{I_{n_{4a}} - I_{n_{4a,0}}\}} (p_j - 1)$ coefficients q_3, and the congruence equation has one solution in the variable $q_{1,j}$, for each of the $\prod_{j \in \{I_{n_{4a}} - I_{n_{4a,0}}\}} p_j$ coefficients q_2 and the $\prod_{j \in \{I_{n_{4a}} - I_{n_{4a,0}}\}} (p_j - 1)$ coefficients q_3. Thus, in total, for groups consisting of $n_{4a,0}$ factors, for which $q_{3,j} = 0 \pmod{p_j}$, $\forall j \in I_{n_{4a,0}}$, we will have $\prod_{j \in I_{n_{4a,0}}} (p_j - 1) \cdot \prod_{j \in \{I_{n_{4a}} - I_{n_{4a,0}}\}} p_j \cdot \prod_{j \in \{I_{n_{4a}} - I_{n_{4a,0}}\}} (p_j - 1) = \prod_{j \in I_{n_{4a}}} (p_j - 1) \cdot \prod_{j \in \{I_{n_{4a}} - I_{n_{4a,0}}\}} p_j = \prod_{j=n_L-n_{4a}+1}^{n_L} (p_j - 1) \cdot \prod_{j \in \{I_{n_{4a}} - I_{n_{4a,0}}\}} p_j$ combinations of coefficients q_i, $i = \overline{1,3}$. If $I_{n_{4a,0}} = I_{n_{4a}}$, that is $n_{4a,0} = n_{4a}$, we have $\prod_{j \in I_{n_{4a}}} (p_j - 1) = \prod_{j=n_L-n_{4a}+1}^{n_L} (p_j - 1)$ coefficients q_1 and one coefficient q_2 and q_3, respectively (namely, the value zero that will be removed from the combinations with other prime factors).

The total number of combinations of coefficients q_i, $i = \overline{1,3}$, for all groups consisting of $n_{4a,0}$ factors, with $n_{4a,0} = \overline{1, n_{4a}}$, for which $q_{3,j} = 0 \pmod{p_j}$, $\forall j \in I_{n_{4a,0}}$, is equal to $\prod_{j=n_L-n_{4a}+1}^{n_L} (p_j - 1) + \sum_{n_{4a,0}=1}^{n_{4a}-1} \left(\sum_{I_{n_{4a,0}} \subset I_{n_{4a}}} \left(\prod_{j=n_L-n_{4a}+1}^{n_L} (p_j - 1) \cdot \prod_{j \in \{I_{n_{4a}} - I_{n_{4a,0}}\}} p_j \right)\right)$, where the sum $\sum_{I_{n_{4a,0}} \subset I_{n_{4a}}} \left(\prod_{j=n_L-n_{4a}+1}^{n_L} (p_j - 1) \cdot \prod_{j \in \{I_{n_{4a}} - I_{n_{4a,0}}\}} p_j \right)$ is over all different sets of distinct indices $I_{n_{4a,0}} \subset I_{n_{4a}}$.

Since the above expression is hard to read, in the following we will give more detailed formulas for $n_{4a} = 2$, $n_{4a} = 3$, and $n_{4a} = 4$, respectively (for $n_{4a} = 1$ the formula is direct), because the smallest number containing at least four factors of type 4. (a) is $5 \cdot 11 \cdot 17 \cdot 23 = 21505$, which is a large enough value for an interleaver length. We specify that the smallest number containing at least five factors of type 4. (a) is $21505 \cdot 29 = 623645$, which is a value too large for an interleaver length.

The number of combinations of coefficients q_i, $i = \overline{1,3}$, resulted for the groups consisting only of $n_{4a,0} = r_0 \in \mathbb{N}^*$ factors from the product $L_{(4a)}$, where $n_{4a,0} \in \{1, 2, \ldots, n_{4a} - 1\}$, with some $n_{4a} = r \in \mathbb{N}^*$, for which $q_{3,j} = 0 \pmod{p_j}$, $\forall j \in I_{n_{4a,0}}$, will be denoted by $C_{L_{(4a)}, n_{4a}=r, n_{4a,0}=r_0}$, where $r_0 < r$. As noted above, for some $n_{4a} \in \mathbb{N}^*$, we have

$$C_{L_{(4a)},n_{4a}=r,n_{4a,0}=r_0} = \sum_{I_{r_0} \subset I_r} \left(\prod_{j=n_L-n_{4a}+1}^{n_L} (p_j - 1) \cdot \prod_{j \in \{I_r - I_{r_0}\}} p_j \right) \tag{4.57}$$

For $r \in \{2, 3, 4\}$, the values $C_{L_{(4a)},n_{4a}=r,n_{4a,0}=r_0}$ are written in detail as

$$C_{L_{(4a)},n_{4a}=2,n_{4a,0}=1} = (p_{n_L-1} - 1) \cdot p_{n_L} \cdot (p_{n_L} - 1)+$$

$$+(p_{n_L} - 1) \cdot p_{n_L-1} \cdot (p_{n_L-1} - 1) \tag{4.58}$$

$$C_{L_{(4a)},n_{4a}=3,n_{4a,0}=1} = (p_{n_L-2} - 1) \cdot p_{n_L-1} \cdot p_{n_L} \cdot (p_{n_L-1} - 1) \cdot (p_{n_L} - 1)+$$

$$+(p_{n_L-1} - 1) \cdot p_{n_L-2} \cdot p_{n_L} \cdot (p_{n_L-2} - 1) \cdot (p_{n_L} - 1)+$$

$$+(p_{n_L} - 1) \cdot p_{n_L-2} \cdot p_{n_L-1} \cdot (p_{n_L-2} - 1) \cdot (p_{n_L-1} - 1) \tag{4.59}$$

$$C_{L_{(4a)},n_{4a}=3,n_{4a,0}=2} = (p_{n_L-2} - 1) \cdot (p_{n_L-1} - 1) \cdot p_{n_L} \cdot (p_{n_L} - 1)+$$

$$+(p_{n_L-2} - 1) \cdot (p_{n_L} - 1) \cdot p_{n_L-1} \cdot (p_{n_L-1} - 1)+$$

$$+(p_{n_L-1} - 1) \cdot (p_{n_L} - 1) \cdot p_{n_L-2} \cdot (p_{n_L-2} - 1) \tag{4.60}$$

$$C_{L_{(4a)},n_{4a}=4,n_{4a,0}=1} =$$

$$= (p_{n_L-3} - 1) \cdot p_{n_L-2} \cdot p_{n_L-1} \cdot p_{n_L} \cdot$$

$$\cdot (p_{n_L-2} - 1) \cdot (p_{n_L-1} - 1) \cdot (p_{n_L} - 1)+$$

$$+(p_{n_L-2} - 1) \cdot p_{n_L-3} \cdot p_{n_L-1} \cdot p_{n_L} \cdot$$

$$\cdot (p_{n_L-3} - 1) \cdot (p_{n_L-1} - 1) \cdot (p_{n_L} - 1)+$$

$$+(p_{n_L-1} - 1) \cdot p_{n_L-3} \cdot p_{n_L-2} \cdot p_{n_L} \cdot$$

$$\cdot (p_{n_L-3} - 1) \cdot (p_{n_L-2} - 1) \cdot (p_{n_L} - 1)+$$

$$+(p_{n_L} - 1) \cdot p_{n_L-3} \cdot p_{n_L-2} \cdot$$

$$\cdot p_{n_L-1} \cdot (p_{n_L-3} - 1) \cdot (p_{n_L-2} - 1) \cdot (p_{n_L-1} - 1) \tag{4.61}$$

$$C_{L_{(4a)},n_{4a}=4,n_{4a,0}=2} =$$

$$= (p_{n_L-3} - 1) \cdot (p_{n_L-2} - 1) \cdot p_{n_L-1} \cdot p_{n_L} \cdot (p_{n_L-1} - 1) \cdot (p_{n_L} - 1)+$$

$$+(p_{n_L-3} - 1) \cdot (p_{n_L-1} - 1) \cdot p_{n_L-2} \cdot p_{n_L} \cdot (p_{n_L-2} - 1) \cdot (p_{n_L} - 1)+$$

$$+(p_{n_L-3}-1)\cdot(p_{n_L}-1)\cdot p_{n_L-2}\cdot p_{n_L-1}\cdot(p_{n_L-2}-1)\cdot(p_{n_L-1}-1)+$$

$$+(p_{n_L-2}-1)\cdot(p_{n_L-1}-1)\cdot p_{n_L-3}\cdot p_{n_L}\cdot(p_{n_L-3}-1)\cdot(p_{n_L}-1)+$$

$$+(p_{n_L-2}-1)\cdot(p_{n_L}-1)\cdot p_{n_L-3}\cdot p_{n_L-1}\cdot(p_{n_L-3}-1)\cdot(p_{n_L-1}-1)+$$

$$+(p_{n_L-1}-1)\cdot(p_{n_L}-1)\cdot p_{n_L-3}\cdot p_{n_L-2}\cdot(p_{n_L-3}-1)\cdot(p_{n_L-2}-1) \qquad (4.62)$$

$$C_{L_{(4a)},n_{4a}=4,n_{4a,0}=3}=$$

$$=(p_{n_L-3}-1)\cdot(p_{n_L-2}-1)\cdot(p_{n_L-1}-1)\cdot p_{n_L}\cdot(p_{n_L}-1)+$$

$$+(p_{n_L-3}-1)\cdot(p_{n_L-2}-1)\cdot(p_{n_L}-1)\cdot p_{n_L-1}\cdot(p_{n_L-1}-1)+$$

$$+(p_{n_L-3}-1)\cdot(p_{n_L-1}-1)\cdot(p_{n_L}-1)\cdot p_{n_L-2}\cdot(p_{n_L-2}-1)+$$

$$+(p_{n_L-2}-1)\cdot(p_{n_L-1}-1)\cdot(p_{n_L}-1)\cdot p_{n_L-3}\cdot(p_{n_L-3}-1) \qquad (4.63)$$

We denote by $C_{L/L_{(4a)}}$ the number of combinations of coefficients q_i, $i=\overline{1,3}$, resulted from the product $L/L_{(4a)}$. Then, the total number of combinations of coefficients q_i, $i=\overline{1,3}$, resulted by considering the groups consisting of $n_{4a,0}$ factors from the product $L_{(4a)}$, for which $q_{3,j}=0 \pmod{p_j}$, $\forall j\in I_{n_{4a,0}}$, with $n_{4a,0}\in\{1,2,\ldots,n_{4a}-1\}$, where $n_{4a}=r\in\mathbb{N}^*$, will be equal to $\sum_{r_0=1}^{n_{4a}-1}\left(C_{L_{(4a)},n_{4a}=r,n_{4a,0}=r_0}\cdot C_{L/L_{(4a)}}\right)$. The previous sum is equal to zero when $n_{4a}=1$. Some of the following theorems give the formulas for the number of true different CPPs when the prime decomposition of L contains prime factors of type 4. (a). For some $n_{4a}=r\in\mathbb{N}^*$, the value $C_{L_{(4a)},n_{4a}=r,n_{4a,0}=r_0}$ is that in (4.57), and for $n_{4a}\in\{2,3,4\}$ the values $C_{L_{(4a)},n_{4a}=r,n_{4a,0}=r_0}$ can be easily computed from relations (4.58)–(4.63). The expression $C_{L/L_{(4a)}}$ will appear explicitly in formulas, according to the prime factors of decomposition of $L/L_{(4a)}$.

In the case of CPPs, there are four types of prime numbers that are considered in stating the conditions in Table 3.2, each with two distinct sets of values of power of the prime number. As the conditions for cases 1. (b), 3. (a), 3. (b) and 4. (b) are the same, there are 23 possible cases of decomposition of the number L in prime factors, which lead to combinations of different conditions on the coefficients q_1, q_2, q_3. Since the formulas resulted for the cases when the decomposition of L contains the factors 2 and 3 with powers 0 or 1 are very similar, we provide these formulas combined into only one, in terms of powers of factors 2 and 3, denoted by $n_2\in\{0,1\}$ and $n_3\in\{0,1\}$, respectively. Three of these 23 cases are for very small values of the number L, being trivial cases. The results regarding the number of all true different CPPs are given in Theorems 4.12–4.27 and are summarized in Table 4.2 at the end of this section.

Table 4.2 Number of all true different CPP-based interleavers

Case	Decomposition of L	$C_{L,CPPs}$	Theorems
(1)	$L = 2^{n_2} \cdot 3^{n_3} \cdot \prod_{j=n_{L1}}^{n_L} p_j^{n_{L,p_j}}$, $n_2 \in \{0, 1\}, n_3 \in \{0, 1\}$, $n_{L1} = n_2 + n_3 + 1, n_L \geq n_{L1}$, $p_j > 3, n_{L,p_j} \geq 1$, when $p_j = 3 \cdot k + 1, k \in \mathbb{N}$ and $n_{L,p_j} > 1$ when $p_j = 3 \cdot k + 2$, $k \in \mathbb{N}^*, j = \overline{n_{L1}, n_L}$	$C_{L,CPPs} = \left(2^{n_3} \cdot \prod_{j=n_{L1}}^{n_L} p_j^{2 \cdot (n_{L,p_j}-1)} \cdot (p_j - 1)\right) \cdot \left(\prod_{j=n_{L1}}^{n_L} p_j^{n_{L,p_j}-1} - 1\right)$	4.12
(2)	$L = 2^{n_{L,2}} \cdot 3^{n_3} \cdot \prod_{j=n_{L1}}^{n_L} p_j^{n_{L,p_j}}$, $n_{L,2} > 1, n_3 \in \{0, 1\}$, $n_{L1} = n_3 + 2, n_L \geq n_{L1}, p_j > 3$, $n_{L,p_j} \geq 1$, when $p_j = 3 \cdot k + 1$, $k \in \mathbb{N}$ and $n_{L,p_j} > 1$ when $p_j = 3 \cdot k + 2, k \in \mathbb{N}^*$, $j = \overline{n_{L1}, n_L}$	$C_{L,CPPs} = \left(2^{n_3} \cdot 2^{2 \cdot n_{L,2}-3} \cdot \prod_{j=n_{L1}}^{n_L} p_j^{2 \cdot (n_{L,p_j}-1)} \cdot (p_j - 1)\right) \cdot \left(2^{n_{L,2}-2} \cdot \prod_{j=n_{L1}}^{n_L} p_j^{n_{L,p_j}-1} - 1\right)$	4.13
(3)	$L = 2^{n_2} \cdot 3^{n_{L,3}} \cdot \prod_{j=n_{L1}}^{n_L} p_j^{n_{L,p_j}}, n_2 \in \{0, 1\}, n_{L,3} > 1$, $n_{L1} = n_2 + 2, n_L \geq n_{L1}, p_j > 3$, $n_{L,p_j} \geq 1$, when $p_j = 3 \cdot k + 1$, $k \in \mathbb{N}$ and $n_{L,p_j} > 1$ when $p_j = 3 \cdot k + 2, k \in \mathbb{N}^*$, $j = \overline{n_{L1}, n_L}$	$C_{L,CPPs} = \left(2 \cdot 3^{2 \cdot (n_{L,3}-1)} \cdot \prod_{j=n_{L1}}^{n_L} p_j^{2 \cdot (n_{L,p_j}-1)} \cdot (p_j - 1)\right) \cdot \left(2 \cdot 3^{n_{L,3}-2} \cdot \prod_{j=n_{L1}}^{n_L} p_j^{n_{L,p_j}-1} - 1\right)$	4.14

(continued)

Table 4.2 (continued)

Case	Decomposition of L	$C_{L,CPPs}$	Theorems
(4)	$L = 2^{n_{L,2}} \cdot 3^{n_{L,3}} \cdot \prod_{j=3}^{n_L} p_j^{n_{L,p_j}}$, $n_{L,2} > 1, n_{L,3} > 1, n_L \geq 3$, $p_j > 3, n_{L,p_j} \geq 1$, when $p_j = 3 \cdot k + 1, k \in \mathbb{N}$ and $n_{L,p_j} > 1$ when $p_j = 3 \cdot k + 2$, $k \in \mathbb{N}^*$, $j = \overline{3, n_L}$	$C_{L,CPPs} = \left(2^{2\cdot(n_{L,2}-1)} \cdot 3^{2\cdot(n_{L,3}-1)} \cdot \prod_{j=3}^{n_L} p_j^{2\cdot(n_{L,p_j}-1)} \cdot (p_j - 1)\right) \cdot \left(2^{n_{L,2}-1} \cdot 3^{n_{L,3}-2} \cdot \prod_{j=3}^{n_L} p_j^{n_{L,p_j}-1} - 1\right)$	4.15
(5)	$L = 2^{n_2} \cdot 3^{n_3} \cdot \prod_{j=n_{L1}}^{n_L - n_{4a}} p_j^{n_{L,p_j}} \cdot \prod_{j=n_L-n_{4a}+1}^{n_L} p_j, n_2 \in \{0, 1\}$, $n_3 \in \{0, 1\}, n_{L1} = n_2 + n_3 + 1$, $n_{4a} \in \mathbb{N}^*, n_L \geq n_{4a} + n_{L1}$, $p_j > 3, n_{L,p_j} \geq 1$, when $p_j = 3 \cdot k + 1, k \in \mathbb{N}$ and $n_{L,p_j} > 1$ when $p_j = 3 \cdot k + 2$, $k \in \mathbb{N}^*, j = \overline{n_{L1}, n_L - n_{4a}}$ and $p_j = 3 \cdot k + 2, k \in \mathbb{N}^*$, $j = \overline{n_L - n_{4a} + 1, n_L}$	$C_{L,CPPs} = 2^{n_3} \cdot \prod_{j=n_L-n_{4a}+1}^{n_L} \left(p_j \cdot (p_j - 1)\right) \cdot \prod_{j=n_{L1}}^{n_L-n_{4a}} \left(p_j^{3\cdot(n_{L,p_j}-1)} \cdot (p_j - 1)\right) + 2^{n_3} \cdot \sum_{r_0=1}^{n_{4a}-1} \left(C_{L_{(4a)},n_{4a}=r,n_{4a,0}=r_0} \cdot \prod_{j=n_{L1}}^{n_L-n_{4a}} \left(p_j^{3\cdot(n_{L,p_j}-1)} \cdot (p_j - 1)\right)\right) + 2^{n_3} \cdot \left(\prod_{j=n_L-n_{4a}+1}^{n_L} (p_j - 1) \cdot \prod_{j=n_{L1}}^{n_L-n_{4a}} \left(p_j^{2\cdot(n_{L,p_j}-1)} \cdot (p_j - 1)\right)\right) \cdot \left(\prod_{j=n_{L1}}^{n_L-n_{4a}} p_j^{n_{L,p_j}-1} - 1\right)$	4.16

(continued)

Table 4.2 (continued)

Case	Decomposition of L	$C_{L,CPPs}$	Theorems
(6)	$L = 2^{n_{L,2}} \cdot 3^{n_3} \cdot \prod_{j=n_{L1}}^{n_L-n_{4a}} p_j^{n_{L,p_j}} \cdot \prod_{j=n_L-n_{4a}+1}^{n_L} p_j, n_{L,2} > 1,$ $n_3 \in \{0,1\}, n_{L1} = n_3 + 2,$ $n_{4a} \in \mathbb{N}^*, n_L \geq n_{4a} + n_{L1},$ $p_j > 3, n_{L,p_j} \geq 1$, when $p_j = 3 \cdot k + 1, k \in \mathbb{N}$ and $n_{L,p_j} > 1$ when $p_j = 3 \cdot k + 2,$ $k \in \mathbb{N}^*, j = \overline{n_{L1}, n_L - n_{4a}}$ and $p_j = 3 \cdot k + 2, k \in \mathbb{N}^*,$ $j = \overline{n_L - n_{4a} + 1, n_L}$	$C_{L,CPPs} = 2^{n_3} \cdot \prod_{j=n_L-n_{4a}+1}^{n_L} \left(p_j \cdot (p_j - 1)\right) \cdot 2^{3 \cdot n_{L,2}-5} \cdot \prod_{j=n_{L1}}^{n_L-n_{4a}} \left(p_j^{3 \cdot (n_{L,p_j}-1)} \cdot (p_j - 1)\right) + 2^{n_3} \cdot \sum_{r_0=1}^{n_{4a}-1} \left(C_{L_{(4a)}, n_{4a}=r, n_{4a,0}=r_0} \cdot 2^{3 \cdot n_{L,2}-5} \cdot \prod_{j=n_{L1}}^{n_L-n_{4a}} \left(p_j^{3 \cdot (n_{L,p_j}-1)} \cdot (p_j - 1)\right)\right) +$ $+ 2^{n_3} \cdot \left(\prod_{j=n_L-n_{4a}+1}^{n_L} (p_j - 1) \cdot 2^{2 \cdot n_{L,2}-3} \cdot \prod_{j=n_{L1}}^{n_L-n_{4a}} \left(p_j^{2 \cdot (n_{L,p_j}-1)} \cdot (p_j - 1)\right)\right) \cdot \left(2^{n_{L,2}-2} \cdot \prod_{j=n_{L1}}^{n_L-n_{4a}} p_j^{n_{L,p_j}-1} - 1\right)$	4.17

(continued)

Table 4.2 (continued)

Case	Decomposition of L	$C_{L,CPPs}$	Theorems
(7)	$L = 2^{n_2} \cdot 3^{n_{L,3}} \cdot \prod_{j=n_{L1}}^{n_L-n_{4a}} p_j^{n_{L,p_j}} \cdot \prod_{j=n_L-n_{4a}+1}^{n_L} p_j, n_2 \in \{0, 1\}$, $n_{L,3} > 1, n_{L1} = n_2 + 2$, $n_{4a} \in \mathbb{N}^*, n_L \geq n_{4a} + n_{L1}$, $p_j > 3, n_{L,p_j} \geq 1$, when $p_j = 3 \cdot k + 1, k \in \mathbb{N}$ and $n_{L,p_j} > 1$ when $p_j = 3 \cdot k + 2$, $k \in \mathbb{N}^*, j = \overline{n_{L1}, n_L - n_{4a}}$ and $p_j = 3 \cdot k + 2, k \in \mathbb{N}^*$, $j = \overline{n_L - n_{4a} + 1, n_L}$	$C_{L,CPPs} = 4 \cdot \left(\prod_{j=n_L-n_{4a}+1}^{n_L} (p_j \cdot (p_j - 1)) \cdot 3^{3 \cdot n_{L,3}-4} \cdot \prod_{j=n_{L1}}^{n_L-n_{4a}} \left(p_j^{3 \cdot (n_{L,p_j}-1)} \cdot (p_j - 1) \right) \right) + 4 \cdot \sum_{r_0=1}^{n_{4a}-1} \left(C_{L_{(4a)}, n_{4a}=r, n_{4a,0}=r_0} \cdot 3^{3 \cdot n_{L,3}-4} \cdot \prod_{j=n_{L1}}^{n_L-n_{4a}} \left(p_j^{3 \cdot (n_{L,p_j}-1)} \cdot (p_j - 1) \right) \right) + $ $+ 2 \cdot \left(\prod_{j=n_L-n_{4a}+1}^{n_L} (p_j - 1) \cdot 3^{2 \cdot (n_{L,3}-1)} \cdot \prod_{j=n_{L1}}^{n_L-n_{4a}} \left(p_j^{2 \cdot (n_{L,p_j}-1)} \cdot (p_j - 1) \right) \right) \cdot \left(2 \cdot 3^{n_{L,3}-2} \cdot \prod_{j=n_{L1}}^{n_L-n_{4a}} p_j^{n_{L,p_j}-1} - 1 \right)$	4.18

(continued)

Table 4.2 (continued)

Case	Decomposition of L	$C_{L,CPPs}$	Theorems
(8)	$L = 2^{n_{L,2}} \cdot 3^{n_{L,3}} \cdot \prod_{j=3}^{n_L-n_{4a}} p_j^{n_{L,p_j}} \cdot \prod_{j=n_L-n_{4a}+1}^{n_L} p_j, n_{L,2} > 1,$ $n_{L,3} > 1, n_{4a} \in \mathbb{N}^*,$ $n_L \geq n_{4a}+3,\ p_j > 3,$ $n_{L,p_j} \geq 1$, when $p_j = 3 \cdot k + 1,$ $k \in \mathbb{N}$ and $n_{L,p_j} > 1$ when $p_j = 3 \cdot k + 2, k \in \mathbb{N}^*,$ $j = \overline{3, n_L - n_{4a}}$ and $p_j = 3 \cdot k + 2,$ $k \in \mathbb{N}^*, j = \overline{n_L - n_{4a} + 1, n_L}$	$C_{L,CPPs} = C_{L,CPPs} = \prod_{j=n_L-n_{4a}+1}^{n_L} (p_j \cdot (p_j - 1)) \cdot$ $2^{3\cdot(n_{L,2}-1)} \cdot 3^{3\cdot n_{L,3}-4} \cdot \prod_{j=3}^{n_L-n_{4a}} \left(p_j^{3\cdot(n_{L,p_j}-1)} \cdot (p_j - 1)\right) +$ $\sum_{r_0=1}^{n_{4a}-1} \Bigg(C_{L_{(4a)},n_{4a}=r,n_{4a,0}=r_0} \cdot 2^{3\cdot(n_{L,2}-1)} \cdot 3^{3\cdot n_{L,3}-4} \cdot$ $\cdot \prod_{j=3}^{n_L-n_{4a}} \left(p_j^{3\cdot(n_{L,p_j}-1)} \cdot (p_j - 1)\right)\Bigg) + \Bigg(\prod_{j=n_L-n_{4a}+1}^{n_L} (p_j - 1) \cdot$ $2^{2\cdot(n_{L,2}-1)} \cdot 3^{2\cdot(n_{L,3}-1)} \cdot \prod_{j=3}^{n_L-n_{4a}} \left(p_j^{2\cdot(n_{L,p_j}-1)} \cdot (p_j - 1)\right)\Bigg) \cdot$ $\left(2^{n_{L,2}-1} \cdot 3^{n_{L,3}-2} \cdot \prod_{j=3}^{n_L-n_{4a}} p_j^{n_{L,p_j}-1} - 1\right)$	4.19
(9)	$L = 2^{n_2} \cdot 3^{n_3} \cdot \prod_{j=n_{L1}}^{n_{4a}+n_{L1}-1} p_j,$ $n_2 \in \{0, 1\}, n_3 \in \{0, 1\},$ $n_{L1} = n_2 + n_3 + 1, n_{4a} \in \mathbb{N}^*,$ $p_j = 3 \cdot k + 2, k \in \mathbb{N}^*,$ $j = \overline{n_{L1}, n_{4a} + n_{L1} - 1}$	$C_{L,CPPs} = 2^{n_3} \cdot \prod_{j=n_{L1}}^{n_{4a}+n_{L1}-1} (p_j \cdot (p_j - 1)) + 2^{n_3} \cdot$ $\sum_{r_0=1}^{n_{4a}-1} C_{L_{(4a)},n_{4a}=r,n_{4a,0}=r_0}$	4.20

(continued)

Table 4.2 (continued)

Case	Decomposition of L	$C_{L,CPPs}$	Theorems
(10)	$L = 2^{n_{L,2}} \cdot 3^{n_3} \cdot \prod_{j=n_{L1}}^{n_{4a}+n_{L1}-1} p_j$, $n_{L,2} > 1, n_3 \in \{0, 1\}$, $n_{L1} = n_3 + 2, n_{4a} \in \mathbb{N}^*$, $p_j = 3 \cdot k + 2, k \in \mathbb{N}^*$, $j = \overline{n_{L1}, n_{4a} + n_{L1} - 1}$	$C_{L,CPPs} = 2^{n_3} \cdot \prod_{j=n_{L1}}^{n_{4a}+n_{L1}-1} (p_j \cdot (p_j - 1)) \cdot 2^{3 \cdot n_{L,2}-5} +$ $+ 2^{n_3} \cdot \sum_{r_0=1}^{n_{4a}-1} \left(C_{L_{(4a)},n_{4a}=r,n_{4a,0}=r_0} \cdot 2^{3 \cdot n_{L,2}-5}\right) +$ $+2^{n_3} \cdot \left(\prod_{j=n_{L1}}^{n_{4a}+n_{L1}-1} (p_j - 1) \cdot 2^{2 \cdot n_{L,2}-3}\right) \cdot \left(2^{\cdot n_{L,2}-2} - 1\right)$	4.21
(11)	$L = 2^{n_2} \cdot 3^{n_{L,3}} \cdot \prod_{j=n_{L1}}^{n_{4a}+n_{L1}-1} p_j$, $n_2 \in \{0, 1\}, n_{L,3} > 1$, $n_{L1} = n_2 + 2, n_{4a} \in \mathbb{N}^*$, $p_j = 3 \cdot k + 2, k \in \mathbb{N}^*$, $j = \overline{n_{L1}, n_{4a} + n_{L1} - 1}$	$C_{L,CPPs} = 4 \cdot \left(\prod_{j=n_{L1}}^{n_{4a}+n_{L1}-1} (p_j \cdot (p_j - 1)) \cdot 3^{3 \cdot n_{L,3}-4}\right) +$ $+ 4 \cdot \sum_{r_0=1}^{n_{4a}-1} \left(C_{L_{(4a)},n_{4a}=r,n_{4a,0}=r_0} \cdot 3^{3 \cdot n_{L,3}-4}\right) +$ $+ 2 \cdot \left(\prod_{j=n_{L1}}^{n_{4a}+n_{L1}-1} (p_j - 1) \cdot 3^{2 \cdot (n_{L,3}-1)}\right) \cdot \left(2 \cdot 3^{\cdot n_{L,3}-2} - 1\right)$	4.22

(continued)

Table 4.2 (continued)

Case	Decomposition of L	$C_{L,CPPs}$	Theorems
(12)	$L = 2^{n_{L,2}} \cdot 3^{n_{L,3}} \cdot \prod_{j=3}^{n_{4a}+2} p_j$, $n_{L,2} > 1, n_{L,3} > 1, n_{4a} \in \mathbb{N}^*$, $p_j = 3 \cdot k + 2, k \in \mathbb{N}^*$, $j = \overline{3, n_{4a} + 2}$	$C_{L,CPPs} = \prod_{j=3}^{n_{4a}+2} (p_j \cdot (p_j - 1)) \cdot 2^{3\cdot(n_{L,2}-1)} \cdot 3^{3\cdot n_{L,3}-4} +$ $+ \sum_{r_0=1}^{n_{4a}-1} \left(C_{L_{(4a)},n_{4a}=r,n_{4a,0}=r_0} \cdot 2^{3\cdot(n_{L,2}-1)} \cdot 3^{3\cdot n_{L,3}-4}\right) +$ $+ \left(\prod_{j=3}^{n_{4a}+2} (p_j - 1) \cdot 2^{2\cdot(n_{L,2}-1)} \cdot 3^{2\cdot(n_{L,3}-1)}\right) \cdot \left(2^{n_{L,2}-1} \cdot 3^{\cdot n_{L,3}-2} - 1\right)$	4.23
(13)	$L = 2^{n_{L,2}} \cdot 3^{n_3}, n_{L,2} > 1$, $n_3 \in \{0, 1\}$	$C_{L,CPPs} = 2^{n_3} \cdot 2^{2\cdot n_{L,2}-3} \cdot \left(2^{n_{L,2}-2} - 1\right)$	4.24
(14)	$L = 2^{n_2} \cdot 3^{n_{L,3}}, n_2 \in \{0, 1\}$, $n_{L,3} > 1$	$C_{L,CPPs} = 2 \cdot 3^{2\cdot(n_{L,3}-1)} \cdot \left(2 \cdot 3^{n_{L,3}-2} - 1\right)$	4.25
(15)	$L = 2^{n_{L,2}} \cdot 3^{n_{L,3}}, n_{L,2} > 1$, $n_{L,3} > 1$	$C_{L,CPPs} = 2^{2\cdot(n_{L,2}-1)} \cdot 3^{2\cdot(n_{L,3}-1)} \cdot \left(2^{n_{L,2}-1} \cdot 3^{n_{L,3}-2} - 1\right)$	4.26
(16)	$L = 2$ or $L = 3$ or $L = 6$	$C_{L,CPPs} = 0$	4.27

Theorem 4.12 *If the decomposition of L is of the form* $L = 2^{n_2} \cdot 3^{n_3} \cdot \prod_{j=n_{L1}}^{n_L} p_j^{n_{L,p_j}}$, *with* $n_2 \in \{0, 1\}$, $n_3 \in \{0, 1\}$, $n_{L1} = n_2 + n_3 + 1$, $n_L \geq n_{L1}$, $p_j > 3$, $n_{L,p_j} \geq 1$, *when* $p_j = 3 \cdot k + 1$, $k \in \mathbb{N}$ *and* $n_{L,p_j} > 1$ *when* $p_j = 3 \cdot k + 2$, $k \in \mathbb{N}^*$, $j = \overline{n_{L1}, n_L}$, *the number of CPPs will be equal to:*

$$C_{L,CPPs} = \left(2^{n_3} \cdot \prod_{j=n_{L1}}^{n_L} p_j^{2\cdot(n_{L,p_j}-1)} \cdot (p_j - 1)\right) \cdot \left(\prod_{j=n_{L1}}^{n_L} p_j^{n_{L,p_j}-1} - 1\right) \tag{4.64}$$

Proof For $n_2 = 0$ and $n_3 = 0$, we have $n_{L1} = 1$, and the factors from the decomposition of L are of type 3. (a) and/or 3. (b) and/or 4. (b). From cases 3. (a), 3. (b) and 4. (b) above, the number of possible combinations for q_1 is equal to $\Phi(L) = \prod_{j=1}^{n_L} p_j^{n_{L,p_j}-1} \cdot (p_j - 1)$ and the total number of coefficients q_2 and q_3, respectively, are equal to $\prod_{j=1}^{n_L} p_j^{n_{L,p_j}-1}$. The value $q_3 = 0$ results only when $q_{3,j} = 0$, $\forall j = \overline{1, n_L}$, i.e., for a single combination of coefficients $q_{3,j}$, $j = \overline{1, n_L}$, that has to be removed. The number of CPPs will be equal to

$$C_{L,CPPs} = \left(\prod_{j=1}^{n_L} p_j^{n_{L,p_j}-1} \cdot (p_j - 1)\right) \cdot \left(\prod_{j=1}^{n_L} p_j^{n_{L,p_j}-1}\right) \cdot$$

$$\cdot \left(\prod_{j=1}^{n_L} p_j^{n_{L,p_j}-1} - 1\right) = \left(\prod_{j=1}^{n_L} p_j^{2\cdot(n_{L,p_j}-1)} \cdot (p_j - 1)\right) \cdot \left(\prod_{j=1}^{n_L} p_j^{n_{L,p_j}-1} - 1\right) \tag{4.65}$$

i.e., the one from (4.64), for $n_2 = 0$ and $n_3 = 0$.

For $n_2 = 1$ and $n_3 = 0$, we have $n_{L1} = 2$, and the factors from the decomposition of L are of type 1. (a) and 3. (a) and/or 3. (b) and/or 4. (b). From the analysis for cases 1. (a) and 3. (a), 3. (b), 4. (b) above, the number of possible combinations for coefficient q_1 is equal to $1 \cdot \Phi(L/2) = \prod_{j=2}^{n_L} p_j^{n_{L,p_j}-1} \cdot (p_j - 1)$ and the total number of coefficients q_2 and q_3, respectively, is equal to $1 \cdot \prod_{j=2}^{n_L} p_j^{n_{L,p_j}-1} = \prod_{j=2}^{n_L} p_j^{n_{L,p_j}-1}$, out of which one value is 0. By removing the value $q_3 = 0$, the number of CPPs will be equal to:

$$C_{L,CPPs} = \left(\prod_{j=2}^{n_L} p_j^{n_{L,p_j}-1} \cdot (p_j - 1)\right) \cdot \left(\prod_{j=2}^{n_L} p_j^{n_{L,p_j}-1}\right) \cdot$$

$$\cdot\left(\prod_{j=2}^{n_L} p_j^{n_{L,p_j}-1}-1\right)=\left(\prod_{j=2}^{n_L} p_j^{2\cdot(n_{L,p_j}-1)}\cdot(p_j-1)\right)\cdot\left(\prod_{j=2}^{n_L} p_j^{n_{L,p_j}-1}-1\right) \tag{4.66}$$

i.e., the one from (4.64), for $n_2 = 1$ and $n_3 = 0$.

For $n_2 = 0$ and $n_3 = 1$, we have $n_{L1} = 2$, and the factors from the decomposition of L are of type 2. (a) and 3. (a) and/or 3. (b) and/or 4. (b). In determining the number of coefficients, we consider that for the two sets of coefficients valid for the factor of type 2. (a), we have only a single value for q_2 and q_3. The number of possible combinations for coefficient q_1 is equal to $2\cdot\prod_{j=2}^{n_L} p_j^{n_{L,p_j}-1}\cdot(p_j-1)$, the number of coefficients q_2 is equal to $\prod_{j=2}^{n_L} p_j^{n_{L,p_j}-1}$ and the number of coefficients q_3 is equal to $\prod_{j=2}^{n_L} p_j^{n_{L,p_j}-1}$, out of which one is 0. By removing the value $q_3 = 0$, the number of CPPs will be equal to:

$$C_{L,CPPs}=\left(2\cdot\prod_{j=2}^{n_L} p_j^{n_{L,p_j}-1}\cdot(p_j-1)\right)\cdot\left(\prod_{j=2}^{n_L} p_j^{n_{L,p_j}-1}\right)\cdot$$
$$\cdot\left(\prod_{j=2}^{n_L} p_j^{n_{L,p_j}-1}-1\right)=\left(2\cdot\prod_{j=2}^{n_L} p_j^{2\cdot(n_{L,p_j}-1)}\cdot(p_j-1)\right)\cdot$$
$$\cdot\left(\prod_{j=2}^{n_L} p_j^{n_{L,p_j}-1}-1\right) \tag{4.67}$$

i.e., the one from (4.64), for $n_2 = 0$ and $n_3 = 1$.

For $n_2 = 1$ and $n_3 = 1$, we have $n_{L1} = 3$, and the factors from the decomposition of L are of type 1. (a) and 2. (a) and 3. (a) and/or 3. (b) and/or 4. (b). The number of possible combinations for coefficient q_1 is equal to $1\cdot 2\cdot\prod_{j=3}^{n_L} p_j^{n_{L,p_j}-1}\cdot(p_j-1)=2\cdot\prod_{j=3}^{n_L} p_j^{n_{L,p_j}-1}\cdot(p_j-1)$, the number of coefficients q_2 and q_3 is equal to $1\cdot 1\cdot\prod_{j=3}^{n_L} p_j^{n_{L,p_j}-1}=\prod_{j=3}^{n_L} p_j^{n_{L,p_j}-1}$, out of which one is 0. After removing the value $q_3 = 0$, the number of CPPs will be equal to:

$$C_{L,CPPs} = \left(2 \cdot \prod_{j=3}^{n_L} p_j^{n_{L,p_j}-1} \cdot (p_j - 1)\right) \cdot \left(\prod_{j=3}^{n_L} p_j^{n_{L,p_j}-1}\right) \cdot$$

$$\cdot \left(\prod_{j=3}^{n_L} p_j^{n_{L,p_j}-1} - 1\right) = \left(2 \cdot \prod_{j=3}^{n_L} p_j^{2\cdot(n_{L,p_j}-1)} \cdot (p_j - 1)\right) \cdot$$

$$\cdot \left(\prod_{j=3}^{n_L} p_j^{n_{L,p_j}-1} - 1\right) \tag{4.68}$$

i.e., the one from (4.64), for $n_2 = 1$ and $n_3 = 1$. ■

From (4.64), we see that the number of CPPs is equal to 0 if the interleaver length is one, two, three or six times a product of prime numbers greater than 3, of the form $3 \cdot k + 1$, $k \in \mathbb{N}$, each of them to the power of 1.

Theorem 4.13 *If the decomposition of L is of the form* $L = 2^{n_{L,2}} \cdot 3^{n_3} \cdot \prod_{j=n_{L1}}^{n_L} p_j^{n_{L,p_j}}$, *with* $n_{L,2} > 1$, $n_3 \in \{0, 1\}$, $n_{L1} = n_3 + 2$, $n_L \geq n_{L1}$, $p_j > 3$, $n_{L,p_j} \geq 1$, *when* $p_j = 3 \cdot k + 1$, $k \in \mathbb{N}$ *and* $n_{L,p_j} > 1$ *when* $p_j = 3 \cdot k + 2$, $k \in \mathbb{N}^*$, $j = \overline{n_{L1}, n_L}$, *the number of CPPs will be equal to:*

$$C_{L,CPPs} = \left(2^{n_3} \cdot 2^{2\cdot n_{L,2}-3} \cdot \prod_{j=n_{L1}}^{n_L} p_j^{2\cdot(n_{L,p_j}-1)} \cdot (p_j - 1)\right) \cdot$$

$$\cdot \left(2^{n_{L,2}-2} \cdot \prod_{j=n_{L1}}^{n_L} p_j^{n_{L,p_j}-1} - 1\right) \tag{4.69}$$

Proof For $n_3 = 0$, we have $n_{L1} = 2$, and the factors from the decomposition of L are of type 1. (b) and 3. (a) and/or 3. (b) and/or 4. (b). From cases 1. (b), 3. (a), 3. (b) and 4. (b) above, the number of possible combinations for coefficient q_1 is equal to $\Phi(L) = 2^{n_{L,2}-1} \cdot \prod_{j=2}^{n_L} p_j^{n_{L,p_j}-1} \cdot (p_j - 1)$ and the total number of coefficients q_2 and q_3, respectively, is equal to $2^{n_{L,2}-1} \cdot \prod_{j=2}^{n_L} p_j^{n_{L,p_j}-1}$. As this case requires from the equivalence conditions that $q_2 < L/2$ and $q_3 < L/2$, a number of $\frac{1}{2} \cdot 2^{n_{L,2}-1} \cdot \prod_{j=2}^{n_L} p_j^{n_{L,p_j}-1} = 2^{n_{L,2}-2} \cdot \prod_{j=2}^{n_L} p_j^{n_{L,p_j}-1}$ possible coefficients q_2 and q_3, respectively, remains, out of which one is zero. By removing the value $q_3 = 0$, the number of CPPs will be equal to:

$$C_{L,CPPs} = \left(2^{n_{L,2}-1} \cdot \prod_{j=2}^{n_L} p_j^{n_{L,p_j}-1} \cdot (p_j - 1)\right) \cdot \left(2^{n_{L,2}-2} \cdot \prod_{j=2}^{n_L} p_j^{n_{L,p_j}-1}\right) \cdot$$

$$\cdot \left(2^{n_{L,2}-2} \cdot \prod_{j=2}^{n_L} p_j^{n_{L,p_j}-1} - 1\right) =$$

$$= \left(2^{2 \cdot n_{L,2}-3} \cdot \prod_{j=2}^{n_L} p_j^{2 \cdot (n_{L,p_j}-1)} \cdot (p_j - 1)\right) \cdot \left(2^{n_{L,2}-2} \cdot \prod_{j=2}^{n_L} p_j^{n_{L,p_j}-1} - 1\right) \quad (4.70)$$

i.e., the one from (4.69), for $n_3 = 0$.

For $n_3 = 1$, we have $n_{L1} = 3$, and the factors from the decomposition of L are of type 1. (b) and 2. (a) and 3. (a) and/or 3. (b) and/or 4. (b). In determining the number of coefficients, we consider that for the two sets of coefficients valid for the factor of type 2. (a) we always have $q_{2,1} = q_{3,1} = 0$. From the two sets we have to keep only one for the coefficients q_2 and q_3. The number of possible combinations for the coefficient q_1 is equal to $2^{n_{L,2}-1} \cdot 2 \cdot \prod_{j=3}^{n_L} p_j^{n_{L,p_j}-1} \cdot (p_j - 1) = 2^{n_{L,2}} \cdot \prod_{j=3}^{n_L} p_j^{n_{L,p_j}-1} \cdot (p_j - 1)$, the number of coefficients q_2 and q_3 is equal to $2^{n_{L,2}-2} \cdot \prod_{j=3}^{n_L} p_j^{n_{L,p_j}-1}$, out of which one is 0. By removing the value $q_3 = 0$, the number of CPPs will be equal to:

$$C_{L,CPPs} = \left(2^{n_{L,2}} \cdot \prod_{j=3}^{n_L} p_j^{n_{L,p_j}-1} \cdot (p_j - 1)\right) \cdot \left(2^{n_{L,2}-2} \cdot \prod_{j=3}^{n_L} p_j^{n_{L,p_j}-1}\right) \cdot$$

$$\cdot \left(2^{n_{L,2}-2} \cdot \prod_{j=3}^{n_L} p_j^{n_{L,p_j}-1} - 1\right) =$$

$$= \left(2^{2 \cdot (n_{L,2}-1)} \cdot \prod_{j=3}^{n_L} p_j^{2 \cdot (n_{L,p_j}-1)} \cdot (p_j - 1)\right) \cdot \left(2^{n_{L,2}-2} \cdot \prod_{j=3}^{n_L} p_j^{n_{L,p_j}-1} - 1\right) \quad (4.71)$$

i.e., the one from (4.69), for $n_3 = 1$. ■

From (4.69), we see that the number of CPPs is equal to 0 if the interleaver length is 4 or 12 times a product of prime numbers greater than 3, of the form $3 \cdot k + 1$, $k \in \mathbb{N}$, each of them to the power of 1.

Theorem 4.14 *If the decomposition of L is of the form* $L = 2^{n_2} \cdot 3^{n_{L,3}} \cdot \prod_{j=n_{L1}}^{n_L} p_j^{n_{L,p_j}}$, *with* $n_2 \in \{0, 1\}$, $n_{L,3} > 1$, $n_{L1} = n_2 + 2$, $n_L \geq n_{L1}$, $p_j > 3$, $n_{L,p_j} \geq 1$, *when* $p_j = 3 \cdot k + 1$, $k \in \mathbb{N}$ *and* $n_{L,p_j} > 1$ *when* $p_j = 3 \cdot k + 2$, $k \in \mathbb{N}^*$, $j = \overline{n_{L1}, n_L}$, *the number of CPPs will be equal to:*

$$C_{L,CPPs} = \left(2 \cdot 3^{2\cdot(n_{L,3}-1)} \cdot \prod_{j=n_{L1}}^{n_L} p_j^{2\cdot(n_{L,p_j}-1)} \cdot (p_j - 1)\right) \cdot \left(2 \cdot 3^{n_{L,3}-2} \cdot \prod_{j=n_{L1}}^{n_L} p_j^{n_{L,p_j}-1} - 1\right) \quad (4.72)$$

Proof For $n_2 = 0$, we have $n_{L1} = 2$, and the factors from the decomposition of L are of type 2. (b) and 3. (a) and/or 3. (b) and/or 4. (b). The number of possible combinations for coefficient q_1 is equal to $2 \cdot 3^{n_{L,3}-1} \cdot \prod_{j=2}^{n_L} p_j^{n_{L,p_j}-1} \cdot (p_j - 1)$, the number of coefficients q_2 is equal to $3^{n_{L,3}-1} \cdot \prod_{j=2}^{n_L} p_j^{n_{L,p_j}-1}$ and the number of coefficients q_3 is equal to $2 \cdot 3^{n_{L,3}-2} \cdot \prod_{j=2}^{n_L} p_j^{n_{L,p_j}-1}$ for $3^{n_{L,3}-1} \cdot \prod_{j=2}^{n_L} p_j^{n_{L,p_j}-1} \cdot (p_j - 1)$ of the $2 \cdot 3^{n_{L,3}-1} \cdot \prod_{j=2}^{n_L} p_j^{n_{L,p_j}-1} \cdot (p_j - 1)$ values of coefficients q_1 and it is equal to $2 \cdot 3^{n_{L,3}-2} \cdot \prod_{j=2}^{n_L} p_j^{n_{L,p_j}-1}$ for the other $3^{n_{L,3}-1} \cdot \prod_{j=2}^{n_L} p_j^{n_{L,p_j}-1} \cdot (p_j - 1)$ values of coefficients q_1, out of which one is 0. By removing the value $q_3 = 0$, the number of CPPs will be equal to:

$$C_{L,CPPs} = \left(3^{n_{L,3}-1} \cdot \prod_{j=2}^{n_L} p_j^{n_{L,p_j}-1} \cdot (p_j - 1)\right) \cdot \left(3^{n_{L,3}-1} \cdot \prod_{j=2}^{n_L} p_j^{n_{L,p_j}-1}\right) \cdot \left(2 \cdot 3^{n_{L,3}-2} \cdot \prod_{j=2}^{n_L} p_j^{n_{L,p_j}-1} - 1\right) + \left(3^{n_{L,3}-1} \cdot \prod_{j=2}^{n_L} p_j^{n_{L,p_j}-1} \cdot (p_j - 1)\right) \cdot \left(3^{n_{L,3}-1} \cdot \prod_{j=2}^{n_L} p_j^{n_{L,p_j}-1}\right) \cdot \left(2 \cdot 3^{n_{L,3}-2} \cdot \prod_{j=2}^{n_L} p_j^{n_{L,p_j}-1} - 1\right) =$$

$$= \left(2 \cdot 3^{2\cdot(n_{L,3}-1)} \cdot \prod_{j=2}^{n_L} p_j^{2\cdot(n_{L,p_j}-1)} \cdot (p_j - 1)\right) \cdot$$

$$\cdot \left(2 \cdot 3^{n_{L,3}-2} \cdot \prod_{j=2}^{n_L} p_j^{n_{L,p_j}-1} - 1\right) \tag{4.73}$$

i.e., the one from (4.72), for $n_2 = 0$.

For $n_2 = 1$, we have $n_{L1} = 3$, and the factors from the decomposition of L are of type 1. (a) and 2. (b) and 3. (a) and/or 3. (b) and/or 4. (b). From the analysis of cases 1. (a), 2. (b) and 3. (a), 3. (b), 4. (b) above, it follows that the number of possible combinations for coefficient q_1 is equal to $1 \cdot 2 \cdot 3^{n_{L,3}-1} \cdot \prod_{j=3}^{n_L} p_j^{n_{L,p_j}-1} \cdot (p_j - 1) = 2 \cdot 3^{n_{L,3}-1} \cdot \prod_{j=3}^{n_L} p_j^{n_{L,p_j}-1} \cdot (p_j - 1)$, the number of coefficients q_2 is equal to $1 \cdot 3^{n_{L,3}-1} \cdot \prod_{j=3}^{n_L} p_j^{n_{L,p_j}-1} = 3^{n_{L,3}-1} \cdot \prod_{j=3}^{n_L} p_j^{n_{L,p_j}-1}$ and the number of coefficients q_3 is equal to $1 \cdot 2 \cdot 3^{n_{L,3}-2} \cdot \prod_{j=3}^{n_L} p_j^{n_{L,p_j}-1} = 2 \cdot 3^{n_{L,3}-2} \cdot \prod_{j=3}^{n_L} p_j^{n_{L,p_j}-1}$ for $3^{n_{L,3}-1} \cdot \prod_{j=3}^{n_L} p_j^{n_{L,p_j}-1} \cdot (p_j - 1)$ values of the $2 \cdot 3^{n_{L,3}-1} \cdot \prod_{j=3}^{n_L} p_j^{n_{L,p_j}-1} \cdot (p_j - 1)$ ones of coefficient q_1 and equal to $1 \cdot 2 \cdot 3^{n_{L,3}-2} \cdot \prod_{j=3}^{n_L} p_j^{n_{L,p_j}-1} = 2 \cdot 3^{n_{L,3}-2} \cdot \prod_{j=3}^{n_L} p_j^{n_{L,p_j}-1}$ for the other $3^{n_{L,3}-1} \cdot \prod_{j=3}^{n_L} p_j^{n_{L,p_j}-1} \cdot (p_j - 1)$ values of coefficient q_1, out of which one is 0. By removing the value $q_3 = 0$, the number of CPPs results equal to:

$$C_{L,CPPs} = \left(3^{n_{L,3}-1} \cdot \prod_{j=3}^{n_L} p_j^{n_{L,p_j}-1} \cdot (p_j - 1)\right) \cdot \left(3^{n_{L,3}-1} \cdot \prod_{j=3}^{n_L} p_j^{n_{L,p_j}-1}\right) \cdot$$

$$\cdot \left(2 \cdot 3^{n_{L,3}-2} \cdot \prod_{j=3}^{n_L} p_j^{n_{L,p_j}-1} - 1\right) + \left(3^{n_{L,3}-1} \cdot \prod_{j=3}^{n_L} p_j^{n_{L,p_j}-1} \cdot (p_j - 1)\right) \cdot$$

$$\cdot \left(3^{n_{L,3}-1} \cdot \prod_{j=3}^{n_L} p_j^{n_{L,p_j}-1}\right) \cdot \left(2 \cdot 3^{n_{L,3}-2} \cdot \prod_{j=3}^{n_L} p_j^{n_{L,p_j}-1} - 1\right) =$$

$$= \left(2 \cdot 3^{2\cdot(n_{L,3}-1)} \cdot \prod_{j=3}^{n_L} p_j^{2\cdot(n_{L,p_j}-1)} \cdot (p_j - 1)\right) \cdot$$

$$\cdot \left(2 \cdot 3^{n_{L,3}-2} \cdot \prod_{j=3}^{n_L} p_j^{n_{L,p_j}-1} - 1\right) \tag{4.74}$$

i.e., the one from (4.72), for $n_2 = 1$. ■

From (4.72) we see that in this case the number of CPPs is always greater than 0.

Theorem 4.15 *If the decomposition of L is of the form* $L = 2^{n_{L,2}} \cdot 3^{n_{L,3}} \cdot \prod_{j=3}^{n_L} p_j^{n_{L,p_j}}$, *with* $n_{L,2} > 1$, $n_{L,3} > 1$, $n_L \geq 3$, $p_j > 3$, $n_{L,p_j} \geq 1$, *when* $p_j = 3 \cdot k + 1$, $k \in \mathbb{N}$ *and* $n_{L,p_j} > 1$ *when* $p_j = 3 \cdot k + 2$, $k \in \mathbb{N}^*$, $j = \overline{3, n_L}$, *the number of CPPs will be equal to:*

$$C_{L,CPPs} = \left(2^{2\cdot(n_{L,2}-1)} \cdot 3^{2\cdot(n_{L,3}-1)} \cdot \prod_{j=3}^{n_L} p_j^{2\cdot(n_{L,p_j}-1)} \cdot (p_j - 1)\right) \cdot$$

$$\cdot \left(2^{n_{L,2}-1} \cdot 3^{n_{L,3}-2} \cdot \prod_{j=3}^{n_L} p_j^{n_{L,p_j}-1} - 1\right) \tag{4.75}$$

Proof In this case, the factors from the decomposition of L are of type 1. (b) and 2. (b) and 3. (a) and/or 3. (b) and/or 4. (b). The numbers of possible combinations for coefficient q_1 is equal to $2^{n_{L,2}-1} \cdot 2 \cdot 3^{n_{L,3}-1} \cdot \prod_{j=3}^{n_L} p_j^{n_{L,p_j}-1} \cdot (p_j - 1) = 2^{n_{L,2}} \cdot 3^{n_{L,3}-1} \cdot \prod_{j=3}^{n_L} p_j^{n_{L,p_j}-1} \cdot (p_j - 1)$, the number of coefficients q_2 is equal to $2^{n_{L,2}-2} \cdot 3^{n_{L,3}-1} \cdot \prod_{j=3}^{n_L} p_j^{n_{L,p_j}-1}$ and the number of coefficients q_3 is equal to $2^{n_{L,2}-2} \cdot 2 \cdot 3^{n_{L,3}-2} \cdot \prod_{j=3}^{n_L} p_j^{n_{L,p_j}-1} = 2^{n_{L,2}-1} \cdot 3^{n_{L,3}-2} \cdot \prod_{j=3}^{n_L} p_j^{n_{L,p_j}-1}$ for $2^{n_{L,2}-1} \cdot 3^{n_{L,3}-1} \cdot \prod_{j=3}^{n_L} p_j^{n_{L,p_j}-1} \cdot (p_j - 1)$ of $2^{n_{L,2}} \cdot 3^{n_{L,3}-1} \cdot \prod_{j=3}^{n_L} p_j^{n_{L,p_j}-1} \cdot (p_j - 1)$ values for coefficient q_1 and equal to $2^{n_{L,2}-2} \cdot 2 \cdot 3^{n_{L,3}-2} \cdot \prod_{j=3}^{n_L} p_j^{n_{L,p_j}-1} = 2^{n_{L,2}-1} \cdot 3^{n_{L,3}-2} \cdot \prod_{j=3}^{n_L} p_j^{n_{L,p_j}-1}$ for the other $2^{n_{L,2}-1} \cdot$

$3^{n_{L,3}-1} \cdot \prod_{j=3}^{n_L} p_j^{n_{L,p_j}-1} \cdot (p_j - 1)$ values of coefficient q_1, out of which one is 0. By removing the value $q_3 = 0$, the number of CPPs will be equal to:

$$C_{L,CPPs} = \left(2^{n_{L,2}-1} \cdot 3^{n_{L,3}-1} \cdot \prod_{j=3}^{n_L} p_j^{n_{L,p_j}-1} \cdot (p_j - 1)\right) \cdot$$

$$\cdot \left(2^{n_{L,2}-2} \cdot 3^{n_{L,3}-1} \cdot \prod_{j=3}^{n_L} p_j^{n_{L,p_j}-1}\right) \cdot \left(2^{n_{L,2}-1} \cdot 3^{n_{L,3}-2} \cdot \prod_{j=3}^{n_L} p_j^{n_{L,p_j}-1} - 1\right) +$$

$$+ \left(2^{n_{L,2}-1} \cdot 3^{n_{L,3}-1} \cdot \prod_{j=3}^{n_L} p_j^{n_{L,p_j}-1} \cdot (p_j - 1)\right) \cdot$$

$$\cdot \left(2^{n_{L,2}-2} \cdot 3^{n_{L,3}-1} \cdot \prod_{j=3}^{n_L} p_j^{n_{L,p_j}-1}\right) \cdot \left(2^{n_{L,2}-1} \cdot 3^{n_{L,3}-2} \cdot \prod_{j=3}^{n_L} p_j^{n_{L,p_j}-1} - 1\right) =$$

$$= \left(2^{2\cdot(n_{L,2}-1)} \cdot 3^{2\cdot(n_{L,3}-1)} \cdot \prod_{j=3}^{n_L} p_j^{2\cdot(n_{L,p_j}-1)} \cdot (p_j - 1)\right) \cdot$$

$$\cdot \left(2^{n_{L,2}-1} \cdot 3^{n_{L,3}-2} \cdot \prod_{j=3}^{n_L} p_j^{n_{L,p_j}-1} - 1\right) \tag{4.76}$$

i.e., the one from (4.75). ■

From (4.75), we see that in this case the number of CPPs is always greater than 0.

Theorem 4.16 *If the decomposition of L is of the form* $L = 2^{n_2} \cdot 3^{n_3} \cdot \prod_{j=n_{L1}}^{n_L - n_{4a}} p_j^{n_{L,p_j}} \cdot \prod_{j=n_L - n_{4a}+1}^{n_L} p_j$, *with* $n_2 \in \{0, 1\}$, $n_3 \in \{0, 1\}$, $n_{L1} = n_2 + n_3 + 1$, $n_{4a} \in \mathbb{N}^*$, $n_L \geq n_{4a} + n_{L1}$, $p_j > 3$, $n_{L,p_j} \geq 1$, *when* $p_j = 3 \cdot k + 1$, $k \in \mathbb{N}$ *and* $n_{L,p_j} > 1$ *when* $p_j = 3 \cdot k + 2$, $k \in \mathbb{N}^*$, $j = \overline{n_{L1}, n_L - n_{4a}}$ *and* $p_j = 3 \cdot k + 2$, $k \in \mathbb{N}^*$, $j = \overline{n_L - n_{4a} + 1, n_L}$, *the number of CPPs will be equal to:*

$$C_{L,CPPs} = 2^{n_3} \cdot \prod_{j=n_L - n_{4a}+1}^{n_L} (p_j \cdot (p_j - 1)) \cdot$$

$$\cdot \prod_{j=n_{L1}}^{n_L-n_{4a}} \left(p_j^{3\cdot(n_{L,p_j}-1)} \cdot (p_j-1) \right) +$$

$$+2^{n_3} \cdot \sum_{r_0=1}^{n_{4a}-1} \left(C_{L_{(4a)},n_{4a}=r,n_{4a,0}=r_0} \cdot \prod_{j=n_{L1}}^{n_L-n_{4a}} \left(p_j^{3\cdot(n_{L,p_j}-1)} \cdot (p_j-1) \right) \right) +$$

$$+2^{n_3} \cdot \left(\prod_{j=n_L-n_{4a}+1}^{n_L} (p_j-1) \cdot \prod_{j=n_{L1}}^{n_L-n_{4a}} \left(p_j^{2\cdot(n_{L,p_j}-1)} \cdot (p_j-1) \right) \right) \cdot$$

$$\cdot \left(\prod_{j=n_{L1}}^{n_L-n_{4a}} p_j^{n_{L,p_j}-1} - 1 \right) \tag{4.77}$$

Proof For $n_2 = 0$ and $n_3 = 0$, we have $n_{L1} = 1$, and the factors from the decomposition of L are of type 4. (a) and 3. (a) and/or 3. (b) and/or 4. (b). We consider the analysis of case 4. (a) above. It follows that for a group of $n_{4a,0}$ prime factors of type 4.a), with $n_{4a,0} < n_{4a}$, for which $q_{3,j} = 0 \pmod{p_j}$, $\forall j \in I_{n_{4a,0}}$, the number of possible combinations for coefficient q_1 is equal to $\prod_{j\in I_{n_{4a,0}}} (p_j-1) \cdot \prod_{j=1}^{n_L-n_{4a}} \left(p_j^{n_{L,p_j}-1} \cdot (p_j-1) \right)$, the number of coefficients q_2 is equal to $\prod_{j\in\{I_{n_{4a}}-I_{n_{4a,0}}\}} p_j \cdot \prod_{j=1}^{n_L-n_{4a}} p_j^{n_{L,p_j}-1}$ and the number of coefficients q_3 is equal to $\prod_{j\in\{I_{n_{4a}}-I_{n_{4a,0}}\}} (p_j-1) \cdot \prod_{j=1}^{n_L-n_{4a}} p_j^{n_{L,p_j}-1}$. In total, for a group of $n_{4a,0}$ prime factors of type 4. (a), with $n_{4a,0} \in \{1, 2, \ldots, n_{4a}-1\}$, the number of CPPs will be equal to:

$$C_{L,CPPs,n_{4a,0}} = \left(\prod_{j\in I_{n_{4a,0}}} (p_j-1) \cdot \prod_{j=1}^{n_L-n_{4a}} \left(p_j^{n_{L,p_j}-1} \cdot (p_j-1) \right) \right) \cdot$$

$$\cdot \left(\prod_{j\in\{I_{n_{4a}}-I_{n_{4a,0}}\}} p_j \cdot \prod_{j=1}^{n_L-n_{4a}} p_j^{n_{L,p_j}-1} \right) \cdot$$

$$\cdot \left(\prod_{j\in\{I_{n_{4a}}-I_{n_{4a,0}}\}} (p_j-1) \cdot \prod_{j=1}^{n_L-n_{4a}} p_j^{n_{L,p_j}-1} \right) =$$

$$= \prod_{j\in I_{n_{4a}}} (p_j - 1) \cdot \prod_{j\in\{I_{n_{4a}} - I_{n_{4a,0}}\}} p_j \cdot \prod_{j=1}^{n_L - n_{4a}} \left(p_j^{3\cdot(n_{L,p_j}-1)} \cdot (p_j - 1) \right) =$$

$$= \prod_{j=n_L-n_{4a}+1}^{n_L} (p_j - 1) \cdot \prod_{j\in\{I_{n_{4a}} - I_{n_{4a,0}}\}} p_j \cdot \prod_{j=1}^{n_L - n_{4a}} \left(p_j^{3\cdot(n_{L,p_j}-1)} \cdot (p_j - 1) \right) \quad (4.78)$$

For a group of n_{4a} prime factors of type 4. (a), we have to remove the case when $q_3 = 0$ and the number of CPPs will be:

$$C_{L,CPPs,n_{4a}} = \left(\prod_{j=n_L-n_{4a}+1}^{n_L} (p_j - 1) \cdot \prod_{j=1}^{n_L-n_{4a}} \left(p_j^{n_{L,p_j}-1} \cdot (p_j - 1) \right) \right) \cdot$$

$$\cdot \left(\prod_{j=1}^{n_L-n_{4a}} p_j^{n_{L,p_j}-1} \right) \cdot \left(\prod_{j=1}^{n_L-n_{4a}} p_j^{n_{L,p_j}-1} - 1 \right) =$$

$$= \left(\prod_{j=n_L-n_{4a}+1}^{n_L} (p_j - 1) \cdot \prod_{j=1}^{n_L-n_{4a}} \left(p_j^{2\cdot(n_{L,p_j}-1)} \cdot (p_j - 1) \right) \right) \cdot$$

$$\cdot \left(\prod_{j=1}^{n_L-n_{4a}} p_j^{n_{L,p_j}-1} - 1 \right) \quad (4.79)$$

When the condition $q_{3,j} = 0 \pmod{p_j}$ is not met for any of the n_{4a} prime factors of type 4. (a), according to condition 4. (a) above, the number of CPPs will be:

$$C_{L,CPPs,0} = \left(\prod_{j=1}^{n_L-n_{4a}} \left(p_j^{n_{L,p_j}-1} \cdot (p_j - 1) \right) \right) \cdot$$

$$\cdot \left(\prod_{j=n_L-n_{4a}+1}^{n_L} p_j \cdot \prod_{j=1}^{n_L-n_{4a}} p_j^{n_{L,p_j}-1} \right) \cdot$$

$$\cdot \left(\prod_{j=n_L-n_{4a}+1}^{n_L} (p_j - 1) \cdot \prod_{j=1}^{n_L-n_{4a}} p_j^{n_{L,p_j}-1} \right) =$$

$$= \prod_{j=n_L-n_{4a}+1}^{n_L} (p_j \cdot (p_j - 1)) \cdot \prod_{j=1}^{n_L-n_{4a}} \left(p_j^{3\cdot(n_{L,p_j}-1)} \cdot (p_j - 1) \right) \quad (4.80)$$

The final number of CPPs results by summing up the quantities in (4.78), for $n_{4a,0} = 1, 2, \ldots, n_{4a} - 1$, (4.79) and (4.80):

$$
\begin{aligned}
C_{L,CPPs} = & \prod_{j=n_L-n_{4a}+1}^{n_L} \left(p_j \cdot (p_j - 1)\right) \cdot \prod_{j=1}^{n_L-n_{4a}} \left(p_j^{3\cdot(n_{L,p_j}-1)} \cdot (p_j - 1)\right) + \\
& + \sum_{n_{4a,0}=1}^{n_{4a}-1} \left(\prod_{j=n_L-n_{4a}+1}^{n_L} (p_j - 1) \cdot \prod_{j\in\{I_{n_{4a}} - I_{n_{4a,0}}\}} p_j \cdot \right. \\
& \left. \cdot \prod_{j=1}^{n_L-n_{4a}} \left(p_j^{3\cdot(n_{L,p_j}-1)} \cdot (p_j - 1)\right)\right) + \\
& + \left(\prod_{j=n_L-n_{4a}+1}^{n_L} (p_j - 1) \cdot \prod_{j=1}^{n_L-n_{4a}} \left(p_j^{2\cdot(n_{L,p_j}-1)} \cdot (p_j - 1)\right)\right) \cdot \\
& \cdot \left(\prod_{j=1}^{n_L-n_{4a}} p_j^{n_{L,p_j}-1} - 1 \right) = \\
= & \prod_{j=n_L-n_{4a}+1}^{n_L} \left(p_j \cdot (p_j - 1)\right) \cdot \prod_{j=1}^{n_L-n_{4a}} \left(p_j^{3\cdot(n_{L,p_j}-1)} \cdot (p_j - 1)\right) + \\
& + \sum_{r_0=1}^{n_{4a}-1} \left(C_{L_{(4a)}, n_{4a}=r, n_{4a,0}=r_0} \cdot \prod_{j=1}^{n_L-n_{4a}} \left(p_j^{3\cdot(n_{L,p_j}-1)} \cdot (p_j - 1)\right)\right) + \\
& + \left(\prod_{j=n_L-n_{4a}+1}^{n_L} (p_j - 1) \cdot \prod_{j=1}^{n_L-n_{4a}} \left(p_j^{2\cdot(n_{L,p_j}-1)} \cdot (p_j - 1)\right)\right) \cdot \\
& \cdot \left(\prod_{j=1}^{n_L-n_{4a}} p_j^{n_{L,p_j}-1} - 1 \right)
\end{aligned} \tag{4.81}
$$

i.e., the one from (4.77), for $n_2 = 0$ and $n_3 = 0$.

For $n_2 = 1$ and $n_3 = 0$, we have $n_{L1} = 2$, and the factors from the decomposition of L are of type 1. (a) and 4. (a) and 3. (a) and/or 3. (b) and/or 4. (b). The number of CPPs results by considering the previous case (for $n_2 = 0$ and $n_3 = 0$) and case 1. (a) above:

$$C_{L,CPPs} = \left(1 \cdot \prod_{j=2}^{n_L - n_{4a}} \left(p_j^{n_{L,p_j}-1} \cdot (p_j - 1)\right)\right) \cdot$$

$$\cdot \left(\prod_{j=n_L-n_{4a}+1}^{n_L} p_j \cdot 1 \cdot \prod_{j=2}^{n_L-n_{4a}} p_j^{n_{L,p_j}-1}\right) \cdot$$

$$\cdot \left(\prod_{j=n_L-n_{4a}+1}^{n_L} (p_j - 1) \cdot 1 \cdot \prod_{j=2}^{n_L-n_{4a}} p_j^{n_{L,p_j}-1}\right) +$$

$$+ \sum_{n_{4a,0}=1}^{n_{4a}-1} \left[\left(\prod_{j \in I_{n_{4a,0}}} (p_j - 1) \cdot 1 \cdot \prod_{j=2}^{n_L-n_{4a}} \left(p_j^{n_{L,p_j}-1} \cdot (p_j - 1)\right)\right) \cdot\right.$$

$$\cdot \left(\prod_{j \in \{I_{n_{4a}} - I_{n_{4a,0}}\}} p_j \cdot 1 \cdot \prod_{j=2}^{n_L-n_{4a}} p_j^{n_{L,p_j}-1}\right) \cdot$$

$$\left.\cdot \left(\prod_{j \in \{I_{n_{4a}} - I_{n_{4a,0}}\}} (p_j - 1) \cdot 1 \cdot \prod_{j=2}^{n_L-n_{4a}} p_j^{n_{L,p_j}-1}\right)\right] + \cdot$$

$$+ \left(\prod_{j=n_L-n_{4a}+1}^{n_L} (p_j - 1) \cdot 1 \cdot \prod_{j=2}^{n_L-n_{4a}} \left(p_j^{n_{L,p_j}-1} \cdot (p_j - 1)\right)\right) \cdot$$

$$\cdot \left(1 \cdot \prod_{j=2}^{n_L-n_{4a}} p_j^{n_{L,p_j}-1}\right) \cdot \left(1 \cdot \prod_{j=2}^{n_L-n_{4a}} p_j^{n_{L,p_j}-1} - 1\right) =$$

$$= \prod_{j=n_L-n_{4a}+1}^{n_L} \left(p_j \cdot (p_j - 1)\right) \cdot \prod_{j=2}^{n_L-n_{4a}} \left(p_j^{3 \cdot (n_{L,p_j}-1)} \cdot (p_j - 1)\right) +$$

$$+ \sum_{n_{4a,0}=1}^{n_{4a}-1} \left(\prod_{j=n_L-n_{4a}+1}^{n_L} (p_j - 1) \cdot \prod_{j \in \{I_{n_{4a}} - I_{n_{4a,0}}\}} p_j \cdot\right.$$

$$\left.\cdot \prod_{j=2}^{n_L-n_{4a}} \left(p_j^{3 \cdot (n_{L,p_j}-1)} \cdot (p_j - 1)\right)\right) +$$

$$+\left(\prod_{j=n_L-n_{4a}+1}^{n_L}(p_j-1)\cdot\prod_{j=2}^{n_L-n_{4a}}\left(p_j^{2\cdot(n_{L,p_j}-1)}\cdot(p_j-1)\right)\right)\cdot$$

$$\cdot\left(\prod_{j=2}^{n_L-n_{4a}}p_j^{n_{L,p_j}-1}-1\right)=$$

$$=\prod_{j=n_L-n_{4a}+1}^{n_L}\left(p_j\cdot(p_j-1)\right)\cdot\prod_{j=2}^{n_L-n_{4a}}\left(p_j^{3\cdot(n_{L,p_j}-1)}\cdot(p_j-1)\right)+$$

$$+\sum_{r_0=1}^{n_{4a}-1}\left(C_{L_{(4a)},n_{4a}=r,n_{4a,0}=r_0}\cdot\prod_{j=2}^{n_L-n_{4a}}\left(p_j^{3\cdot(n_{L,p_j}-1)}\cdot(p_j-1)\right)\right)+$$

$$+\left(\prod_{j=n_L-n_{4a}+1}^{n_L}(p_j-1)\cdot\prod_{j=2}^{n_L-n_{4a}}\left(p_j^{2\cdot(n_{L,p_j}-1)}\cdot(p_j-1)\right)\right)\cdot$$

$$\cdot\left(\prod_{j=2}^{n_L-n_{4a}}p_j^{n_{L,p_j}-1}-1\right)\tag{4.82}$$

i.e., the one from (4.77), for $n_2=1$ and $n_3=0$.

For $n_2=0$ and $n_3=1$, we have $n_{L1}=2$, and the factors from the decomposition of L are of type 2. (a) and 4. (a) and 3. (a) and/or 3. (b) and/or 4. (b). We consider the analysis from the proof of this theorem for $n_2=0$ and $n_3=0$ and the conditions from case 2. (a) above. For a group of $n_{4a,0}$ prime factors of type 4. (a), with $n_{4a,0}\in\{1,2,\ldots,n_{4a}-1\}$, the number of CPPs is equal to:

$$C_{L,CPPs,n_{4a,0}}=\left(\prod_{j\in I_{n_{4a,0}}}(p_j-1)\cdot 2\cdot\prod_{j=2}^{n_L-n_{4a}}\left(p_j^{n_{L,p_j}-1}\cdot(p_j-1)\right)\right)\cdot$$

$$\cdot\left(\prod_{j\in\{I_{n_{4a}}-I_{n_{4a,0}}\}}p_j\cdot\prod_{j=2}^{n_L-n_{4a}}p_j^{n_{L,p_j}-1}\right)\cdot$$

$$\cdot\left(\prod_{j\in\{I_{n_{4a}}-I_{n_{4a,0}}\}}(p_j-1)\cdot\prod_{j=2}^{n_L-n_{4a}}p_j^{n_{L,p_j}-1}\right)=$$

$$= \prod_{j\in I_{n_{4a}}} (p_j - 1)\cdot 2 \cdot \prod_{j\in\{I_{n_{4a}} - I_{n_{4a,0}}\}} p_j \cdot \prod_{j=2}^{n_L - n_{4a}} \left(p_j^{3\cdot(n_{L,p_j}-1)} \cdot (p_j - 1)\right) =$$

$$= \prod_{j=n_L-n_{4a}+1}^{n_L} (p_j - 1)\cdot 2 \cdot \prod_{j\in\{I_{n_{4a}} - I_{n_{4a,0}}\}} p_j \cdot \prod_{j=2}^{n_L - n_{4a}} \left(p_j^{3\cdot(n_{L,p_j}-1)} \cdot (p_j - 1)\right) \quad (4.83)$$

For a group of n_{4a} prime factors of type 4. (a), we have to remove the case when $q_3 = 0$ and the number of CPPs will be equal to:

$$C_{L,CPPs,n_{4a}} = \left(\prod_{j=n_L-n_{4a}+1}^{n_L} (p_j - 1)\cdot 2 \cdot \prod_{j=2}^{n_L - n_{4a}} \left(p_j^{n_{L,p_j}-1} \cdot (p_j - 1)\right)\right) \cdot$$

$$\cdot \left(\prod_{j=2}^{n_L - n_{4a}} p_j^{n_{L,p_j}-1} \right) \cdot \left(\prod_{j=2}^{n_L - n_{4a}} p_j^{n_{L,p_j}-1} - 1 \right) =$$

$$= \left(\prod_{j=n_L-n_{4a}+1}^{n_L} (p_j - 1)\cdot 2 \cdot \prod_{j=2}^{n_L - n_{4a}} \left(p_j^{2\cdot(n_{L,p_j}-1)} \cdot (p_j - 1)\right)\right) \cdot$$

$$\cdot \left(\prod_{j=2}^{n_L - n_{4a}} p_j^{n_{L,p_j}-1} - 1 \right) \quad (4.84)$$

When the condition $q_{3,j} = 0 \pmod{p_j}$ is not met for any of the n_{4a} prime factors of type 4. (a), the number of CPPs will be:

$$C_{L,CPPs,0} = \left(2 \cdot \prod_{j=2}^{n_L - n_{4a}} \left(p_j^{n_{L,p_j}-1} \cdot (p_j - 1)\right)\right) \cdot$$

$$\cdot \left(\prod_{j=n_L-n_{4a}+1}^{n_L} p_j \cdot \prod_{j=2}^{n_L - n_{4a}} p_j^{n_{L,p_j}-1} \right) \cdot$$

$$\cdot \left(\prod_{j=n_L-n_{4a}+1}^{n_L} (p_j - 1) \cdot \prod_{j=2}^{n_L - n_{4a}} p_j^{n_{L,p_j}-1} \right) =$$

$$= \prod_{j=n_L-n_{4a}+1}^{n_L} \left(p_j \cdot (p_j - 1)\right) \cdot 2 \cdot \prod_{j=2}^{n_L-n_{4a}} \left(p_j^{3\cdot(n_{L,p_j}-1)} \cdot (p_j - 1)\right) \tag{4.85}$$

The number of CPPs results by summing up the quantities in (4.83), for $n_{4a,0} = 1, 2, \ldots, n_{4a} - 1$, (4.84) and (4.85):

$$C_{L,CPPs} = \prod_{j=n_L-n_{4a}+1}^{n_L} \left(p_j \cdot (p_j - 1)\right) \cdot$$

$$\cdot 2 \cdot \prod_{j=2}^{n_L-n_{4a}} \left(p_j^{3\cdot(n_{L,p_j}-1)} \cdot (p_j - 1)\right) +$$

$$+ \sum_{n_{4a,0}=1}^{n_{4a}-1} \left(\prod_{j=n_L-n_{4a}+1}^{n_L} (p_j - 1) \cdot 2 \cdot \prod_{j\in\{I_{n_{4a}}-I_{n_{4a,0}}\}} p_j \cdot \right.$$

$$\left. \cdot \prod_{j=2}^{n_L-n_{4a}} \left(p_j^{3\cdot(n_{L,p_j}-1)} \cdot (p_j - 1)\right) \right) +$$

$$+ \left(\prod_{j=n_L-n_{4a}+1}^{n_L} (p_j - 1) \cdot 2 \cdot \prod_{j=2}^{n_L-n_{4a}} \left(p_j^{2\cdot(n_{L,p_j}-1)} \cdot (p_j - 1)\right) \right) \cdot$$

$$\cdot \left(\prod_{j=2}^{n_L-n_{4a}} p_j^{n_{L,p_j}-1} - 1 \right) =$$

$$= 2 \cdot \prod_{j=n_L-n_{4a}+1}^{n_L} \left(p_j \cdot (p_j - 1)\right) \cdot \prod_{j=2}^{n_L-n_{4a}} \left(p_j^{3\cdot(n_{L,p_j}-1)} \cdot (p_j - 1)\right) +$$

$$+2 \cdot \sum_{r_0=1}^{n_{4a}-1} \left(C_{L_{(4a)},n_{4a}=r,n_{4a,0}=r_0} \cdot \prod_{j=2}^{n_L-n_{4a}} \left(p_j^{3\cdot(n_{L,p_j}-1)} \cdot (p_j - 1)\right) \right) +$$

$$+2 \cdot \left(\prod_{j=n_L-n_{4a}+1}^{n_L} (p_j - 1) \cdot \prod_{j=2}^{n_L-n_{4a}} \left(p_j^{2\cdot(n_{L,p_j}-1)} \cdot (p_j - 1)\right) \right) \cdot$$

$$\cdot\left(\prod_{j=2}^{n_L-n_{4a}} p_j^{n_{L,p_j}-1}-1\right) \tag{4.86}$$

i.e., the one from (4.77), for $n_2 = 0$ and $n_3 = 1$.

For $n_2 = 1$ and $n_3 = 1$, we have $n_{L1} = 3$, and the factors from the decomposition of L are of type 1. (a) and 2. (a) and 4. (a) and 3. (a) and/or 3. (b) and/or 4. (b). In determining the number of CPPs, we consider the proof of this theorem for $n_2 = 0$ and $n_3 = 1$ and the analysis for case 1. (a) above. The number of CPPs will be:

$$C_{L,CPPs} = \left(2\cdot\prod_{j=3}^{n_L-n_{4a}}\left(p_j^{n_{L,p_j}-1}\cdot(p_j-1)\right)\right)\cdot$$

$$\cdot\left(\prod_{j=n_L-n_{4a}+1}^{n_L} p_j\cdot\prod_{j=3}^{n_L-n_{4a}} p_j^{n_{L,p_j}-1}\right)\cdot$$

$$\cdot\left(\prod_{j=n_L-n_{4a}+1}^{n_L}(p_j-1)\cdot\prod_{j=3}^{n_L-n_{4a}} p_j^{n_{L,p_j}-1}\right)+$$

$$+\sum_{n_{4a,0}=1}^{n_{4a}-1}\left[\left(\prod_{j\in I_{n_{4a,0}}}(p_j-1)\cdot 2\cdot\prod_{j=3}^{n_L-n_{4a}}\left(p_j^{n_{L,p_j}-1}\cdot(p_j-1)\right)\right)\cdot\right.$$

$$\cdot\left(\prod_{j\in\{I_{n_{4a}}-I_{n_{4a,0}}\}} p_j\cdot\prod_{j=3}^{n_L-n_{4a}} p_j^{n_{L,p_j}-1}\right)\cdot$$

$$\left.\cdot\left(\prod_{j\in\{I_{n_{4a}}-I_{n_{4a,0}}\}}(p_j-1)\cdot\prod_{j=3}^{n_L-n_{4a}} p_j^{n_{L,p_j}-1}\right)\right]+$$

$$+\left(\prod_{j=n_L-n_{4a}+1}^{n_L}(p_j-1)\cdot 2\cdot\prod_{j=3}^{n_L-n_{4a}}\left(p_j^{n_{L,p_j}-1}\cdot(p_j-1)\right)\right)\cdot$$

$$\cdot\left(\prod_{j=3}^{n_L-n_{4a}} p_j^{n_{L,p_j}-1}\right)\cdot\left(\prod_{j=3}^{n_L-n_{4a}} p_j^{n_{L,p_j}-1}-1\right)=$$

$$= \prod_{j=n_L-n_{4a}+1}^{n_L} \left(p_j \cdot (p_j - 1)\right) \cdot 2 \cdot \prod_{j=3}^{n_L-n_{4a}} \left(p_j^{3\cdot(n_{L,p_j}-1)} \cdot (p_j - 1)\right) +$$

$$+ \sum_{n_{4a,0}=1}^{n_{4a}-1} \left(\prod_{j=n_L-n_{4a}+1}^{n_L} (p_j - 1) \cdot \prod_{j \in \{I_{n_{4a}} - I_{n_{4a,0}}\}} p_j \cdot \right.$$

$$\left. \cdot 2 \cdot \prod_{j=3}^{n_L-n_{4a}} \left(p_j^{3\cdot(n_{L,p_j}-1)} \cdot (p_j - 1)\right)\right) +$$

$$+ \left(\prod_{j=n_L-n_{4a}+1}^{n_L} (p_j - 1) \cdot 2 \cdot \prod_{j=3}^{n_L-n_{4a}} \left(p_j^{2\cdot(n_{L,p_j}-1)} \cdot (p_j - 1)\right)\right) \cdot$$

$$\cdot \left(\prod_{j=3}^{n_L-n_{4a}} p_j^{n_{L,p_j}-1} - 1 \right) =$$

$$= 2 \cdot \prod_{j=n_L-n_{4a}+1}^{n_L} \left(p_j \cdot (p_j - 1)\right) \cdot \prod_{j=3}^{n_L-n_{4a}} \left(p_j^{3\cdot(n_{L,p_j}-1)} \cdot (p_j - 1)\right) +$$

$$+2 \cdot \sum_{r_0=1}^{n_{4a}-1} \left(C_{L_{(4a)}, n_{4a}=r, n_{4a,0}=r_0} \cdot \prod_{j=3}^{n_L-n_{4a}} \left(p_j^{3\cdot(n_{L,p_j}-1)} \cdot (p_j - 1)\right)\right) +$$

$$+2 \cdot \left(\prod_{j=n_L-n_{4a}+1}^{n_L} (p_j - 1) \cdot \prod_{j=3}^{n_L-n_{4a}} \left(p_j^{2\cdot(n_{L,p_j}-1)} \cdot (p_j - 1)\right)\right) \cdot$$

$$\cdot \left(\prod_{j=3}^{n_L-n_{4a}} p_j^{n_{L,p_j}-1} - 1 \right) \quad (4.87)$$

i.e., the one from (4.77), for $n_2 = 1$ and $n_3 = 1$. ■

From (4.77) we see that in this case the number of CPPs is always greater than 0.

Theorem 4.17 *If the decomposition of L is of the form* $L = 2^{n_{L,2}} \cdot 3^{n_3} \cdot \prod_{j=n_{L1}}^{n_L-n_{4a}} p_j^{n_{L,p_j}} \cdot \prod_{j=n_L-n_{4a}+1}^{n_L} p_j$, *with* $n_{L,2} > 1$, $n_3 \in \{0, 1\}$, $n_{L1} = n_3 + 2$, $n_{4a} \in \mathbb{N}^*$, $n_L \geq n_{4a} + n_{L1}$,

$p_j > 3$, $n_{L,p_j} \geq 1$, *when* $p_j = 3 \cdot k + 1$, $k \in \mathbb{N}$ *and* $n_{L,p_j} > 1$ *when* $p_j = 3 \cdot k + 2$, $k \in \mathbb{N}^*$, $j = \overline{n_{L1}, n_L - n_{4a}}$ *and* $p_j = 3 \cdot k + 2$, $k \in \mathbb{N}^*$, $j = \overline{n_L - n_{4a} + 1, n_L}$, *the number of CPPs will be equal to:*

$$C_{L,CPPs} = 2^{n_3} \cdot \prod_{j=n_L-n_{4a}+1}^{n_L} (p_j \cdot (p_j - 1)) \cdot 2^{3 \cdot n_{L,2} - 5} \cdot$$

$$\cdot \prod_{j=n_{L1}}^{n_L - n_{4a}} \left(p_j^{3 \cdot (n_{L,p_j} - 1)} \cdot (p_j - 1) \right) +$$

$$+ 2^{n_3} \cdot \sum_{r_0=1}^{n_{4a}-1} \left(C_{L_{(4a)}, n_{4a}=r, n_{4a,0}=r_0} \cdot 2^{3 \cdot n_{L,2} - 5} \cdot \right.$$

$$\left. \cdot \prod_{j=n_{L1}}^{n_L - n_{4a}} \left(p_j^{3 \cdot (n_{L,p_j} - 1)} \cdot (p_j - 1) \right) \right) +$$

$$+ 2^{n_3} \cdot \left(\prod_{j=n_L-n_{4a}+1}^{n_L} (p_j - 1) \cdot 2^{2 \cdot n_{L,2} - 3} \cdot \prod_{j=n_{L1}}^{n_L - n_{4a}} \left(p_j^{2 \cdot (n_{L,p_j} - 1)} \cdot (p_j - 1) \right) \right) \cdot$$

$$\cdot \left(2^{\cdot n_{L,2} - 2} \cdot \prod_{j=n_{L1}}^{n_L - n_{4a}} p_j^{n_{L,p_j} - 1} - 1 \right) \quad (4.88)$$

Proof For $n_3 = 0$, we have $n_{L1} = 2$, and the factors from the decomposition of L are of type 1. (b) and 4. (a) and 3. (a) and/or 3. (b) and/or 4. (b). We consider the proof of Theorem 4.16 for $n_2 = 0$ and $n_3 = 0$ and the additional case 1. (b) above. The number of CPPs will be equal to:

$$C_{L,CPPs} = \left(2^{n_{L,2} - 1} \cdot \prod_{j=2}^{n_L - n_{4a}} \left(p_j^{n_{L,p_j} - 1} \cdot (p_j - 1) \right) \right) \cdot$$

$$\cdot \left(\prod_{j=n_L-n_{4a}+1}^{n_L} p_j \cdot 2^{n_{L,2} - 2} \cdot \prod_{j=2}^{n_L - n_{4a}} p_j^{n_{L,p_j} - 1} \right) \cdot$$

$$\cdot \left(\prod_{j=n_L-n_{4a}+1}^{n_L} (p_j - 1) \cdot 2^{n_{L,2} - 2} \cdot \prod_{j=2}^{n_L - n_{4a}} p_j^{n_{L,p_j} - 1} \right) +$$

$$+\sum_{n_{4a,0}=1}^{n_{4a}-1}\left[\left(\prod_{j\in I_{n_{4a,0}}}(p_j-1)\cdot 2^{n_{L,2}-1}\cdot\prod_{j=2}^{n_L-n_{4a}}\left(p_j^{n_{L,p_j}-1}\cdot(p_j-1)\right)\right)\cdot\right.$$

$$\cdot\left(\prod_{j\in\{I_{n_{4a}}-I_{n_{4a,0}}\}}p_j\cdot 2^{n_{L,2}-2}\cdot\prod_{j=2}^{n_L-n_{4a}}p_j^{n_{L,p_j}-1}\right)\cdot$$

$$\left.\cdot\left(\prod_{j\in\{I_{n_{4a}}-I_{n_{4a,0}}\}}(p_j-1)\cdot 2^{n_{L,2}-2}\cdot\prod_{j=2}^{n_L-n_{4a}}p_j^{n_{L,p_j}-1}\right)\right]+$$

$$+\left(\prod_{j=n_L-n_{4a}+1}^{n_L}(p_j-1)\cdot 2^{n_{L,2}-1}\cdot\prod_{j=2}^{n_L-n_{4a}}\left(p_j^{n_{L,p_j}-1}\cdot(p_j-1)\right)\right)\cdot$$

$$\cdot\left(2^{\cdot n_{L,2}-2}\cdot\prod_{j=2}^{n_L-n_{4a}}p_j^{n_{L,p_j}-1}\right)\cdot\left(2^{\cdot n_{L,2}-2}\cdot\prod_{j=2}^{n_L-n_{4a}}p_j^{n_{L,p_j}-1}-1\right)=$$

$$=\prod_{j=n_L-n_{4a}+1}^{n_L}\left(p_j\cdot(p_j-1)\right)\cdot 2^{3\cdot n_{L,2}-5}\cdot\prod_{j=2}^{n_L-n_{4a}}\left(p_j^{3\cdot(n_{L,p_j}-1)}\cdot(p_j-1)\right)+$$

$$+\sum_{n_{4a,0}=1}^{n_{4a}-1}\left(\prod_{j=n_L-n_{4a}+1}^{n_L}(p_j-1)\cdot\prod_{j\in\{I_{n_{4a}}-I_{n_{4a,0}}\}}p_j\cdot 2^{3\cdot n_{L,2}-5}\cdot\right.$$

$$\left.\cdot\prod_{j=2}^{n_L-n_{4a}}\left(p_j^{3\cdot(n_{L,p_j}-1)}\cdot(p_j-1)\right)\right)+$$

$$+\left(\prod_{j=n_L-n_{4a}+1}^{n_L}(p_j-1)\cdot 2^{2\cdot n_{L,2}-3}\cdot\prod_{j=2}^{n_L-n_{4a}}\left(p_j^{2\cdot(n_{L,p_j}-1)}\cdot(p_j-1)\right)\right)\cdot$$

$$\cdot\left(2^{\cdot n_{L,2}-2}\cdot\prod_{j=2}^{n_L-n_{4a}}p_j^{n_{L,p_j}-1}-1\right)=$$

$$=\prod_{j=n_L-n_{4a}+1}^{n_L}\left(p_j\cdot(p_j-1)\right)\cdot 2^{3\cdot n_{L,2}-5}.$$

$$
\cdot \prod_{j=2}^{n_L-n_{4a}} \left(p_j^{3\cdot(n_{L,p_j}-1)} \cdot (p_j-1) \right) +
$$

$$
+ \sum_{r_0=1}^{n_{4a}-1} \left(C_{L_{(4a)},n_{4a}=r,n_{4a,0}=r_0} \cdot 2^{3\cdot n_{L,2}-5} \cdot \right.
$$

$$
\left. \cdot \prod_{j=2}^{n_L-n_{4a}} \left(p_j^{3\cdot(n_{L,p_j}-1)} \cdot (p_j-1) \right) \right) +
$$

$$
+ \left(\prod_{j=n_L-n_{4a}+1}^{n_L} (p_j-1) \cdot 2^{2\cdot n_{L,2}-3} \cdot \prod_{j=2}^{n_L-n_{4a}} \left(p_j^{2\cdot(n_{L,p_j}-1)} \cdot (p_j-1) \right) \right) \cdot
$$

$$
\cdot \left(2^{\cdot n_{L,2}-2} \cdot \prod_{j=2}^{n_L-n_{4a}} p_j^{n_{L,p_j}-1} - 1 \right) \tag{4.89}
$$

i.e., the one from (4.88), for $n_3 = 0$.

For $n_3 = 1$, we have $n_{L1} = 3$, and the factors from the decomposition of L are of type 1. (b) and 2. (a) and 4. (a) and 3. (a) and/or 3. (b) and/or 4. (b). We consider the proof of Theorem 4.16 for $n_2 = 0$ and $n_3 = 1$ and the additional case 1. (b) above. The number of CPPs is equal to:

$$
C_{L,CPPs} = \left(2^{n_{L,2}-1} \cdot 2 \cdot \prod_{j=3}^{n_L-n_{4a}} \left(p_j^{n_{L,p_j}-1} \cdot (p_j-1) \right) \right) \cdot
$$

$$
\cdot \left(\prod_{j=n_L-n_{4a}+1}^{n_L} p_j \cdot 2^{n_{L,2}-2} \cdot \prod_{j=3}^{n_L-n_{4a}} p_j^{n_{L,p_j}-1} \right) \cdot
$$

$$
\cdot \left(\prod_{j=n_L-n_{4a}+1}^{n_L} (p_j-1) \cdot 2^{n_{L,2}-2} \cdot \prod_{j=3}^{n_L-n_{4a}} p_j^{n_{L,p_j}-1} \right) +
$$

$$
+ \sum_{n_{4a,0}=1}^{n_{4a}-1} \left[\left(\prod_{j\in I_{n_{4a,0}}} (p_j-1) \cdot 2^{n_{L,2}-1} \cdot 2 \cdot \prod_{j=3}^{n_L-n_{4a}} \left(p_j^{n_{L,p_j}-1} \cdot (p_j-1) \right) \right) \cdot \right.
$$

$$
\cdot \left(\prod_{j\in\{I_{n_{4a}}-I_{n_{4a,0}}\}} p_j \cdot 2^{n_{L,2}-2} \cdot \prod_{j=3}^{n_L-n_{4a}} p_j^{n_{L,p_j}-1} \right) \cdot
$$

$$\cdot\left(\prod_{j\in\{I_{n_{4a}}-I_{n_{4a,0}}\}}(p_j-1)\cdot 2^{n_{L,2}-2}\cdot\prod_{j=3}^{n_L-n_{4a}}p_j^{n_{L,p_j}-1}\right)\Bigg]+$$

$$+\left(\prod_{j=n_L-n_{4a}+1}^{n_L}(p_j-1)\cdot 2^{n_{L,2}-1}\cdot 2\cdot\prod_{j=3}^{n_L-n_{4a}}\left(p_j^{n_{L,p_j}-1}\cdot(p_j-1)\right)\right)\cdot$$

$$\cdot\left(2^{\cdot n_{L,2}-2}\cdot\prod_{j=3}^{n_L-n_{4a}}p_j^{n_{L,p_j}-1}\right)\cdot\left(2^{\cdot n_{L,2}-2}\cdot\prod_{j=3}^{n_L-n_{4a}}p_j^{n_{L,p_j}-1}-1\right)=$$

$$=\prod_{j=n_L-n_{4a}+1}^{n_L}\left(p_j\cdot(p_j-1)\right)\cdot 2^{3\cdot n_{L,2}-4}\cdot\prod_{j=3}^{n_L-n_{4a}}\left(p_j^{3\cdot(n_{L,p_j}-1)}\cdot(p_j-1)\right)+$$

$$+\sum_{n_{4a,0}=1}^{n_{4a}-1}\left(\prod_{j=n_L-n_{4a}+1}^{n_L}(p_j-1)\cdot\prod_{j\in\{I_{n_{4a}}-I_{n_{4a,0}}\}}p_j\cdot 2^{3\cdot n_{L,2}-4}\cdot\right.$$

$$\left.\cdot\prod_{j=3}^{n_L-n_{4a}}\left(p_j^{3\cdot(n_{L,p_j}-1)}\cdot(p_j-1)\right)\right)+$$

$$+\left(\prod_{j=n_L-n_{4a}+1}^{n_L}(p_j-1)\cdot 2^{2\cdot(n_{L,2}-1)}\cdot\prod_{j=3}^{n_L-n_{4a}}\left(p_j^{2\cdot(n_{L,p_j}-1)}\cdot(p_j-1)\right)\right)\cdot$$

$$\cdot\left(2^{\cdot n_{L,2}-2}\cdot\prod_{j=3}^{n_L-n_{4a}}p_j^{n_{L,p_j}-1}-1\right)=$$

$$=\prod_{j=n_L-n_{4a}+1}^{n_L}\left(p_j\cdot(p_j-1)\right)\cdot 2^{3\cdot n_{L,2}-4}\cdot$$

$$\cdot\prod_{j=3}^{n_L-n_{4a}}\left(p_j^{3\cdot(n_{L,p_j}-1)}\cdot(p_j-1)\right)+$$

$$+\sum_{r_0=1}^{n_{4a}-1}\left(C_{L_{(4a)},n_{4a}=r,n_{4a,0}=r_0}\cdot 2^{3\cdot n_{L,2}-4}\cdot\right.$$

$$\cdot \prod_{j=3}^{n_L-n_{4a}} \left(p_j^{3\cdot(n_{L,p_j}-1)} \cdot (p_j-1) \right) \Bigg) +$$

$$+ \left(\prod_{j=n_L-n_{4a}+1}^{n_L} (p_j-1) \cdot 2^{2\cdot(n_{L,2}-1)} \cdot \prod_{j=3}^{n_L-n_{4a}} \left(p_j^{2\cdot(n_{L,p_j}-1)} \cdot (p_j-1) \right) \right) \cdot$$

$$\cdot \left(2^{\cdot n_{L,2}-2} \cdot \prod_{j=3}^{n_L-n_{4a}} p_j^{n_{L,p_j}-1} - 1 \right) \tag{4.90}$$

i.e., the one from (4.88), for $n_3 = 1$. ■

From (4.88) we see that in this case the number of CPPs is always greater than 0.

Theorem 4.18 *If the decomposition of L is of the form* $L = 2^{n_2} \cdot 3^{n_{L,3}} \cdot \prod_{j=n_{L1}}^{n_L-n_{4a}} p_j^{n_{L,p_j}} \cdot \prod_{j=n_L-n_{4a}+1}^{n_L} p_j$, *with* $n_2 \in \{0, 1\}$, $n_{L,3} > 1$, $n_{L1} = n_2 + 2$, $n_{4a} \in \mathbb{N}^*$, $n_L \geq n_{4a} + n_{L1}$, $p_j > 3$, $n_{L,p_j} \geq 1$, *when* $p_j = 3 \cdot k + 1$, $k \in \mathbb{N}$ *and* $n_{L,p_j} > 1$ *when* $p_j = 3 \cdot k + 2$, $k \in \mathbb{N}^*$, $j = \overline{n_{L1}, n_L - n_{4a}}$ *and* $p_j = 3 \cdot k + 2$, $k \in \mathbb{N}^*$, $j = \overline{n_L - n_{4a} + 1, n_L}$, *the number of CPPs will be equal to:*

$$C_{L,CPPs} = 4 \cdot \left(\prod_{j=n_L-n_{4a}+1}^{n_L} (p_j \cdot (p_j-1)) \cdot 3^{3\cdot n_{L,3}-4} \cdot \right.$$

$$\cdot \prod_{j=n_{L1}}^{n_L-n_{4a}} \left(p_j^{3\cdot(n_{L,p_j}-1)} \cdot (p_j-1) \right) \Bigg) +$$

$$+4 \cdot \sum_{r_0=1}^{n_{4a}-1} \left(C_{L_{(4a)},n_{4a}=r,n_{4a,0}=r_0} \cdot 3^{3\cdot n_{L,3}-4} \cdot \right.$$

$$\cdot \prod_{j=n_{L1}}^{n_L-n_{4a}} \left(p_j^{3\cdot(n_{L,p_j}-1)} \cdot (p_j-1) \right) \Bigg) +$$

$$+2 \cdot \left(\prod_{j=n_L-n_{4a}+1}^{n_L} (p_j-1) \cdot 3^{2\cdot(n_{L,3}-1)} \cdot \right.$$

$$\cdot \prod_{j=n_{L1}}^{n_L - n_{4a}} \left(p_j^{2\cdot(n_{L,p_j}-1)} \cdot (p_j - 1) \right) \Bigg) \cdot$$

$$\cdot \left(2 \cdot 3^{n_{L,3}-2} \cdot \prod_{j=n_{L1}}^{n_L - n_{4a}} p_j^{n_{L,p_j}-1} - 1 \right) \tag{4.91}$$

Proof For $n_2 = 0$, we have $n_{L1} = 2$, and the factors from the decomposition of L are of type 2. (b) and 4. (a) and 3. (a) and/or 3. (b) and/or 4. (b). We consider the analysis from the proof of the Theorem 4.16 for $n_2 = 0$ and $n_3 = 0$ and the conditions from the case 2. (b) above. For a group of $n_{4a,0}$ prime factors of the type 4. (a), with $n_{4a,0} \in \{1, 2, \ldots, n_{4a} - 1\}$, the number of CPPs is equal to:

$$C_{L,CPPs,n_{4a,0}} = 2 \cdot \Bigg(\prod_{j \in I_{n_{4a,0}}} (p_j - 1) \cdot 3^{n_{L,3}-1} \cdot$$

$$\cdot \prod_{j=2}^{n_L - n_{4a}} \left(p_j^{n_{L,p_j}-1} \cdot (p_j - 1) \right) \Bigg) \cdot$$

$$\cdot \left(\prod_{j \in \{I_{n_{4a}} - I_{n_{4a,0}}\}} p_j \cdot 3^{n_{L,3}-1} \cdot \prod_{j=2}^{n_L - n_{4a}} p_j^{n_{L,p_j}-1} \right) \cdot$$

$$\cdot \left(\prod_{j \in \{I_{n_{4a}} - I_{n_{4a,0}}\}} (p_j - 1) \cdot 2 \cdot 3^{n_{L,3}-2} \cdot \prod_{j=2}^{n_L - n_{4a}} p_j^{n_{L,p_j}-1} \right) =$$

$$= \prod_{j \in I_{n_{4a}}} (p_j - 1) \cdot 3^{3\cdot n_{L,3}-4} \cdot \prod_{j \in \{I_{n_{4a}} - I_{n_{4a,0}}\}} p_j \cdot$$

$$\cdot \prod_{j=2}^{n_L - n_{4a}} \left(p_j^{3\cdot(n_{L,p_j}-1)} \cdot (p_j - 1) \right) =$$

$$= \prod_{j=n_L - n_{4a}+1}^{n_L} (p_j - 1) \cdot 3^{3\cdot n_{L,3}-4} \cdot \prod_{j \in \{I_{n_{4a}} - I_{n_{4a,0}}\}} p_j \cdot$$

$$\cdot \prod_{j=2}^{n_L - n_{4a}} \left(p_j^{3\cdot(n_{L,p_j}-1)} \cdot (p_j - 1) \right) \tag{4.92}$$

For a group of n_{4a} prime factors of the type 4. (a), we have to remove the case when $q_3 = 0$ and the number of CPPs will be equal to:

$$C_{L,CPPs,n_{4a}} = 2 \cdot \left(\prod_{j=n_L-n_{4a}+1}^{n_L} (p_j - 1) \cdot 3^{n_{L,3}-1} \cdot \prod_{j=2}^{n_L-n_{4a}} \left(p_j^{n_{L,p_j}-1} \cdot (p_j - 1) \right) \right) \cdot$$
$$\cdot \left(3^{n_{L,3}-1} \cdot \prod_{j=2}^{n_L-n_{4a}} p_j^{n_{L,p_j}-1} \right) \cdot \left(2 \cdot 3^{n_{L,3}-2} \cdot \prod_{j=2}^{n_L-n_{4a}} p_j^{n_{L,p_j}-1} - 1 \right) =$$
$$= 2 \cdot \left(\prod_{j=n_L-n_{4a}+1}^{n_L} (p_j - 1) \cdot 3^{2\cdot(n_{L,3}-1)} \cdot \prod_{j=2}^{n_L-n_{4a}} \left(p_j^{2\cdot(n_{L,p_j}-1)} \cdot (p_j - 1) \right) \right) \cdot$$
$$\cdot \left(2 \cdot 3^{n_{L,3}-2} \cdot \prod_{j=2}^{n_L-n_{4a}} p_j^{n_{L,p_j}-1} - 1 \right) \tag{4.93}$$

When the condition $q_{3,j} = 0 \pmod{p_j}$ is not met for any of the n_{4a} prime factors of the type 4. (a), the number of CPPs will be:

$$C_{L,CPPs,0} = 2 \cdot \left(3^{n_{L,3}-1} \cdot \prod_{j=2}^{n_L-n_{4a}} \left(p_j^{n_{L,p_j}-1} \cdot (p_j - 1) \right) \right) \cdot$$
$$\cdot \left(\prod_{j=n_L-n_{4a}+1}^{n_L} p_j \cdot 3^{n_{L,3}-1} \cdot \prod_{j=2}^{n_L-n_{4a}} p_j^{n_{L,p_j}-1} \right) \cdot$$
$$\cdot \left(\prod_{j=n_L-n_{4a}+1}^{n_L} (p_j - 1) \cdot 2 \cdot 3^{n_{L,3}-2} \cdot \prod_{j=2}^{n_L-n_{4a}} p_j^{n_{L,p_j}-1} \right) =$$
$$= 4 \cdot \left(\prod_{j=n_L-n_{4a}+1}^{n_L} (p_j \cdot (p_j - 1)) \cdot 3^{3\cdot n_{L,3}-4} \cdot \right.$$

$$\cdot \prod_{j=2}^{n_L-n_{4a}} \left(p_j^{3\cdot(n_{L,p_j}-1)} \cdot (p_j-1) \right) \Bigg) \tag{4.94}$$

The number of CPPs results by summing up the quantities in (4.92), for $n_{4a,0} = 1, 2, \ldots, n_{4a}-1$, (4.93) and (4.94):

$$C_{L,CPPs} = 4 \cdot \left(\prod_{j=n_L-n_{4a}+1}^{n_L} \left(p_j \cdot (p_j-1)\right) \cdot 3^{3\cdot n_{L,3}-4} \cdot \right.$$

$$\left. \cdot \prod_{j=2}^{n_L-n_{4a}} \left(p_j^{3\cdot(n_{L,p_j}-1)} \cdot (p_j-1) \right) \right) +$$

$$+4 \cdot \sum_{n_{4a,0}=1}^{n_{4a}-1} \left(\prod_{j=n_L-n_{4a}+1}^{n_L} (p_j-1) \cdot 3^{3\cdot n_{L,3}-4} \cdot \prod_{j\in\{I_{n_{4a}}-I_{n_{4a,0}}\}} p_j \cdot \right.$$

$$\left. \cdot \prod_{j=2}^{n_L-n_{4a}} \left(p_j^{3\cdot(n_{L,p_j}-1)} \cdot (p_j-1) \right) \right) +$$

$$+2 \cdot \left(\prod_{j=n_L-n_{4a}+1}^{n_L} (p_j-1) \cdot 3^{2\cdot(n_{L,3}-1)} \cdot \prod_{j=2}^{n_L-n_{4a}} \left(p_j^{2\cdot(n_{L,p_j}-1)} \cdot (p_j-1) \right) \right) \cdot$$

$$\cdot \left(2 \cdot 3^{n_{L,3}-2} \cdot \prod_{j=2}^{n_L-n_{4a}} p_j^{n_{L,p_j}-1} - 1 \right) =$$

$$= 4 \cdot \left(\prod_{j=n_L-n_{4a}+1}^{n_L} \left(p_j \cdot (p_j-1)\right) \cdot 3^{3\cdot n_{L,3}-4} \cdot \right.$$

$$\left. \cdot \prod_{j=2}^{n_L-n_{4a}} \left(p_j^{3\cdot(n_{L,p_j}-1)} \cdot (p_j-1) \right) \right) +$$

$$+4 \cdot \sum_{r_0=1}^{n_{4a}-1} \left(C_{L_{(4a)},n_{4a}=r,n_{4a,0}=r_0} \cdot 3^{3\cdot n_{L,3}-4} \cdot \right.$$

$$\cdot \prod_{j=2}^{n_L-n_{4a}} \left(p_j^{3\cdot(n_{L,p_j}-1)} \cdot (p_j-1) \right) \Bigg) +$$

$$+2 \cdot \left(\prod_{j=n_L-n_{4a}+1}^{n_L} (p_j-1) \cdot 3^{2\cdot(n_{L,3}-1)} \cdot \prod_{j=2}^{n_L-n_{4a}} \left(p_j^{2\cdot(n_{L,p_j}-1)} \cdot (p_j-1) \right) \right) \cdot$$

$$\cdot \left(2 \cdot 3^{n_{L,3}-2} \cdot \prod_{j=2}^{n_L-n_{4a}} p_j^{n_{L,p_j}-1} - 1 \right) \tag{4.95}$$

i.e., the one from (4.91), for $n_2 = 0$.

For $n_2 = 1$, we have $n_{L1} = 3$, and the factors from the decomposition of L are of type 1. (a) and 2. (b) and 4. (a) and 3. (a) and/or 3. (b) and/or 4. (b). In determining the number of CPPs we consider the proof of this theorem for $n_2 = 0$ and the analysis for case 1. (a) above. The number of CPPs will be:

$$C_{L,CPPs} = 2 \cdot \left(3^{n_{L,3}-1} \cdot \prod_{j=3}^{n_L-n_{4a}} \left(p_j^{n_{L,p_j}-1} \cdot (p_j-1) \right) \right) \cdot$$

$$\cdot \left(\prod_{j=n_L-n_{4a}+1}^{n_L} p_j \cdot 3^{n_{L,3}-1} \cdot \prod_{j=3}^{n_L-n_{4a}} p_j^{n_{L,p_j}-1} \right) \cdot$$

$$\cdot \left(\prod_{j=n_L-n_{4a}+1}^{n_L} (p_j-1) \cdot 2 \cdot 3^{n_{L,3}-2} \cdot \prod_{j=3}^{n_L-n_{4a}} p_j^{n_{L,p_j}-1} \right) +$$

$$+2 \cdot \sum_{n_{4a,0}=1}^{n_{4a}-1} \left[\left(\prod_{j \in I_{n_{4a,0}}} (p_j-1) \cdot 3^{n_{L,3}-1} \cdot \prod_{j=3}^{n_L-n_{4a}} \left(p_j^{n_{L,p_j}-1} \cdot (p_j-1) \right) \right) \cdot \right.$$

$$\cdot \left(\prod_{j \in \{I_{n_{4a}} - I_{n_{4a,0}}\}} p_j \cdot 3^{n_{L,3}-1} \cdot \prod_{j=3}^{n_L-n_{4a}} p_j^{n_{L,p_j}-1} \right) \cdot$$

$$\left. \cdot \left(\prod_{j \in \{I_{n_{4a}} - I_{n_{4a,0}}\}} (p_j-1) \cdot 2 \cdot 3^{n_{L,3}-2} \cdot \prod_{j=3}^{n_L-n_{4a}} p_j^{n_{L,p_j}-1} \right) \right] +$$

$$+2\cdot\left(\prod_{j=n_L-n_{4a}+1}^{n_L}(p_j-1)\cdot 3^{n_{L,3}-1}\cdot\prod_{j=3}^{n_L-n_{4a}}\left(p_j^{n_{L,p_j}-1}\cdot(p_j-1)\right)\right)\cdot$$

$$\cdot\left(3^{n_{L,3}-1}\cdot\prod_{j=3}^{n_L-n_{4a}}p_j^{n_{L,p_j}-1}\right)\cdot\left(2\cdot 3^{n_{L,3}-2}\cdot\prod_{j=3}^{n_L-n_{4a}}p_j^{n_{L,p_j}-1}-1\right)=$$

$$=4\cdot\left(\prod_{j=n_L-n_{4a}+1}^{n_L}(p_j\cdot(p_j-1))\cdot 3^{3\cdot n_{L,3}-4}\cdot\right.$$

$$\left.\cdot\prod_{j=3}^{n_L-n_{4a}}\left(p_j^{3\cdot(n_{L,p_j}-1)}\cdot(p_j-1)\right)\right)+$$

$$+4\cdot\sum_{n_{4a,0}=1}^{n_{4a}-1}\left(\prod_{j\in I_{n_{4a}}}(p_j-1)\cdot 3^{3\cdot n_{L,3}-4}\cdot\prod_{j\in\{I_{n_{4a}}-I_{n_{4a,0}}\}}p_j\cdot\right.$$

$$\left.\cdot\prod_{j=3}^{n_L-n_{4a}}\left(p_j^{3\cdot(n_{L,p_j}-1)}\cdot(p_j-1)\right)\right)+$$

$$+2\cdot\left(\prod_{j=n_L-n_{4a}+1}^{n_L}(p_j-1)\cdot 3^{2\cdot(n_{L,3}-1)}\cdot\right.$$

$$\left.\cdot\prod_{j=3}^{n_L-n_{4a}}\left(p_j^{2\cdot(n_{L,p_j}-1)}\cdot(p_j-1)\right)\right)\cdot\left(2\cdot 3^{n_{L,3}-2}\cdot\prod_{j=3}^{n_L-n_{4a}}p_j^{n_{L,p_j}-1}-1\right)=$$

$$=4\cdot\left(\prod_{j=n_L-n_{4a}+1}^{n_L}(p_j\cdot(p_j-1))\cdot 3^{3\cdot n_{L,3}-4}\cdot\right.$$

$$\left.\cdot\prod_{j=3}^{n_L-n_{4a}}\left(p_j^{3\cdot(n_{L,p_j}-1)}\cdot(p_j-1)\right)\right)+$$

$$+4\cdot\sum_{r_0=1}^{n_{4a}-1}\left(C_{L_{(4a)},n_{4a}=r,n_{4a,0}=r_0}\cdot 3^{3\cdot n_{L,3}-4}\cdot\right.$$

$$\cdot \prod_{j=3}^{n_L-n_{4a}} \left(p_j^{3\cdot(n_{L,p_j}-1)} \cdot (p_j - 1) \right) \Bigg) +$$

$$+2 \cdot \left(\prod_{j=n_L-n_{4a}+1}^{n_L} (p_j - 1) \cdot 3^{2\cdot(n_{L,3}-1)} \cdot \prod_{j=3}^{n_L-n_{4a}} \left(p_j^{2\cdot(n_{L,p_j}-1)} \cdot (p_j - 1) \right) \right) \cdot$$

$$\cdot \left(2 \cdot 3^{n_{L,3}-2} \cdot \prod_{j=3}^{n_L-n_{4a}} p_j^{n_{L,p_j}-1} - 1 \right) \tag{4.96}$$

i.e., the one from (4.91), for $n_2 = 1$. ■

From (4.91) we see that in this case the number of CPPs is always greater than 0.

Theorem 4.19 *If the decomposition of L is of the form* $L = 2^{n_{L,2}} \cdot 3^{n_{L,3}} \cdot \prod_{j=3}^{n_L-n_{4a}} p_j^{n_{L,p_j}} \cdot \prod_{j=n_L-n_{4a}+1}^{n_L} p_j$, *with* $n_{L,2} > 1$, $n_{L,3} > 1$, $n_{4a} \in \mathbb{N}^*$, $n_L \geq n_{4a} + 3$, $p_j > 3$, $n_{L,p_j} \geq 1$, *when* $p_j = 3 \cdot k + 1$, $k \in \mathbb{N}$ *and* $n_{L,p_j} > 1$ *when* $p_j = 3 \cdot k + 2$, $k \in \mathbb{N}^*$, $j = \overline{3, n_L - n_{4a}}$ *and* $p_j = 3 \cdot k + 2$, $k \in \mathbb{N}^*$, $j = \overline{n_L - n_{4a} + 1, n_L}$, *the number of CPPs will be equal to:*

$$C_{L,CPPs} = \prod_{j=n_L-n_{4a}+1}^{n_L} (p_j \cdot (p_j - 1)) \cdot 2^{3\cdot(n_{L,2}-1)} \cdot 3^{3\cdot n_{L,3}-4} \cdot$$

$$\cdot \prod_{j=3}^{n_L-n_{4a}} \left(p_j^{3\cdot(n_{L,p_j}-1)} \cdot (p_j - 1) \right) +$$

$$+ \sum_{r_0=1}^{n_{4a}-1} \Bigg(C_{L_{(4a)},n_{4a}=r,n_{4a,0}=r_0} \cdot 2^{3\cdot(n_{L,2}-1)} \cdot 3^{3\cdot n_{L,3}-4} \cdot$$

$$\cdot \prod_{j=3}^{n_L-n_{4a}} \left(p_j^{3\cdot(n_{L,p_j}-1)} \cdot (p_j - 1) \right) \Bigg) +$$

$$+ \left(\prod_{j=n_L-n_{4a}+1}^{n_L} (p_j - 1) \cdot 2^{2\cdot(n_{L,2}-1)} \cdot 3^{2\cdot(n_{L,3}-1)} \cdot \right.$$

$$\left. \cdot \prod_{j=3}^{n_L-n_{4a}} \left(p_j^{2\cdot(n_{L,p_j}-1)} \cdot (p_j - 1) \right) \right) \cdot$$

$$\cdot\left(2^{n_{L,2}-1}\cdot 3^{n_{L,3}-2}\cdot\prod_{j=3}^{n_L-n_{4a}} p_j^{n_{L,p_j}-1}-1\right) \tag{4.97}$$

Proof We consider the proof of Theorem 4.18 for $n_2=0$ and the additional case 1. (b) above. The number of CPPs is equal to:

$$C_{L,CPPs}=\left(2\cdot 2^{n_{L,2}-1}\cdot 3^{n_{L,3}-1}\cdot\prod_{j=3}^{n_L-n_{4a}}\left(p_j^{n_{L,p_j}-1}\cdot(p_j-1)\right)\right)\cdot$$

$$\cdot\left(\prod_{j=n_L-n_{4a}+1}^{n_L} p_j\cdot 2^{n_{L,2}-2}\cdot 3^{n_{L,3}-1}\cdot\prod_{j=3}^{n_L-n_{4a}} p_j^{n_{L,p_j}-1}\right)\cdot$$

$$\cdot\left(\prod_{j=n_L-n_{4a}+1}^{n_L}(p_j-1)\cdot 2^{n_{L,2}-2}\cdot 2\cdot 3^{n_{L,3}-2}\cdot\prod_{j=3}^{n_L-n_{4a}} p_j^{n_{L,p_j}-1}\right)+$$

$$+2\cdot\sum_{n_{4a,0}=1}^{n_{4a}-1}\left[\left(\prod_{j\in I_{n_{4a,0}}}(p_j-1)\cdot 2^{n_{L,2}-1}\cdot 3^{n_{L,3}-1}\cdot\right.\right.$$

$$\left.\cdot\prod_{j=3}^{n_L-n_{4a}}\left(p_j^{n_{L,p_j}-1}\cdot(p_j-1)\right)\right)\cdot$$

$$\cdot\left(\prod_{j\in\{I_{n_{4a}}-I_{n_{4a,0}}\}} p_j\cdot 2^{n_{L,2}-2}\cdot 3^{n_{L,3}-1}\cdot\prod_{j=3}^{n_L-n_{4a}} p_j^{n_{L,p_j}-1}\right)\cdot$$

$$\left.\cdot\left(\prod_{j\in\{I_{n_{4a}}-I_{n_{4a,0}}\}}(p_j-1)\cdot 2^{n_{L,2}-2}\cdot 2\cdot 3^{n_{L,3}-2}\cdot\prod_{j=3}^{n_L-n_{4a}} p_j^{n_{L,p_j}-1}\right)\right]+$$

$$+2\cdot\left(\prod_{j=n_L-n_{4a}+1}^{n_L}(p_j-1)\cdot 2^{n_{L,2}-1}\cdot 3^{n_{L,3}-1}\cdot\right.$$

$$\left.\cdot\prod_{j=3}^{n_L-n_{4a}}\left(p_j^{n_{L,p_j}-1}\cdot(p_j-1)\right)\right)\cdot\left(2^{n_{L,2}-2}\cdot 3^{n_{L,3}-1}\cdot\prod_{j=3}^{n_L-n_{4a}} p_j^{n_{L,p_j}-1}\right)\cdot$$

$$\cdot\left(2^{n_{L,2}-2}\cdot 2\cdot 3^{n_{L,3}-2}\cdot\prod_{j=3}^{n_L-n_{4a}} p_j^{n_{L,p_j}-1}-1\right)=$$

$$=\prod_{j=n_L-n_{4a}+1}^{n_L}\left(p_j\cdot(p_j-1)\right)\cdot 2^{3\cdot(n_{L,2}-1)}\cdot 3^{3\cdot n_{L,3}-4}\cdot$$

$$\cdot\prod_{j=3}^{n_L-n_{4a}}\left(p_j^{3\cdot(n_{L,p_j}-1)}\cdot(p_j-1)\right)+$$

$$+4\cdot\sum_{n_{4a,0}=1}^{n_{4a}-1}\left(\prod_{j=n_L-n_{4a}+1}^{n_L}(p_j-1)\cdot\prod_{j\in\{I_{n_{4a}}-I_{n_{4a,0}}\}} p_j\cdot 2^{3\cdot(n_{L,2}-1)}\cdot 3^{3\cdot n_{L,3}-4}\cdot\right.$$

$$\left.\cdot\prod_{j=3}^{n_L-n_{4a}}\left(p_j^{3\cdot(n_{L,p_j}-1)}\cdot(p_j-1)\right)\right)+$$

$$+\left(\prod_{j=n_L-n_{4a}+1}^{n_L}(p_j-1)\cdot 2^{2\cdot(n_{L,2}-1)}\cdot 3^{2\cdot(n_{L,3}-1)}\cdot\right.$$

$$\left.\cdot\prod_{j=3}^{n_L-n_{4a}}\left(p_j^{2\cdot(n_{L,p_j}-1)}\cdot(p_j-1)\right)\right)\cdot$$

$$\cdot\left(2^{n_{L,2}-1}\cdot 3^{n_{L,3}-2}\cdot\prod_{j=3}^{n_L-n_{4a}} p_j^{n_{L,p_j}-1}-1\right)=$$

$$=\prod_{j=n_L-n_{4a}+1}^{n_L}\left(p_j\cdot(p_j-1)\right)\cdot 2^{3\cdot(n_{L,2}-1)}\cdot 3^{3\cdot n_{L,3}-4}\cdot$$

$$\cdot\prod_{j=3}^{n_L-n_{4a}}\left(p_j^{3\cdot(n_{L,p_j}-1)}\cdot(p_j-1)\right)+$$

$$+\sum_{r_0=1}^{n_{4a}-1}\left(C_{L_{(4a)},n_{4a}=r,n_{4a,0}=r_0}\cdot 2^{3\cdot(n_{L,2}-1)}\cdot 3^{3\cdot n_{L,3}-4}\cdot\right.$$

$$\cdot \prod_{j=3}^{n_L-n_{4a}} \left(p_j^{3\cdot(n_{L,p_j}-1)} \cdot (p_j-1) \right) \Bigg) +$$

$$+ \left(\prod_{j=n_L-n_{4a}+1}^{n_L} (p_j-1) \cdot 2^{2\cdot(n_{L,2}-1)} \cdot 3^{2\cdot(n_{L,3}-1)} \cdot \right.$$

$$\left. \cdot \prod_{j=3}^{n_L-n_{4a}} \left(p_j^{2\cdot(n_{L,p_j}-1)} \cdot (p_j-1) \right) \right) \cdot$$

$$\cdot \left(2^{n_{L,2}-1} \cdot 3^{n_{L,3}-2} \cdot \prod_{j=3}^{n_L-n_{4a}} p_j^{n_{L,p_j}-1} - 1 \right) \tag{4.98}$$

■

From (4.97) we see that in this case the number of CPPs is always greater than 0.

Theorem 4.20 *If the decomposition of L is of the form* $L = 2^{n_2} \cdot 3^{n_3} \cdot \prod_{j=n_{L1}}^{n_{4a}+n_{L1}-1} p_j$, *with* $n_2 \in \{0, 1\}$, $n_3 \in \{0, 1\}$, $n_{L1} = n_2 + n_3 + 1$, $n_{4a} \in \mathbb{N}^*$, $p_j = 3 \cdot k + 2$, $k \in \mathbb{N}^*$, $j = \overline{n_{L1}, n_{4a} + n_{L1} - 1}$, *the number of CPPs will be equal to:*

$$C_{L,CPPs} = 2^{n_3} \cdot \prod_{j=n_{L1}}^{n_{4a}+n_{L1}-1} \left(p_j \cdot (p_j-1)\right) + 2^{n_3} \cdot \sum_{r_0=1}^{n_{4a}-1} C_{L_{(4a)},n_{4a}=r,n_{4a,0}=r_0} \tag{4.99}$$

Proof This is a particular case of Theorem 4.16, therefore we can use Eq. (4.77) in which the products $\prod_{j=n_{L1}}^{n_L-n_{4a}} \left(p_j^{3\cdot(n_{L,p_j}-1)} \cdot (p_j-1) \right)$, $\prod_{j=n_{L1}}^{n_L-n_{4a}} \left(p_j^{2\cdot(n_{L,p_j}-1)} \cdot (p_j-1) \right)$ and $\prod_{j=n_{L1}}^{n_L-n_{4a}} p_j^{n_{L,p_j}-1}$ are replaced by 1, and $n_L = n_{4a} + n_{L1} - 1$. ■

From (4.99) we see that in this case the number of CPPs is always greater than 0.

Theorem 4.21 *If the decomposition of L is of the form* $L = 2^{n_{L,2}} \cdot 3^{n_3} \cdot \prod_{j=n_{L1}}^{n_{4a}+n_{L1}-1} p_j$, *with* $n_{L,2} > 1$, $n_3 \in \{0, 1\}$, $n_{L1} = n_3 + 2$, $n_{4a} \in \mathbb{N}^*$, $p_j = 3 \cdot k + 2$, $k \in \mathbb{N}^*$, $j = \overline{n_{L1}, n_{4a} + n_{L1} - 1}$, *the number of CPPs will be equal to:*

$$C_{L,CPPs} = 2^{n_3} \cdot \prod_{j=n_{L1}}^{n_{4a}+n_{L1}-1} \left(p_j \cdot (p_j-1)\right) \cdot 2^{3\cdot n_{L,2}-5} +$$

$$+2^{n_3} \cdot \sum_{r_0=1}^{n_{4a}-1} \left(C_{L_{(4a)}, n_{4a}=r, n_{4a,0}=r_0} \cdot 2^{3 \cdot n_{L,2}-5} \right) +$$

$$+2^{n_3} \cdot \left(\prod_{j=n_{L1}}^{n_{4a}+n_{L1}-1} (p_j - 1) \cdot 2^{2 \cdot n_{L,2}-3} \right) \cdot \left(2^{\cdot n_{L,2}-2} - 1 \right) \tag{4.100}$$

Proof This is a particular case of Theorem 4.17, therefore we can use Eq. (4.88), in which the products $\prod_{j=n_{L1}}^{n_L-n_{4a}} \left(p_j^{3 \cdot (n_{L,p_j}-1)} \cdot (p_j - 1) \right)$, $\prod_{j=n_{L1}}^{n_L-n_{4a}} \left(p_j^{2 \cdot (n_{L,p_j}-1)} \cdot (p_j - 1) \right)$ and $\prod_{j=n_{L1}}^{n_L-n_{4a}} p_j^{n_{L,p_j}-1}$ are replaced by 1, and $n_L = n_{4a} + n_{L1} - 1$. ■

From (4.100) we see that in this case the number of CPPs is always greater than 0.

Theorem 4.22 *If the decomposition of L is of the form* $L = 2^{n_2} \cdot 3^{n_{L,3}} \cdot \prod_{j=n_{L1}}^{n_{4a}+n_{L1}-1} p_j$, *with* $n_2 \in \{0, 1\}$, $n_{L,3} > 1$, $n_{L1} = n_2 + 2$, $n_{4a} \in \mathbb{N}^*$, $p_j = 3 \cdot k + 2$, $k \in \mathbb{N}^*$, $j = \overline{n_{L1}, n_{4a} + n_{L1} - 1}$, *the number of CPPs will be equal to:*

$$C_{L,CPPs} = 4 \cdot \left(\prod_{j=n_{L1}}^{n_{4a}+n_{L1}-1} \left(p_j \cdot (p_j - 1) \right) \cdot 3^{3 \cdot n_{L,3}-4} \right) +$$

$$+4 \cdot \sum_{r_0=1}^{n_{4a}-1} \left(C_{L_{(4a)}, n_{4a}=r, n_{4a,0}=r_0} \cdot 3^{3 \cdot n_{L,3}-4} \right) +$$

$$+2 \cdot \left(\prod_{j=n_{L1}}^{n_{4a}+n_{L1}-1} (p_j - 1) \cdot 3^{2 \cdot (n_{L,3}-1)} \right) \cdot \left(2 \cdot 3^{\cdot n_{L,3}-2} - 1 \right) \tag{4.101}$$

Proof This is a particular case of Theorem 4.18, therefore we can use Eq. (4.91), in which the products $\prod_{j=n_{L1}}^{n_L-n_{4a}} \left(p_j^{3 \cdot (n_{L,p_j}-1)} \cdot (p_j - 1) \right)$, $\prod_{j=n_{L1}}^{n_L-n_{4a}} \left(p_j^{2 \cdot (n_{L,p_j}-1)} \cdot (p_j - 1) \right)$ and $\prod_{j=n_{L1}}^{n_L-n_{4a}} p_j^{n_{L,p_j}-1}$ are replaced by 1, and $n_L = n_{4a} + n_{L1} - 1$. ■

From (4.101) we see that in this case the number of CPPs is always greater than 0.

Theorem 4.23 *If the decomposition of L is of the form* $L = 2^{n_{L,2}} \cdot 3^{n_{L,3}} \cdot \prod_{j=3}^{n_{4a}+2} p_j$, *with* $n_{L,2} > 1$, $n_{L,3} > 1$, $n_{4a} \in \mathbb{N}^*$, $p_j = 3 \cdot k + 2$, $k \in \mathbb{N}^*$, $j = \overline{3, n_{4a} + 2}$, *the number of CPPs will be equal to:*

$$C_{L,CPPs} = \prod_{j=3}^{n_{4a}+2} \left(p_j \cdot (p_j - 1)\right) \cdot 2^{3\cdot(n_{L,2}-1)} \cdot 3^{3\cdot n_{L,3}-4} +$$

$$+ \sum_{r_0=1}^{n_{4a}-1} \left(C_{L_{(4a)},n_{4a}=r,n_{4a,0}=r_0} \cdot 2^{3\cdot(n_{L,2}-1)} \cdot 3^{3\cdot n_{L,3}-4}\right) +$$

$$+ \left(\prod_{j=3}^{n_{4a}+2} (p_j - 1) \cdot 2^{2\cdot(n_{L,2}-1)} \cdot 3^{2\cdot(n_{L,3}-1)}\right) \cdot \left(2^{n_{L,2}-1} \cdot 3^{n_{L,3}-2} - 1\right) \tag{4.102}$$

Proof This is a particular case of Theorem 4.19, therefore we can use Eq. (4.97), in which the products $\prod_{j=n_{L1}}^{n_L-n_{4a}} \left(p_j^{3\cdot(n_{L,p_j}-1)} \cdot (p_j - 1)\right)$, $\prod_{j=n_{L1}}^{n_L-n_{4a}} \left(p_j^{2\cdot(n_{L,p_j}-1)} \cdot (p_j - 1)\right)$ and $\prod_{j=n_{L1}}^{n_L-n_{4a}} p_j^{n_{L,p_j}-1}$ are replaced by 1, and $n_L = n_{4a} + n_{L1} - 1$. ∎

From (4.102) we see that in this case the number of CPPs is always greater than 0.

Theorem 4.24 *If the decomposition of L is of the form* $L = 2^{n_{L,2}} \cdot 3^{n_3}$, *with* $n_{L,2} > 1$, $n_3 \in \{0, 1\}$, *the number of CPPs will be equal to:*

$$C_{L,CPPs} = 2^{n_3} \cdot 2^{2\cdot n_{L,2}-3} \cdot \left(2^{n_{L,2}-2} - 1\right) \tag{4.103}$$

Proof This is a particular case of Theorem 4.13, therefore we can use Eq. (4.69), in which the products $\prod_{j=n_{L1}}^{n_L-n_{4a}} \left(p_j^{2\cdot(n_{L,p_j}-1)} \cdot (p_j - 1)\right)$ and $\prod_{j=n_{L1}}^{n_L-n_{4a}} p_j^{n_{L,p_j}-1}$ are replaced by 1. ∎

From (4.103) we see that the number of CPPs is equal to 0 if the interleaver length is 4 or 12.

Theorem 4.25 *If the decomposition of L is of the form* $L = 2^{n_2} \cdot 3^{n_{L,3}}$, *with* $n_2 \in \{0, 1\}$, $n_{L,3} > 1$, *the number of CPPs will be equal to:*

$$C_{L,CPPs} = 2 \cdot 3^{2\cdot(n_{L,3}-1)} \cdot \left(2 \cdot 3^{n_{L,3}-2} - 1\right) \tag{4.104}$$

Proof This is a particular case of Theorem 4.14, therefore we can use Eq. (4.72), in which the products $\prod_{j=n_{L1}}^{n_L-n_{4a}} \left(p_j^{2\cdot(n_{L,p_j}-1)} \cdot (p_j-1)\right)$ and $\prod_{j=n_{L1}}^{n_L-n_{4a}} p_j^{n_{L,p_j}-1}$ are replaced by 1. ■

From (4.104) we see that in this case the number of CPPs is always greater than 0.

Theorem 4.26 *If the decomposition of L is of the form* $L = 2^{n_{L,2}} \cdot 3^{n_{L,3}}$, *with* $n_{L,2} > 1$, $n_{L,3} > 1$, *the number of CPPs will be equal to:*

$$C_{L,CPPs} = 2^{2\cdot(n_{L,2}-1)} \cdot 3^{2\cdot(n_{L,3}-1)} \cdot \left(2^{n_{L,2}-1} \cdot 3^{n_{L,3}-2} - 1\right) \tag{4.105}$$

Proof This is a particular case of Theorem 4.15, therefore we can use Eq. (4.75), in which the products $\prod_{j=3}^{n_{4a}} \left(p_j^{2\cdot(n_{L,p_j}-1)} \cdot (p_j-1)\right)$ and $\prod_{j=3}^{n_{4a}} p_j^{n_{L,p_j}-1}$ are replaced by 1. ■

From (4.105) we see that in this case the number of CPPs is always greater than 0.

Theorem 4.27 *If* $L = 2$ *or* $L = 3$ *or* $L = 6$, *the number of CPPs is equal to zero.*

Proof Since $L = 2$ is even, it is required that $q_3 < L/2 = 1$. As q_3 can not be 0, there is no true CPP in this case, i.e. $C_{2,CPPs} = 0$.

When $L = 3$, from the equivalence conditions, it is required that $q_3 < L/3 = 1$. As q_3 cannot be 0, there is no true CPP in this case, i.e. $C_{3,CPPs} = 0$.

When $L = 6$, from the equivalence conditions of CPPs, it is required that $q_3 < L/6 = 1$. As q_3 cannot be 0, there is no trueCPP, i.e. $C_{6,CPPs} = 0$. ■

We bring together the conclusions from Theorems 4.12–4.27 and conclude that the number of CPPs is equal to 0 if the interleaver length is of the form:

$$L = 2^{n_{L,2}} \cdot 3^{n_{L,3}} \cdot \prod_{j=3}^{n_L} p_j, \text{ with } n_{L,2} = \overline{0,2}, n_{L,3} = \overline{0,1}, p_j > 3,$$

$$\text{with } p_j = 3 \cdot k + 1, k \in \mathbb{N}, j = \overline{3, n_L} \tag{4.106}$$

Such lengths have to be avoided in designing CPP-based interleavers.

By comparing Eqs. (4.56) and (4.106), it can be concluded that for any interleaver length for which the number of CPPs is 0, the number of QPPs is also 0. But there are lengths for which the number of QPPs is 0 and the number of CPPs is greater than 0. Such lengths are generated by multiplying by one, two or four products of prime numbers greater than 2, each to the power of 1. In each product, there should be at least a prime number of the form $3 \cdot k + 2$, $k \in \mathbb{N}^*$, so that:

$$L = 2^{n_{L,2}} \cdot 3^{n_{L,3}} \cdot \prod_{j=3}^{n_L - n_{4a}} p_j \cdot \prod_{j=n_L-n_{4a}+1}^{n_L} p_j, \text{ with } n_{L,2} = \overline{0,2}, n_{L,3} = \overline{0,1},$$

$$p_j > 3, \text{ with } p_j = 3 \cdot k + 1, k \in \mathbb{N}, \text{ if } j = \overline{3, n_L - n_{4a}} \text{ and}$$

$$p_j = 3 \cdot k + 2, k \in \mathbb{N}^*, \text{ if } j = \overline{n_L - n_{4a} + 1, n_L}, n_L \geq n_{4a} \geq 1 \tag{4.107}$$

For the lengths of the type in (4.107) the CPP-based interleavers can be used instead of QPP-based ones. The number of lengths for which the number of QPPs is 0 is significantly greater than the number of lengths for which the number of CPPs is equal to 0. For example, from 2 to 1,00,000 there are 7098 lengths for which the number of QPPs is 0 and only 2264 lengths for which the number of CPPs is 0.

In the following, we give an example of checking the formula for the number of true different CPPs, for a more comprehensive case, i.e., when the prime-factor decomposition of the interleaver length contains almost all types of prime factors that appear in the theorems presented in the paper. The chosen length of the interleaver corresponds to the decomposition from Theorem 4.16, when $n_2 = 1$ and $n_3 = 1$.

Example 4.1 Let $L = 16170 = 2 \cdot 3 \cdot 5 \cdot 7^2 \cdot 11$. According to Theorem 4.16, we have $n_2 = 1$, $n_3 = 1$, $n_{L1} = 3$, $n_{4a} = 2$. We mention that the two factors of type 4. (a) are 5 and 11, and the only factor of type 3. (b) is 7^2.

The quantity $C_{N_{(4a)}, n_{4a}=2, n_{4a,0}=1}$ used in (4.77), is given in (4.58). It is equal to

$$C_{N_{(4a)}, n_{4a}=2, n_{4a,0}=1} = (5-1) \cdot 11 \cdot (11-1) + (11-1) \cdot 5 \cdot (5-1) = 640$$

According to (4.77), the number of all true different CPPs is equal to:

$$C_{16170,CPPs} = 2 \cdot 5 \cdot (5-1) \cdot 11 \cdot (11-1) \cdot 7^{3 \cdot (2-1)} \cdot (7-1) +$$

$$+ 2 \cdot 640 \cdot 7^{3 \cdot (2-1)} \cdot (7-1) +$$

$$+ 2 \cdot (5-1) \cdot (11-1) \cdot 7^{2 \cdot (2-1)} \cdot (7-1) \cdot (7^{2-1} - 1) = 11830560$$

In the following, we detail the coefficients of the 11830560 true different CPPs.

Because $6 \mid L$, it is required that $q_3 < L/6 = 2695$ and $q_2 < L/2 = 8085$.

Because of the factor of type 3. (b) from the decomposition of 16170, coefficients q_3 and q_2 have to be multiples of 7, and coefficient q_1 has to be relatively prime with 7. Because of the factor of type 2. (a) from the decomposition of 16170, coefficient q_2 also has to be multiple of 3. Therefore, q_3 can take values in the set $\{7, 14, 21, \ldots, 2688\}$ (with cardinality 384), q_2 can take values in the set $\{0, 21, 42, \ldots, 8064\}$ (with cardinality 385), and q_1 can take values in the set $\{1, 2, 3, 4, 5, 6, 8, 9, \ldots, 16169\}$ (with cardinality $16170 \cdot \frac{6}{7} = 13860$).

Because of the factor of type 2. (a), the condition $(q_1 + q_3) \neq 0 \pmod 3$ must be fulfilled. Because of the factor of type 1. (a), the condition $(q_1 + q_2 + q_3) \neq 0 \pmod 2$ must be fulfilled. Thus, when q_2 is even, $(q_1 + q_3)$ is required to be odd, and when q_2 is odd, $(q_1 + q_3)$ is required to be even.

Because of the two factors of type 4. (a) (5 and 11), the analysis has to be performed for four different cases: (a) q_3 is a multiple of $5 \cdot 11 = 55$; (b) q_3 is a multiple of 5, but not a multiple of 11; (c) q_3 is a multiple of 11, but not a multiple of 5; and (d) q_3 is neither a multiple of 5, nor a multiple of 11.

(a) When q_3 is multiple of $5 \cdot 11 = 55$, q_2 has to be multiple of $5 \cdot 11 = 55$, and q_1 has to be relatively prime with 5 and 11. This situation happens for six values from the set of 384 possible values for q_3 (namely, $\{385, 770, 1155, 1540, 1925, 2310\}$), for 7 values from the set of 385 possible values for q_2 (namely, $\{0, 1155, 2310, 3465, 4620, 5775, 6930\}$) and for $16170 \cdot \frac{6}{7} \cdot \frac{4}{5} \cdot \frac{10}{11} = 10080$ values from the set of 13860 possible values for q_1 (namely, $\{1, 2, 3, 4, 6, 8, 9, 12, 13, \ldots, 16169\}$).

From the six previous values for q_3, two are multiples of 3 (namely, 1155 and 2310), two are multiples of 3 plus 1 (namely 385 and 1540) and two are multiples of 3 plus 2 (namely, 770 and 1925). From the 10080 values for q_1, there are $10080/3 = 3360$ values multiples of 3, multiples of 3 plus 1, and multiples of 3 plus 2, respectively. Therefore, the condition $(q_1 + q_3) \neq 0 \pmod 3$ is fulfilled for $2 \cdot 3360 \cdot 2 \cdot 3 = 40320$ pairs of values for the coefficients (q_3, q_1). Of the seven values for q_2, four values are even and three are odd. From each of the three groups of two values for q_3, one is even and one is odd and from each of three groups of 3360 values for q_1, 1680 are even and 1680 are odd. It follows that from the 40320 pairs of values for the coefficients (q_3, q_1), the sum $(q_1 + q_3)$ is even for $40320/2 = 20160$ pairs of values and odd for $40320/2 = 20160$ pairs. Therefore, when q_3 is multiple of $5 \cdot 11 = 55$, we will have $4 \cdot 20160 + 3 \cdot 20160 = 141120$ sets of coefficients (q_3, q_2, q_1) leading to true different CPPs modulo 16170.

(b) When q_3 is a multiple of 5, but not a multiple of 11, q_2 has to be a multiple of 5, q_1 has to be relatively prime with 5, and the three coefficients must fulfill the condition $q_2^2 = 3q_1q_3 \pmod{11}$. q_3 is a multiple of 5, but not a multiple of 11, for 70 values from the set of 384 possible values for q_3 (namely, $\{35, 70, 105, \ldots, 350, 420, 455, \ldots, 735, 805, \ldots, 2660\}$). q_2 is a multiple of 5 for 77 values from the set of 385 possible values for q_2 (namely, $\{0, 105, 210, \ldots, 7980\}$). For each two coefficients q_3 and q_2, from the sets previously mentioned, the congruence equation $q_2^2 = 3q_1q_3 \pmod{11}$ has only one solution modulo 11, in variable q_1 and, thus, $16170/11 = 1470$ solutions modulo 16170. Of these, we have to remove those that are multiples of 5 or multiples of 7. In this way, $1470 - 1470/5 - 1470/7 + 1470/35 = 1008$ valid solutions remain. For example, for $q_3 = 35$ and $q_2 = 105$, the possible values for q_1 are from the set $\{6, 17, 39, 61, \ldots, 16154\}$.

From the 70 possible solutions for q_3, 23 values are multiples of 3 and multiples of 3 plus 1, respectively, and 24 values are multiples of 3 plus 2. From the 1008 possible values for q_1, $1008/3 = 336$ values are multiples of 3, multiples of 3 plus 1, and multiples of 3 plus 2, respectively. Therefore, the condition $(q_1 + q_3) \neq 0 \pmod 3$ is fulfilled for $2 \cdot 23 \cdot 336 \cdot 2 + 24 \cdot 336 \cdot 2 = 47040$ pairs of values of the coefficients (q_3, q_1). From the 77 values for q_2, 39 values are even and 38 values are odd. From each of the three groups of 336 values for q_1, 168 values are even and 168 values are odd. From the two groups of 23 values for q_3, one contains 11 even values and 12 odd values and the other group contains 12 even values and 11 odd values. The group of 24 values for q_3 contains 12 even and odd values, respectively. It follows that from the 47040 pairs of values for the coefficients (q_3, q_1), the sum $(q_1 + q_3)$ is even and odd for $2 \cdot 11 \cdot 168 \cdot 2 + 2 \cdot 12 \cdot 168 \cdot 2 + 2 \cdot 12 \cdot 168 \cdot 2 = 23520$ pairs of values, respectively. Therefore, the condition $(q_1 + q_2 + q_3) \neq 0 \pmod 2$ is fulfilled for $39 \cdot 23520 + 38 \cdot 23520 = 1811040$ sets of coefficients (q_3, q_2, q_1). Thus, when q_3 is a multiple of 5, but not a multiple of 11, we will have 1811040 true different CPPs modulo 16170.

(c) When q_3 is a multiple of 11, but not a multiple of 5, q_2 has to be a multiple of 11, q_1 has to be relatively prime with 11, and the three coefficients must fulfill the condition $q_2^2 = 3q_1q_3 \pmod 5$. q_3 is a multiple of 11, but not a multiple of 5, for 28 values from the set of 384 possible values for q_3 (namely, $\{77, 154, 231, 308, 462, 539, 616, 693, \ldots, 2387, 2464, 2541, 2618\}$). q_2 is a multiple of 11 for 35 values from the set of 385 possible values for q_2 (namely, $\{0, 231, 462, \ldots, 7854\}$). For each two coefficients q_3 and q_2, fixed in the sets previously mentioned, the congruence equation $q_2^2 = 3q_1q_3 \pmod 5$ has only one solution modulo 5, in variable q_1, and thus, $16170/5 = 3234$ solutions modulo 16170. From these, we have to remove those that are multiples of 11 or multiples of 7. In this way, $3234 - 3234/11 - 3234/7 + 3234/77 = 2520$ valid solutions remain. For example, for $q_3 = 77$ and $q_2 = 231$, the possible values for q_1 are from the set $\{1, 6, 16, 26, \ldots, 16166\}$.

From the 28 possible values for q_3, nine values are multiples of 3 and multiples of 3 plus 1, respectively, and 10 values are multiples of 3 plus 2. From the 2520 possible values for q_1, $2520/3 = 840$ values are multiples of 3, multiples of 3 plus 1, and multiples of 3 plus 2, respectively. Therefore, the condition $(q_1 + q_3) \neq 0 \pmod 3$ is fulfilled by $2 \cdot 9 \cdot 840 \cdot 2 + 10 \cdot 840 \cdot 2 = 47040$ pairs of values of coefficients (q_3, q_1). From the 35 values for q_2, 18 values are even and 17 values are odd. From each of the three groups of 840 values for q_1, 420 values are even and 420 are odd. From the two groups of nine values for q_3, one contains four even and five odd values and the other contains five even and four odd values. The group of 10 values for q_3 contains five even and five odd values. It follows that from the 47040 pairs of values of coefficients (q_3, q_1), the sum $(q_1 + q_3)$ is even and odd for $2 \cdot 4 \cdot 420 \cdot 2 + 2 \cdot 5 \cdot 420 \cdot 2 + 2 \cdot 5 \cdot 420 \cdot 2 = 23520$ pairs of values (q_3, q_1), respectively. Therefore, the condition $(q_1 + q_2 + q_3) \neq 0 \pmod 2$ is fulfilled for $18 \cdot 23520 + 17 \cdot 23520 = 823200$ sets of coefficients. Thus, when q_3 is a multiple of 11, but not a multiple of 5, we have 823200 true different CPPs modulo 16170.

(d) When q_3 is neither a multiple of 5 nor a multiple of 11, the three coefficients have to fulfill the conditions $q_2^2 = 3q_1q_3 \pmod 5$ and $q_2^2 = 3q_1q_3 \pmod{11}$. q_3 is neither a multiple of 5 nor of 11, but is a multiple of 7, for $384 - 6 - 70 - 28 = 280$ values from the initial set of 384 possible values for q_3 (namely, $\{7, 14, 21, 28, 42, 49, 56, 63, 84, \ldots, 2688\}$). q_2 should have only 3 or 7 as prime factors; thus, it can take 385 values, as it was stated at the beginning of this example. For each two coefficients q_3 and q_2 from the sets previously mentioned, the congruence equations $q_2^2 = 3q_1q_3 \pmod 5$ and $q_2^2 = 3q_1q_3 \pmod{11}$ have a single solution modulo 55, in variable q_1, and thus, $16170/55 = 294$ solutions modulo 16170. From these, we have to remove those that are multiples of 7. In this way $294 - 294/7 = 252$ valid solutions remain. For example, for $q_3 = 7$ and $q_2 = 21$, the possible values for q_1 are from the set $\{76, 131, 186, \ldots, 16136\}$.

Out of the 280 possible values for q_3, 94 are multiples of 3 and multiples of 3 plus 1, respectively and 92 values are multiples of 3 plus 2. Out of the 252 possible values for q_1, $252/3 = 84$ values are multiples of 3, multiples of 3 plus 1, and multiples of 3 plus 2, respectively. Therefore, the condition $(q_1 + q_3) \neq 0 \pmod 3$ is fulfilled by $2 \cdot 94 \cdot 84 \cdot 2 + 92 \cdot 84 \cdot 2 = 47040$ pairs of values of coefficients (q_3, q_1). Out of the 385 values for q_2, 193 are even and 192 are odd. Out of each of the three groups of 84 values for q_1, 42 are even and 42 are odd. Out of the two groups of 94 values for q_3, one contains 46 even values and 48 odd values and the other one contains 48 even values and 46 odd values. The group of 92 values for q_3 contains 46 even and odd values, respectively. It follows that, from the 47040 pairs of values of coefficients (q_3, q_1), the sum $(q_1 + q_3)$ is even and odd for $2 \cdot 46 \cdot 42 \cdot 2 + 2 \cdot 48 \cdot 42 \cdot 2 + 2 \cdot 46 \cdot 42 \cdot 2 = 23520$ pairs of values (q_3, q_1), respectively. Therefore the condition $(q_1 + q_2 + q_3) \neq 0 \pmod 2$ is fulfilled for $193 \cdot 23520 + 192 \cdot 23520 = 9055200$ sets of coefficients (q_3, q_2, q_1). Thus, when q_3 is neither a multiple of 11 nor of 5, we will have 9055200 true different CPPs modulo 16170.

The total number of true different CPPs modulo 16170 results by summing up the number of CPPs from the four cases. It is equal to $141120 + 1811040 + 823200 + 9055200 = 11830560$, which is the one determined by the formula in Theorem 4.16. ■

4.6 Determining the Number of True Different Permutation Polynomial-Based Interleavers Under Zhao and Fan Sufficient Conditions for Degrees 3, 4 and 5

The method used to determine the number of different true PPs under Zhao and Fan sufficient conditions (denoted ZF PPs) is that given in Sect. 4.3. However the equivalence conditions for PPs fulfilling Zhao and Fan sufficient conditions at point (c) can be different from those for general PPs.

The form of NPs of degree up to d is given in (4.48) from Theorem 4.8. We have checked that (4.48) includes all NPs of degree d or less, for $d = 5$, which is of interest, using the method from Sect. 4.2 for QNPs and CNPs. The number of all NPs of degree up to d is equal to $\prod_{k=1}^{d} \gcd(k!, L)$, as it was shown in Theorem 4.7. We notice that always one of the NPs has all coefficients null, named trivial NP.

Let

$$z(x) = \sum_{k=1}^{d} z_k \cdot x^k \pmod{L} \tag{4.108}$$

be a NP modulo L of degree d.

We define a *NP valid under Zhao and Fan sufficient conditions* (denoted a ZF NP) a NP $z(x)$ (denoted $z_{ZF}(x)$) fulfilling the condition that if $\pi(x)$ is a ZF PP then $\pi(x) + z_{ZF}(x)$ is also a ZF PP. From Table 3.6 we can determine the conditions on the coefficients of a valid NP under Zhao and Fan sufficient conditions. Since Zhao and Fan sufficient conditions are also necessary for the prime $p = 2$ we have to take into account only the conditions for primes $p > 2$ (i.e. the last row in Table 3.6). This means:

$$z_1 \neq (-q_1) \pmod{p},\ z_2 = z_3 = \cdots = z_d = 0 \pmod{p},\ \forall p \mid L,\ p > 2. \tag{4.109}$$

In the following, we explain how the coefficients q_k in (4.108), with $k = \overline{1, d}$, take values according to (4.48), for $d = 2, d = 3, d = 4$ or $d = 5$. Since $\gcd(1!, L) = 1$, it follows that $\tau_1 = 0$ and, consequently, there are no non-trivial NPs of degree 1 and therefore the index k can take values from 2 to d in the sum in (4.48). As the NPs obtained for $d = 5$ in (4.48) include all NPs of degree up to 5, we will detail the Eq. (4.48) only for $d = 5$. $z_{d=5}(x)$ denotes $z(x)$ for $d = 5$. Thus, we have

$$z_{d=5}(x) = \frac{L}{\gcd(2!, L)} \cdot \tau_2 \cdot x \cdot (x-1) +$$

$$+ \frac{L}{\gcd(3!, L)} \cdot \tau_3 \cdot x \cdot (x-1) \cdot (x-2) +$$

$$+ \frac{L}{\gcd(4!, L)} \cdot \tau_4 \cdot x \cdot (x-1) \cdot (x-2) \cdot (x-3) +$$

$$+ \frac{L}{\gcd(5!, L)} \cdot \tau_5 \cdot x \cdot (x-1) \cdot (x-2) \cdot (x-3) \cdot (x-4) =$$

$$= \frac{L}{\gcd(5!, L)} \cdot \tau_5 \cdot x^5 + \left(\frac{L}{\gcd(4!, L)} \cdot \tau_4 - \frac{L}{\gcd(5!, L)} \cdot 10 \cdot \tau_5 \right) \cdot x^4 +$$

$$+ \left(\frac{L}{\gcd(3!, L)} \cdot \tau_3 - \frac{L}{\gcd(4!, L)} \cdot 6 \cdot \tau_4 + \frac{L}{\gcd(5!, L)} \cdot 35 \cdot \tau_5 \right) \cdot x^3 +$$

$$+ \left(\frac{L}{\gcd(2!, L)} \cdot \tau_2 - \frac{L}{\gcd(3!, L)} \cdot 3 \cdot \tau_3 + \frac{L}{\gcd(4!, L)} \cdot 11 \cdot \tau_4 - \right.$$

$$\left. - \frac{L}{\gcd(5!, L)} \cdot 50 \cdot \tau_5 \right) \cdot x^2 + \left(\frac{L}{\gcd(3!, L)} \cdot 2 \cdot \tau_3 - \frac{L}{\gcd(2!, L)} \cdot \tau_2 - \right.$$

$$-\frac{L}{\gcd(4!, L)} \cdot 6 \cdot \tau_4 + \frac{L}{\gcd(5!, L)} \cdot 24 \cdot \tau_5 \Bigg) \cdot x,$$

$$\text{with } 0 \leq \tau_2 \leq \gcd(2!, L) - 1, 0 \leq \tau_3 \leq \gcd(3!, L) - 1,$$

$$0 \leq \tau_4 \leq \gcd(4!, L) - 1 \text{ and } 0 \leq \tau_5 \leq \gcd(5!, L) - 1. \tag{4.110}$$

In (4.110), τ_2 can take any value between 0 and $\gcd(2!, L) - 1$, τ_3 can take any value between 0 and $\gcd(3!, L) - 1$, τ_4 can take any value between 0 and $\gcd(4!, L) - 1$, and τ_5 can take any value between 0 and $\gcd(5!, L) - 1$. Non-trivial NPs of degree 5 result if $\tau_5 > 0$, non-trivial NPs of degree 4 if $\tau_5 = 0$ and $\tau_4 > 0$, non-trivial NPs of degree 3 (CNPs) if $\tau_5 = 0$, $\tau_4 = 0$ and $\tau_3 > 0$, and non-trivial NPs of degree 2 (QNPs) if $\tau_5 = 0$, $\tau_4 = 0$, $\tau_3 = 0$ and $\tau_2 > 0$. As the coefficients of a non-trivial NP have to be integers, and the coefficient of the maximum degree term has to be positive, there are non-trivial NPs of degree 5, only if $2 \mid L$ or/and $3 \mid L$ or/and $5 \mid L$, non-trivial NPs of degree 4, only if $2 \mid L$ or/and $3 \mid L$, non-trivial CNPs only if $2 \mid L$ or/and $3 \mid L$, and non-trivial QNPs only if $2 \mid L$. With the above considerations, it follows that the number of non-trivial NPs of degree 5 is equal to $\gcd(2!, L) \cdot \gcd(3!, L) \cdot \gcd(4!, L) \cdot (\gcd(5!, L) - 1)$, the number of non-trivial NPs of degree 4 is equal to $\gcd(2!, L) \cdot \gcd(3!, L) \cdot (\gcd(4!, L) - 1)$, the number of non-trivial CNPs is equal to $\gcd(2!, L) \cdot (\gcd(3!, L) - 1)$, and the number of non-trivial QNPs is equal to $\gcd(2!, L) - 1$. Thus, according to the value of L, the number of non-trivial QNPs may be 0 or 1 (as it also results from Theorem 4.4), the number of non-trivial CNPs may be 0, 2 or 10 (as it also results from Theorem 4.6), the number of non-trivial NPs of degree 4 may be 0, 4, 6, 12, 28, 60, 132 or 276, and the number of non-trivial NPs of degree 5 may be 0, 4, 8, 18, 48, 72, 126, 224, 304, 360, 1248, 1584, 2088, 6624, 8496 or 34272.

We have to impose additional constraints for $z_{n=5}(x)$ to be a valid ZF NP only if $3 \mid L$ and/or $5 \mid L$ because for every prime $p > 5$, with $p \mid L$, the five coefficients of $z_{n=5}(x)$ will be divisible by p, and thus they fulfill the conditions (4.109).

We denote by g_k the value $\gcd(k!, L)$, for $k \in \mathbb{N}^*, k < L$. Let the prime factorization of g_k, for $k = 3, 4, 5$, be of the form

$$g_k = \gcd(k!, L) = 2^{n_{g_k,2}} \cdot 3^{n_{g_k,3}} \cdot 5^{n_{g_k,5}},$$

$$\text{where } 0 \leq n_{g_k,2} \leq 3, 0 \leq n_{g_k,3} \leq 1, \text{ and } 0 \leq n_{g_k,5} \leq 1. \tag{4.111}$$

Obviously, for $k = 3$ or $k = 4$ we have $n_{g_k,5} = 0$ in (4.111).

We write the prime factorization of L as

$$L = 2^{n_{L,2}} \cdot 3^{n_{L,3}} \cdot 5^{n_{L,5}} \cdot \prod_{j=4}^{n_L} p_j^{n_{L,p_j}},$$

$$\text{with } n_{L,p} \geq 0 \text{ for } p = 2, 3, 5, \text{ and } n_{L,p_j} \geq 1 \, \forall j = \overline{4, n_L}. \tag{4.112}$$

We note that if $n_{L,3} = 0$ then $3 \nmid L$ and we not have to impose additional constraints for prime $p = 3$ for $z_{d=k}(x), k = 3, 4, 5$, to be a valid ZF NP. Also if $n_{L,5} = 0$ then $5 \nmid L$ and we not have to impose additional constraints for prime $p = 5$ for $z_{d=5}(x)$.

If $n_{L,3} > 1$ we have $3 \mid \dfrac{L}{\gcd(k!, L)}$, $\forall k = 2, 3, 4, 5$, and from (4.110) it results that 3 divides all coefficients of $z_{d=5}(x)$, thus fulfilling the conditions for prime $p = 3$ for a valid ZF NP. If $n_{L,5} > 1$ we have $5 \mid \dfrac{L}{\gcd(5!, L)}$, $\forall k = 2, 3, 4, 5$, and from (4.110) it results that 5 divides all coefficients of $z_{d=5}(x)$, thus fulfilling the conditions for prime $p = 5$ for a valid ZF NP.

From those above it results that we have to impose additional constraints for $z_{d=5}(x)$ to be a valid ZF NP only if $n_{L,3} = 1$ and/or $n_{L,5} = 1$. We analyze these cases below.

If $\pi(x)$ is a ZF PP modulo L of degree d, and $z_{ZF}(x)$ is a non-trivial ZF NP modulo L of degree less than or equal to d, then $\pi(x) + z_{ZF}(x)$ is also a ZF PP modulo L of degree d, with coefficients different from those of $\pi(x)$. From (4.48), we note that the coefficient of the maximum degree term of a NP (denoted by d_{NP}) is any multiple of $\dfrac{L}{\gcd(d_{NP}!, L)}$, i.e. $\dfrac{L}{\gcd(d_{NP}!, L)} \cdot \tau_{NP}$, with $0 \le \tau_{NP} \le \gcd(d_{NP}!, L) - 1$. It follows that any PP modulo L of degree $d \ge d_{NP}$ is equivalent to a PP having the coefficient of the d_{NP} degree term $q_{d_{NP}} < \dfrac{L}{\gcd(d_{NP}!, L)}$. We assume that the coefficient of the d_{NP} degree term of the initial PP is in the range $\left[\dfrac{L}{\gcd(d_{NP}!, L)} \cdot \tau_{NP}, \dfrac{L}{\gcd(d_{NP}!, L)} \cdot (\tau_{NP} + 1)\right)$ for a fixed τ_{NP} with $0 \le \tau_{NP} \le \gcd(d_{NP}!, L) - 1$. We obtain the equivalent PP by subtracting the NP of the maximum degree d_{NP} with the leading coefficient $\dfrac{L}{\gcd(d_{NP}!, L)} \cdot \tau_{NP}$ from the initial PP. However the coefficient of the maximum degree term of a ZF NP (denoted by d_{ZF-NP}) is not any multiple of $\dfrac{L}{\gcd(d_{ZF-NP}!, L)}$ because this coefficient, i.e. $\dfrac{L}{\gcd(d_{ZF-NP}!, L)} \cdot \tau_{ZF-NP}$, have to be a multiple of any prime greater than 2 which is a divisor of L. Then, for $\pi(x)$ in (3.1) a ZF PP, we can consider only those PP coefficients for which $q_k < \dfrac{L}{\gcd(k!, L)} \cdot \tau_k, \forall k = \overline{2, d}$, where τ_k is the least positive integer such that $p \mid \left(\dfrac{L}{\gcd(k!, L)} \cdot \tau_k\right)$, for every prime $p \mid L$ with $2 < p \le k$. In this way we get the conditions on the maximum values of the coefficients for a ZF CPP, a ZF PP of fourth degree (denoted by ZF 4-PP) and a ZF PP of fifth degree (denoted by ZF 5-PP) in Sects. 4.6.1–4.6.3, respectively. Thus, the number of ZF NPs for a certain degree will also give the number of combinations of maximum values for the coefficient of ZF PPs we are interested in.

We detail below the conditions for a ZF NP of degree three, four, and five.

If $n_{L,3} = 1$ then $3 \nmid \dfrac{L}{\gcd(k!, L)}$, $\forall k = 3, 4, 5$. In this case for $z_{d=k}(x)$, $\forall k = 3, 4, 5$, to be a valid ZF NP we have to impose the condition that $3 \mid \tau_k$, $\forall k = 3, 4, 5$. Similarly, if $n_{L,5} = 1$ then $5 \nmid \dfrac{L}{\gcd(5!, L)}$. In this case for $z_{d=5}(x)$ to be a valid ZF NP we have to impose the condition that $5 \mid \tau_5$.

Thus, when $n_{L,3} = 1$ any ZF PP modulo L is equivalent to a ZF PP having the coefficient of the third degree term $q_3 < 3 \cdot \dfrac{L}{\gcd(3!, L)}$.

For similar reasons when $n_{L,3} = 1$ any ZF PP modulo L is equivalent to a ZF PP having the coefficient of the fourth degree term $q_4 < 3 \cdot \dfrac{L}{\gcd(4!, L)}$.

When $n_{L,3} = 1$ and $n_{L,5} = 0$ any ZF PP modulo L is equivalent to a ZF PP having the coefficient of the fifth degree term $q_5 < 3 \cdot \dfrac{L}{\gcd(5!, L)}$.

When $n_{L,3} = 0$ and $n_{L,5} = 1$ any ZF PP modulo L is equivalent to a ZF PP having the coefficient of the fifth degree term $q_5 < 5 \cdot \dfrac{L}{\gcd(5!, L)}$.

When $n_{L,3} = 1$ and $n_{L,5} = 1$ any ZF PP modulo L is equivalent to a ZF PP having the coefficient of the fifth degree term $q_5 < 15 \cdot \dfrac{L}{\gcd(5!, L)}$.

We have not to impose additional constraints for the coefficients z_1 and z_2 of the NP $z_{d=5}(x)$ because when $3 \mid \tau_k \cdot \dfrac{L}{\gcd(k!, L)}$, $\forall k = 3, 4, 5$ and $5 \mid \tau_5 \cdot \dfrac{L}{\gcd(5!, L)}$, then $p \mid z_1$ and $p \mid z_2$, for $p = 3$ or 5. Thus $z_2 = z_1 = 0 \pmod{p}$, $p = 3$ or 5, and the conditions from (4.109) for primes 3 and 5 are fulfilled.

From the Zhao and Fan sufficient conditions on the coefficients of a PP from Sect. 3.10, there are three types of prime factors, as shown in Table 3.6. These conditions are considered in the following and for each type of prime factor, the number of ZF PPs is determined. The notation ZF stands for Zhao and Fan sufficient conditions.

We mention that as we want to determine the number of true ZF PPs, the coefficient of the term of maximum degree has to be different from zero. Therefore, whenever appropriate, we will mention when a coefficient is zero as a consequence of the conditions we imposed. This zero value will be eliminated in counting.

Case ZF1. (a) If $p = 2$ and $n_{L,2} = 1$, the coefficients $q_{i,1} \in \mathbb{Z}_2 = \{0, 1\}$, $\forall i = \overline{1, d}$. In this case, there are two distinct permutations. The ZF NPs of degree d result by imposing $L = 2$. Since $\gcd(k!, L) = 2$, $\forall k = \overline{2, d}$, it follows that $\dfrac{L}{\gcd(k!, L)} = 1$, $\forall k = \overline{2, d}$, and we can consider $q_{i,1} = 0$, $\forall i = \overline{2, d}$. We take into account the condition $(q_{1,1} + q_{2,1} + \cdots + q_{d,1}) \neq 0 \pmod{2}$, resulting $q_{1,1} = 1$ and, therefore, we have only a single combination of coefficients $q_{i,1} \in \mathbb{Z}_2$, $i = \overline{1, d}$, that has to be considered in combinations with other prime factors.

Case ZF1. (b) If $p = 2$ and $n_{L,2} > 1$, the coefficients $q_{i,1} \in \mathbb{Z}_{2^{n_{L,2}}}$, $\forall i = \overline{1, d}$. The condition $q_{1,1} \neq 0 \pmod{2}$ is met for $\Phi(2^{n_{L,2}}) = 2^{n_{L,2}-1}$ coefficients. In the following, we will consider thee particular cases for $d = 3$, $d = 4$ and $d = 5$.

For $d = 3$, the conditions 1. (b) in Table 3.6 become $q_{2,1} = q_{3,1} = 0 \pmod 2$. By considering ZF NPs of third degree (ZF CNPs), we have $q_{2,1}, q_{3,1} < 2^{n_{L,2}-1}$ and there are $2^{n_{L,2}-2}$ coefficients $q_{2,1}$ and $q_{3,1}$, respectively, out of which one is zero.

For $d = 4$, the conditions 1. (b) from Table 3.6 become $(q_{2,1} + q_{4,1}) = 0 \pmod 2$ and $q_{3,1} = 0 \pmod 2$. By considering ZF NPs of fourth degree, we have $q_{2,1}, q_{3,1} < 2^{n_{L,2}-1}$, $q_{4,1} < \dfrac{2^{n_{L,2}}}{\gcd(24, 2^{n_{L,2}})} = 2^{n_{L,2}-\min(3,n_{L,2})}$.

If $n_{L,2} \geq 4$, the condition $(q_{2,1} + q_{4,1}) = 0 \pmod 2$ will be met by $2^{n_{L,2}-2}$ and $2^{n_{L,2}-4}$ even coefficients $q_{2,1}$, and $q_{4,1}$, respectively, out of which one is zero and by $2^{n_{L,2}-2}$ and $2^{n_{L,2}-4}$ odd coefficients $q_{2,1}$ and $q_{4,1}$, respectively. If $n_{L,2} = 2$ or $n_{L,2} = 3$ the only possible value for $q_{4,1}$ is 0, and the possible values for $q_{2,1}$ are 0 for $n_{L,2} = 2$, and 0 and 2, respectively, for $n_{L,2} = 3$. The condition $q_{3,1} = 0 \pmod 2$ is met for $2^{n_{L,2}-2}$ coefficients $q_{3,1}$.

For $d = 5$, the conditions 1. (b) from Table 3.6 become $(q_{2,1} + q_{4,1}) = 0 \pmod 2$ and $(q_{3,1} + q_{5,1}) = 0 \pmod 2$. By considering the ZF NPs of fifth degree, we have $q_{2,1}, q_{3,1} < 2^{n_{L,2}-1}$, $q_{4,1} < \dfrac{2^{n_{L,2}}}{\gcd(24, 2^{n_{L,2}})} = 2^{n_{L,2}-\min(3,n_{L,2})}$, $q_{5,1} < \dfrac{2^{n_{L,2}}}{\gcd(120, 2^{n_{L,2}})} = 2^{n_{L,2}-\min(3,n_{L,2})}$.

If $n_{L,2} \geq 4$, the condition $(q_{2,1} + q_{4,1}) = 0 \pmod 2$ will be met by $2^{n_{L,2}-2}$ and $2^{n_{L,2}-4}$ even coefficients $q_{2,1}$ and $q_{4,1}$, respectively and by $2^{n_{L,2}-2}$ and $2^{n_{L,2}-4}$ odd coefficients $q_{2,1}$ and $q_{4,1}$, respectively.

Similarly, the condition $(q_{3,1} + q_{5,1}) = 0 \pmod 2$ will be met by $2^{n_{L,2}-2}$ and $2^{n_{L,2}-4}$ even coefficients $q_{3,1}$ and $q_{5,1}$, respectively, out of which one is zero, and by $2^{n_{L,2}-2}$ and $2^{n_{L,2}-4}$ odd coefficients $q_{3,1}$ and $q_{5,1}$, respectively. If $n_{L,2} = 2$ or $n_{L,2} = 3$, the single possible value for $q_{4,1}$ and $q_{5,1}$ is 0, and the possible values for $q_{2,1}$ and $q_{3,1}$ will be 0 for $n_{L,2} = 2$, and 0 and 2, respectively, for $n_{L,2} = 3$.

Case ZF2. If $p_j > 2$ and $n_{L,p_j} \geq 1$, the coefficients $q_{i,j} \in \mathbb{Z}_{p_j^{n_{L,p_j}}}$, $\forall i = \overline{1,d}$. The condition $q_{1,j} \neq 0 \pmod{p_j}$ is met for $\Phi(p_j^{n_{L,p_j}}) = p_j^{n_{L,p_j}-1} \cdot (p-1)$ coefficients. The condition $q_{i,j} = 0 \pmod{p_j}$, $\forall i = \overline{2,d}$ is met for $\dfrac{p_j^{n_{L,p_j}}}{p_j} = p_j^{n_{L,p_j}-1}$ coefficients, out of which one is zero. We also impose here the equivalence conditions for degrees $d = 3$, $d = 4$ and $d = 5$.

For $d = 3$, ZF CNPs depend on whether the prime factor p_j is equal to 3. If $p_j = 3$, we have $q_{3,j} < 3^{n_{L,3}-1}$ if $n_{L,3} \geq 2$ and $q_{3,j} < 3$ if $n_{L,3} = 1$. Therefore, there are $3^{n_{L,3}-1}$ coefficients $q_{2,j}$ and $3^{n_{L,3}-2}$ coefficients $q_{3,j}$ if $n_{L,3} \geq 2$, out of which one is zero. If $n_{L,3} = 1$, $q_{2,j}$ and $q_{3,j}$ can be only zero. Summarizing, for $n_{L,3} \geq 1$ there are $3^{n_{L,3}-1}$ coefficients $q_{2,j}$ and $3^{\max\{0; n_{L,3}-2\}}$ coefficients $q_{3,j}$, out of which one is zero. If $p_j \neq 3$, there are no ZF NPs of degree less than or equal to three and, therefore, there are $p_j^{n_{L,p_j}-1}$ coefficients $q_{2,j}$ and $q_{3,j}$, respectively, out of which one is zero.

For $d = 4$, the ZF NPs of fourth degree depend on whether $p_j = 3$. If $p_j = 3$, we have $q_{3,j}, q_{4,j} < 3^{n_{L,3}-1}$ if $n_{L,3} \geq 2$ and $q_{3,j}, q_{4,j} < 3$ if $n_{L,3} = 1$. Thus, there will be $3^{n_{L,3}-1}$ coefficients $q_{2,j}$ and if $n_{L,3} \geq 2$, $3^{n_{L,3}-2}$ coefficients $q_{3,j}$ and $q_{4,j}$, respectively, out of which one is zero. Summarizing, for $n_{L,3} \geq 1$ there are $3^{n_{L,3}-1}$ coefficients $q_{2,j}$ and $3^{\max\{0;n_{L,3}-2\}}$ coefficients $q_{3,j}$ and $q_{4,j}$, respectively, out of which one is zero. If $n_{L,3} = 1$, $q_{2,j}$, $q_{3,j}$ and $q_{4,j}$ can take only the value zero. If $p_j \neq 3$, there are no ZF NPs of degree less than or equal to four and, therefore, there will be $p_j^{n_{L,p_j}-1}$ coefficients $q_{2,j}$, $q_{3,j}$, and $q_{4,j}$, respectively, out of which one is zero.

For $d = 5$, the ZF NPs of fifth degree depend on whether $p_j = 3$, as well as whether $p_j = 5$.

If $p_j = 3$, we have $q_{3,j}, q_{4,j}, q_{5,j} < 3^{n_{L,3}-1}$ and, thus, there will be $3^{n_{L,3}-1}$ coefficients $q_{2,j}$ and $3^{n_{L,3}-2}$ (if $n_{L,3} \geq 2$) coefficients $q_{3,j}$, $q_{4,j}$ and $q_{5,j}$, respectively, out of which one is zero. If $n_{L,3} = 1$, $q_{2,j}$, $q_{3,j}$, $q_{4,j}$ and $q_{5,j}$ can take only the value zero. Summarizing, for $n_{L,3} \geq 1$ there are $3^{n_{L,3}-1}$ coefficients $q_{2,j}$ and $3^{\max\{0;n_{L,3}-2\}}$ coefficients $q_{3,j}$, $q_{4,j}$ and $q_{5,j}$, respectively, out of which one is zero.

If $p_j = 5$, we have $q_{5,j} < 5^{n_{L,5}-1}$ if $n_{L,5} \geq 2$ and $q_{5,j} < 5$ if $n_{L,5} = 1$. Thus, there will be $5^{n_{L,5}-1}$ coefficients $q_{2,j}$, $q_{3,j}$ and $q_{4,j}$ respectively, and $5^{n_{L,5}-2}$ (if $n_{L,5} \geq 2$) coefficients $q_{5,j}$, out of which one is zero. If $n_{L,5} = 1$, $q_{5,j}$ can take only the value zero. Summarizing, for $n_{L,5} \geq 1$ there are $5^{n_{L,5}-1}$ coefficients $q_{2,j}$, $q_{3,j}$ and $q_{4,j}$ respectively, and $5^{\max\{0;n_{L,5}-2\}}$ coefficients $q_{5,j}$, respectively, out of which one is zero.

If $p_j \neq 3$ and $p_j \neq 5$, there are no ZF NPs of degree less than or equal to five and, therefore, there will be $p_j^{n_{L,p_j}-1}$ coefficients $q_{2,j}$, $q_{3,j}$, $q_{4,j}$ and $q_{5,j}$, respectively, out of which one is zero.

In the following, we apply the method described in Sect. 4.3 to determine the number of true different ZF PPs of degrees 3, 4 and 5.

4.6.1 Determining the Number of True Different Cubic Permutation Polynomial-Based Interleavers Under Zhao and Fan Sufficient Conditions

From the equivalence conditions for ZF CPPs, given in Theorems 4.3 and 4.6 with the additional constraints from the beginning of this section, we must have:

- $q_2 < L/2$ and $q_3 < L/2$, when $2 \mid L$ and $3 \nmid L$;
- $q_3 < L/3$, when $9 \mid L$ and $2 \nmid L$;
- $q_2 < L/2$ and $q_3 < L/2$, when $6 \mid L$ and $9 \nmid L$;
- $q_2 < L/2$ and $q_3 < L/6$, when $18 \mid L$.

For CPPs, we distinguish six cases for prime decomposition of L. The results regarding the number of true different ZF CPPs are given in Theorems 4.28–4.33 below and are summarized in Table 4.3 at the end of this subsection.

Theorem 4.28 *If* $2 \nmid L$ *and* $3 \nmid L$*, i.e.* $L = \prod_{j=1}^{n_L} p_j^{n_{L,p_j}}$*, with* $p_j > 3$ *and* $n_{L,p_j} \geq 1$*,* $\forall j = \overline{1, n_L}$*, the number of ZF CPPs will be equal to:*

$$C_{L,CPPs,ZF} = \left(\prod_{j=1}^{n_L} p_j^{2 \cdot (n_{L,p_j} - 1)} \cdot (p_j - 1) \right) \cdot \left(\prod_{j=1}^{n_L} p_j^{n_{L,p_j} - 1} - 1 \right) \quad (4.113)$$

Proof From case ZF2 at the beginning of Sect. 4.6, it follows that the number of possible combinations for the coefficient q_1 is equal to $\prod_{j=1}^{n_L} \left(p_j^{n_{L,p_j} - 1} \cdot (p_j - 1) \right)$ and the number of coefficients q_2 and q_3, respectively, is equal to $\prod_{j=1}^{n_L} p_j^{n_{L,p_j} - 1}$. The value $q_3 = 0$ results only when $q_{3,j} = 0$, $\forall j = \overline{1, n_L}$, i.e. for a single combination of coefficients $q_{3,j}$, $j = \overline{1, n_L}$, that has to be removed. The number of ZF CPPs will be equal to that in (4.113). ■

From (4.113) we see that the number of ZF CPPs is equal to 0 if L is a product of prime numbers greater than three, each of them to the power of 1.

Theorem 4.29 *If* $2 \mid L$*,* $4 \nmid L$ *and* $3 \nmid L$*, i.e.* $L = 2 \cdot \prod_{j=2}^{n_L} p_j^{n_{L,p_j}}$*, with* $p_j > 3$ *and* $n_{L,p_j} \geq 1$*,* $\forall j = \overline{2, n_L}$*, the number of ZF CPPs will be equal to:*

$$C_{L,CPPs,ZF} = \left(\prod_{j=2}^{n_L} p_j^{2 \cdot (n_{L,p_j} - 1)} \cdot (p_j - 1) \right) \cdot \left(\prod_{j=2}^{n_L} p_j^{n_{L,p_j} - 1} - 1 \right) \quad (4.114)$$

Proof From the analysis for cases ZF1. (a) and ZF2 at the beginning of Sect. 4.6, the number of possible combinations for coefficient q_1 is equal to $1 \cdot \Phi(L/2) = \prod_{j=2}^{n_L} \left(p_j^{n_{L,p_j} - 1} \cdot (p_j - 1) \right)$, and the number of coefficients q_2 and q_3, respectively, is equal to $1 \cdot \prod_{j=2}^{n_L} p_j^{n_{L,p_j} - 1}$, out of which one value is zero. By removing the value $q_3 = 0$, the number of ZF CPPs will be equal to:

$$C_{L,CPPs,ZF} = \left(\prod_{j=2}^{n_L} p_j^{n_{L,p_j} - 1} \cdot (p_j - 1) \right) \cdot \left(\prod_{j=2}^{n_L} p_j^{n_{L,p_j} - 1} \right) \cdot$$

$$\cdot\left(\prod_{j=2}^{n_L} p_j^{n_{L,p_j}-1}-1\right)=\left(\prod_{j=2}^{n_L} p_j^{2\cdot(n_{L,p_j}-1)}\cdot(p_j-1)\right)\cdot\left(\prod_{j=2}^{n_L} p_j^{n_{L,p_j}-1}-1\right) \tag{4.115}$$

■

From (4.114) we see that the number of ZF CPPs is equal to 0 if L is two times a product of prime numbers greater than three, each of them to the power of 1.

Theorem 4.30 *If* $4 \mid L$ *and* $3 \nmid L$, *i.e.* $L = 2^{n_{L,2}} \cdot \prod_{j=2}^{n_L} p_j^{n_{L,p_j}}$, *with* $n_{L,2} > 1,\ p_j > 3$ *and* $n_{L,p_j} \geq 1,\ \forall j = \overline{2, n_L}$, *the number of ZF CPPs will be equal to:*

$$C_{L,CPPs,ZF} = 2^{2\cdot n_{L,2}-3}\cdot\left(\prod_{j=2}^{n_L} p_j^{2\cdot(n_{L,p_j}-1)}\cdot(p_j-1)\right)\cdot$$

$$\cdot\left(2^{n_{L,2}-2}\cdot\prod_{j=2}^{n_L} p_j^{n_{L,p_j}-1}-1\right) \tag{4.116}$$

Proof From cases ZF1. (a) and ZF2 at the beginning of Sect. 4.6, it follows that the number of possible combinations for coefficient q_1 is equal to $\Phi(L) = 2^{n_{L,2}-1}\cdot\prod_{j=2}^{n_L}\left(p_j^{n_{L,p_j}-1}\cdot(p_j-1)\right)$, and the number of coefficients q_2 and q_3, respectively, is equal to $2^{n_{L,2}-2}\cdot\prod_{j=2}^{n_L} p_j^{n_{L,p_j}-1}$, out of which one value is zero. By removing the value $q_3 = 0$, the number of ZF CPPs will be equal to:

$$C_{L,CPPs,ZF} = \left(2^{n_{L,2}-1}\cdot\prod_{j=2}^{n_L} p_j^{n_{L,p_j}-1}\cdot(p_j-1)\right)\cdot$$

$$\cdot\left(2^{n_{L,2}-2}\cdot\prod_{j=2}^{n_L} p_j^{n_{L,p_j}-1}\right)\cdot\left(2^{n_{L,2}-2}\cdot\prod_{j=2}^{n_L} p_j^{n_{L,p_j}-1}-1\right)=$$

$$=2^{2\cdot n_{L,2}-3}\cdot\left(\prod_{j=2}^{n_L} p_j^{2\cdot(n_{L,p_j}-1)}\cdot(p_j-1)\right)\cdot\left(2^{n_{L,2}-2}\cdot\prod_{j=2}^{n_L} p_j^{n_{L,p_j}-1}-1\right) \tag{4.117}$$

■

From (4.116) we see that the number of ZF CPPs is equal to 0 if L is four times a product of prime numbers greater than three, each of them to the power of 1.

Theorem 4.31 *If* $2 \nmid L$ *and* $3 \mid L$, *i.e.* $L = 3^{n_{L,3}} \cdot \prod_{j=2}^{n_L} p_j^{n_{L,p_j}}$, *with* $n_{L,3} \geq 1$, $p_j > 3$ *and* $n_{L,p_j} \geq 1$, $\forall j = \overline{2, n_L}$, *the number of ZF CPPs will be equal to:*

$$C_{L,CPPs,ZF} = 2 \cdot 3^{2\cdot(n_{L,3}-1)} \cdot \left(\prod_{j=2}^{n_L} p_j^{2\cdot(n_{L,p_j}-1)} \cdot (p_j - 1) \right) \cdot \left(3^{\max\{0; n_{L,3}-2\}} \cdot \prod_{j=2}^{n_L} p_j^{n_{L,p_j}-1} - 1 \right) \tag{4.118}$$

Proof The number of possible combinations for coefficient q_1 is equal to $2 \cdot 3^{n_{L,3}-1} \cdot \prod_{j=2}^{n_L} \left(p_j^{n_{L,p_j}-1} \cdot (p_j - 1) \right)$, the number of coefficients q_2 is equal to $3^{n_{L,3}-1} \cdot \prod_{j=2}^{n_L} p_j^{n_{L,p_j}-1}$, and the number of coefficients q_3 is equal to $3^{\max\{0; n_{L,3}-2\}} \cdot \prod_{j=2}^{n_L} p_j^{n_{L,p_j}-1}$, out of which one value is zero. By removing the value $q_3 = 0$, the number of ZF CPPs will be equal to:

$$C_{L,CPPs,ZF} = \left(2 \cdot 3^{n_{L,3}-1} \cdot \prod_{j=2}^{n_L} p_j^{n_{L,p_j}-1} \cdot (p_j - 1) \right) \cdot \left(3^{n_{L,3}-1} \cdot \prod_{j=2}^{n_L} p_j^{n_{L,p_j}-1} \right) \cdot \left(3^{\max\{0; n_{L,3}-2\}} \cdot \prod_{j=2}^{n_L} p_j^{n_{L,p_j}-1} - 1 \right) =$$

$$= 2 \cdot 3^{2\cdot(n_{L,3}-1)} \cdot \left(\prod_{j=2}^{n_L} p_j^{2\cdot(n_{L,p_j}-1)} \cdot (p_j - 1) \right) \cdot \left(3^{\max\{0; n_{L,3}-2\}} \cdot \prod_{j=2}^{n_L} p_j^{n_{L,p_j}-1} - 1 \right) \tag{4.119}$$

■

From (4.118) we see that the number of ZF CPPs is equal to 0, if L is three or nine times a product of prime numbers greater than three, each of them to the power of 1.

Theorem 4.32 *If* $2 \mid L$, $4 \nmid L$ *and* $3 \mid L$, *i.e.* $L = 2 \cdot 3^{n_{L,3}} \cdot \prod_{j=3}^{n_L} p_j^{n_{L,p_j}}$, *with* $n_{L,3} \geq 1$, $p_j > 3$ *and* $n_{L,p_j} \geq 1$, $\forall j = \overline{3, n_L}$, *the number of ZF CPPs will be equal to:*

$$C_{L,CPPs,ZF} = 2 \cdot 3^{2 \cdot (n_{L,3}-1)} \cdot \left(\prod_{j=3}^{n_L} p_j^{2 \cdot (n_{L,p_j}-1)} \cdot (p_j - 1) \right) \cdot \left(3^{\max\{0; n_{L,3}-2\}} \cdot \prod_{j=3}^{n_L} p_j^{n_{L,p_j}-1} - 1 \right) \tag{4.120}$$

Proof The proof for Theorem 4.32 is similar to that of Theorem 4.31, by replacing L by $L/2$. Therefore, the number of ZF CPPs will be equal to:

$$C_{L,CPPs,ZF} = \left(2 \cdot 3^{n_{L,3}-1} \cdot \prod_{j=3}^{n_L} p_j^{n_{L,p_j}-1} \cdot (p_j - 1) \right) \cdot \left(3^{n_{L,3}-1} \cdot \prod_{j=3}^{n_L} p_j^{n_{L,p_j}-1} \right) \cdot \left(3^{\max\{0; n_{L,3}-2\}} \cdot \prod_{j=3}^{n_L} p_j^{n_{L,p_j}-1} - 1 \right) =$$

$$= 2 \cdot 3^{2 \cdot (n_{L,3}-1)} \cdot \left(\prod_{j=3}^{n_L} p_j^{2 \cdot (n_{L,p_j}-1)} \cdot (p_j - 1) \right) \cdot \left(3^{\max\{0; n_{L,3}-2\}} \cdot \prod_{j=3}^{n_L} p_j^{n_{L,p_j}-1} - 1 \right) \tag{4.121}$$

■

From (4.120) we see that the number of ZF CPPs is equal to 0 if L is 6 or 18 times a product of prime numbers greater than three, each of them to the power of 1.

Theorem 4.33 *If* $4 \mid L$ *and* $3 \mid L$, *i.e.* $L = 2^{n_{L,2}} \cdot 3^{n_{L,3}} \cdot \prod_{j=3}^{n_L} p_j^{n_{L,p_j}}$, *with* $n_{L,2} > 1$, $n_{L,3} \geq 1$, $p_j > 3$ *and* $n_{L,p_j} \geq 1$, $\forall j = \overline{3, n_L}$, *the number of ZF CPPs will be equal to:*

$$C_{L,CPPs,ZF} = 2^{2 \cdot (n_{L,2}-1)} \cdot 3^{2 \cdot (n_{L,3}-1)} \cdot \left(\prod_{j=3}^{n_L} p_j^{2 \cdot (n_{L,p_j}-1)} \cdot (p_j - 1) \right) \cdot$$

$$\cdot\left(2^{n_{L,2}-2}\cdot 3^{\max\{0;n_{L,3}-2\}}\cdot\prod_{j=3}^{n_L}p_j^{n_{L,p_j}-1}-1\right) \tag{4.122}$$

Proof The number of possible combinations for coefficient q_1 is equal to $\Phi(L)=2^{n_{L,2}-1}\cdot 2\cdot 3^{n_{L,3}-1}\cdot\prod_{j=3}^{n_L}\left(p_j^{n_{L,p_j}-1}\cdot(p_j-1)\right)$, the number of coefficients q_2 is equal to $2^{n_{L,2}-2}\cdot 3^{n_{L,3}-1}\cdot\prod_{j=3}^{n_L}p_j^{n_{L,p_j}-1}$ and the number of coefficients q_3 is equal to $2^{n_{L,2}-2}\cdot 3^{\max\{0;n_{L,3}-2\}}\cdot\prod_{j=3}^{n_L}p_j^{n_{L,p_j}-1}$, out of which one is zero. By removing the value $q_3=0$, the number of ZF CPPs will be equal to:

$$\begin{aligned}C_{L,CPPs,ZF}&=\left(2^{n_{L,2}-1}\cdot 2\cdot 3^{n_{L,3}-1}\cdot\prod_{j=3}^{n_L}p_j^{n_{L,p_j}-1}\cdot(p_j-1)\right)\cdot\\&\quad\cdot\left(2^{n_{L,2}-2}\cdot 3^{n_{L,3}-1}\cdot\prod_{j=3}^{n_L}p_j^{n_{L,p_j}-1}\right)\cdot\\&\quad\cdot\left(2^{n_{L,2}-2}\cdot 3^{\max\{0;n_{L,3}-2\}}\cdot\prod_{j=3}^{n_L}p_j^{n_{L,p_j}-1}-1\right)=\\&=2^{2\cdot(n_{L,2}-1)}\cdot 3^{2\cdot(n_{L,3}-1)}\cdot\left(\prod_{j=3}^{n_L}p_j^{2\cdot(n_{L,p_j}-1)}\cdot(p_j-1)\right)\cdot\\&\quad\cdot\left(2^{n_{L,2}-2}\cdot 3^{\max\{0;n_{L,3}-2\}}\cdot\prod_{j=3}^{n_L}p_j^{n_{L,p_j}-1}-1\right)\end{aligned} \tag{4.123}$$

■

From (4.122) we see that the number of ZF CPPs is equal to 0, if L is 12 or 36 times a product of prime numbers greater than three, each of them to the power of 1.

From Theorems 4.28–4.33 above, we conclude that under Zhao and Fan sufficient conditions the number of CPPs is 0, when the interleaver length is

$$L=2^{n_{L,2}}\cdot 3^{n_{L,3}}\cdot\prod_{j=3}^{n_L}p_j,\; n_{L,2}=\overline{0,2},\; n_{L,3}=\overline{0,2},\; p_j>3,\forall j=\overline{3,n_L} \tag{4.124}$$

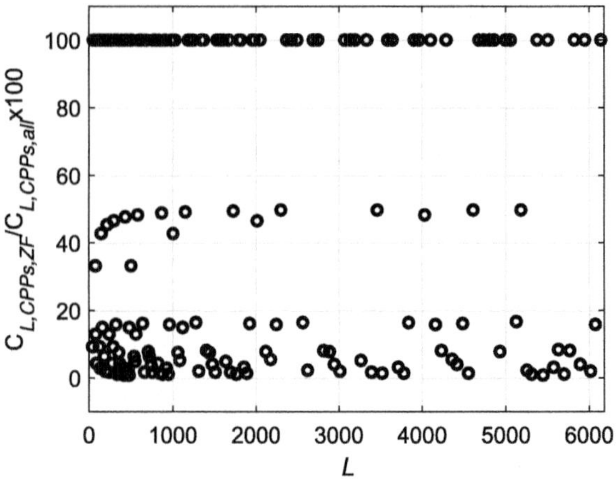

Fig. 4.1 The ratio (in percentage) between the number of true different ZF CPPs and the number of all true different CPPs, for all lengths of LTE standard

Such lengths have to be avoided in designing CPP-based interleavers, under Zhao and Fan sufficient conditions.

Because all the lengths from the LTE standard (3GPP 2008) are multiples of 8, from (4.124) it results that for all these lengths there exist CPPs, under Zhao and Fan sufficient conditions.

The necessary and sufficient conditions that the coefficients of CPPs must fulfill are given in Table 3.2. By comparing them with those in Table 3.6, we notice that Zhao and Fan sufficient conditions become necessary when 3 at any power, or a prime number greater than 2, of the form $3 \cdot k + 2$, $k \in \mathbb{N}$, at the power 1, are not prime factors in the decomposition of the interleaver length. LTE standard contains 58 such lengths. However, for all lengths for which in their prime factorization there exists the prime 3 at power of 1, the number of all true different CPPs is equal to the number of true different ZF CPPs. There exist 25 such lengths among the LTE lengths. Thus for 83 lengths from the LTE standard the number of all true different CPPs is equal to the number of true different ZF CPPs.

Figure 4.1 shows the ratio (in percentage) between the number of true different ZF CPPs and the number of all true different CPPs, for all the 188 lengths of the LTE standard. For 105 lengths (others than the 83 ones for which the number of all true different CPPs is equal to the number of true different ZF CPPs), this percentage is less than 50%.

In the following, we give an example with related comments for the most comprehensive case, i.e. for Theorems 4.33. The example provides information about all the values the coefficients of polynomials of degree 3 can take, so that they are CPPs modulo the considered length.

Example 4.2 (*Example for Theorem* 4.33) Let the length be $L = 2100 = 2^2 \cdot 3 \cdot 5^2 \cdot 7$. According to (4.122), the number of CPPs will be equal to:

$$C_{2100,CPPs,ZF} = 2^{2\cdot(2-1)} \cdot 3^{2\cdot(1-1)} \cdot 5^{2\cdot(2-1)} \cdot (5-1) \cdot 7^{2\cdot(1-1)} \cdot (7-1)\cdot$$

$$\cdot\left(2^{2-2} \cdot 3^{\max\{0;1-2\}} \cdot 5^{2-1} \cdot 7^{1-1} - 1\right) =$$

$$= 4 \cdot 1 \cdot 25 \cdot 4 \cdot 1 \cdot 6 \cdot \left(1 \cdot 3^0 \cdot 5 \cdot 1 - 1\right) = 2400 \cdot (5 - 1) = 9600.$$

Further, we detail the coefficients of the 9600 true different ZF CPPs. As $6 \mid L$ and $9 \nmid L$, it is required that $q_3 < 3 \cdot L/6 = L/2 = 1050$ and $q_2 < L/2 = 1050$. The coefficients q_2 and q_3 have to be multiples of $2 \cdot 3 \cdot 5 \cdot 7 = 210$, and coefficient q_1 has to be relatively prime with 2100. Therefore, coefficient q_3 can take 4 values: 210, 420, 630 and 840, q_2 can take 5 values: 0, 210, 420, 630 and 840, and coefficient q_1 can take $\Phi(2100) = 2100 \cdot \frac{1}{2} \cdot \frac{2}{3} \cdot \frac{4}{5} \cdot \frac{6}{7} = 480$ values. The total number of ZF CPPs is equal to $4 \cdot 5 \cdot 480 = 9600$, just the number found by the formula we determined. We note that the number of all true different CPP is also 9600. However the equivalence conditions for CPPs, in this case, are: $q_3 < L/6 = 350$ and $q_2 < L/2 = 1050$. We note that every ZF CPP having coefficient $q_3 > 350$ (i.e. 420, 630, or 840) has a equivalent CPP having coefficient $q_3 < 350$ and coefficient $q_2 < 1050$. For example, ZF CPP $1 \cdot x + 0 \cdot x^2 + 420 \cdot x^3 \pmod{2100}$ is equivalent to CPP $351 \cdot x + 0 \cdot x^2 + 70 \cdot x^3 \pmod{2100}$ only by NP $z(x) = 350 \cdot x + 0 \cdot x^2 + 1750 \cdot x^3 \pmod{2100}$. This NP is not a ZF NP since $3 \nmid 350$ and $3 \nmid 1750$. ■

4.6.2 *Determining the Number of True Different Fourth Degree Permutation Polynomial-Based Interleavers Under Zhao and Fan Sufficient Conditions*

As it is shown at the beginning of Sect. 4.6, from the equivalence conditions for ZF 4-PPs we must have:

- $q_2 < L/2$, $q_3 < L/2$ and $q_4 < L/2$, when $2 \mid L$, $3 \nmid L$ and $4 \nmid L$;
- $q_2 < L/2$, $q_3 < L/2$ and $q_4 < L/4$, when $3 \nmid L$, $4 \mid L$ and $8 \nmid L$;
- $q_2 < L/2$, $q_3 < L/2$ and $q_4 < L/2$, when $4 \nmid L$, $6 \mid L$, and $9 \nmid L$;
- $q_2 < L/2$, $q_3 < L/2$ and $q_4 < L/8$, when $3 \nmid L$ and $8 \mid L$;
- $q_3 < L/3$ and $q_4 < L/3$, when $2 \nmid L$ and $9 \mid L$;
- $q_2 < L/2$, $q_3 < L/2$ and $q_4 < L/4$, when $8 \nmid L$, $9 \nmid L$, and $12 \mid L$;
- $q_2 < L/2$, $q_3 < L/6$ and $q_4 < L/6$, when $4 \nmid L$ and $18 \mid L$;
- $q_2 < L/2$, $q_3 < L/6$ and $q_4 < L/12$, when $8 \nmid L$ and $36 \mid L$;
- $q_2 < L/2$, $q_3 < L/2$ and $q_4 < L/8$, when $9 \nmid L$ and $24 \mid L$;
- $q_2 < L/2$, $q_3 < L/6$ and $q_4 < L/24$, when $72 \mid L$.

Table 4.3 Number of true different CPP-based interleavers under Zhao and Fan sufficient conditions

Case	Decomposition of L	$C_{L,CPPs,ZF}$	Theorems
(1)	$L=\prod_{j=1}^{n_L} p_j^{n_{L,p_j}}$, $p_j>3$ and $n_{L,p_j}\geq 1, \forall j=\overline{1,n_L}$	$C_{L,CPPs,ZF}=\left(\prod_{j=1}^{n_L} p_j^{2\cdot(n_{L,p_j}-1)}\cdot(p_j-1)\right)\cdot\left(\prod_{j=1}^{n_L} p_j^{n_{L,p_j}-1}-1\right)$	4.28
(2)	$L=2\cdot\prod_{j=2}^{n_L} p_j^{n_{L,p_j}}$, $p_j>3$ and $n_{L,p_j}\geq 1, \forall j=\overline{2,n_L}$	$C_{L,CPPs,ZF}=\left(\prod_{j=2}^{n_L} p_j^{2\cdot(n_{L,p_j}-1)}\cdot(p_j-1)\right)\cdot\left(\prod_{j=2}^{n_L} p_j^{n_{L,p_j}-1}-1\right)$	4.29
(3)	$L=2^{n_{L,2}}\cdot\prod_{j=2}^{n_L} p_j^{n_{L,p_j}}$, $n_{L,2}>1$, $p_j>3$ and $n_{L,p_j}\geq 1, \forall j=\overline{2,n_L}$	$C_{L,CPPs,ZF}=2^{2\cdot n_{L,2}-3}\cdot\left(\prod_{j=2}^{n_L} p_j^{2\cdot(n_{L,p_j}-1)}\cdot(p_j-1)\right)\cdot\left(2^{n_{L,2}-2}\cdot\prod_{j=2}^{n_L} p_j^{n_{L,p_j}-1}-1\right)$	4.30
(4)	$L=3^{n_{L,3}}\cdot\prod_{j=2}^{n_L} p_j^{n_{L,p_j}}$, $n_{L,3}\geq 1$, $p_j>3$ and $n_{L,p_j}\geq 1, \forall j=\overline{2,n_L}$	$C_{L,CPPs,ZF}=2\cdot 3^{2\cdot(n_{L,3}-1)}\cdot\left(\prod_{j=2}^{n_L} p_j^{2\cdot(n_{L,p_j}-1)}\cdot(p_j-1)\right)\cdot\left(3^{\max\{0;n_{L,3}-2\}}\cdot\prod_{j=2}^{n_L} p_j^{n_{L,p_j}-1}-1\right)$	4.31
(5)	$L=2\cdot 3^{n_{L,3}}\cdot\prod_{j=3}^{n_L} p_j^{n_{L,p_j}}$, $n_{L,3}\geq 1$, $p_j>3$ and $n_{L,p_j}\geq 1$, $\forall j=\overline{3,n_L}$	$C_{L,CPPs,ZF}=2\cdot 3^{2\cdot(n_{L,3}-1)}\cdot\left(\prod_{j=3}^{n_L} p_j^{2\cdot(n_{L,p_j}-1)}\cdot(p_j-1)\right)\cdot\left(3^{\max\{0;n_{L,3}-2\}}\cdot\prod_{j=3}^{n_L} p_j^{n_{L,p_j}-1}-1\right)$	4.32
(6)	$L=2^{n_{L,2}}\cdot 3^{n_{L,3}}\cdot\prod_{j=3}^{n_L} p_j^{n_{L,p_j}}$, $n_{L,2}>1, n_{L,3}\geq 1$, $p_j>3$ and $n_{L,p_j}\geq 1, \forall j=\overline{3,n_L}$	$C_{L,CPPs,ZF}=2^{2\cdot(n_{L,2}-1)}\cdot 3^{2\cdot(n_{L,3}-1)}\cdot\left(\prod_{j=3}^{n_L} p_j^{2\cdot(n_{L,p_j}-1)}\cdot(p_j-1)\right)\cdot\left(2^{n_{L,2}-2}\cdot 3^{\max\{0;n_{L,3}-2\}}\cdot\prod_{j=3}^{n_L} p_j^{n_{L,p_j}-1}-1\right)$	4.33

For ZF PPs of fourth degree, there are eight cases for the prime decomposition of L. The results regarding the number of true different ZF 4-PPs are given in Theorems 4.34–4.41 below and are summarized in Table 4.4 at the end of this subsection.

Table 4.4 Number of true different 4-PP-based interleavers under Zhao and Fan sufficient conditions

Case	Decomposition of L	$C_{L,4-PPs,ZF}$	Theorems
(1)	$L=\prod_{j=1}^{n_L} p_j^{n_{L,p_j}}$, $p_j>3$ and $n_{L,p_j}\geq 1, \forall j=\overline{1,n_L}$	$C_{L,4-PPs,ZF}=\left(\prod_{j=1}^{n_L} p_j^{3\cdot(n_{L,p_j}-1)}\cdot(p_j-1)\right)\cdot\left(\prod_{j=1}^{n_L} p_j^{n_{L,p_j}-1}-1\right)$	4.34
(2)	$L=2\cdot\prod_{j=2}^{n_L} p_j^{n_{L,p_j}}$, $p_j>3$ and $n_{L,p_j}\geq 1, \forall j=\overline{2,n_L}$	$C_{L,4-PPs,ZF}=\left(\prod_{j=2}^{n_L} p_j^{3\cdot(n_{L,p_j}-1)}\cdot(p_j-1)\right)\cdot\left(\prod_{j=2}^{n_L} p_j^{n_{L,p_j}-1}-1\right)$	4.35
(3)	$L=3^{n_{L,3}}\cdot\prod_{j=2}^{n_L} p_j^{n_{L,p_j}}$, $n_{L,3}\geq 1$, $p_j>3$ and $n_{L,p_j}\geq 1$, $\forall j=\overline{2,n_L}$	$C_{L,4-PPs,ZF}=2\cdot 3^{2\cdot(n_{L,3}-1)+\max\{0;n_{L,3}-2\}}\cdot\prod_{j=2}^{n_L}\left(p_j^{3\cdot(n_{L,p_j}-1)}\cdot(p_j-1)\right)\cdot\left(3^{\max\{0;n_{L,3}-2\}}\cdot\prod_{j=2}^{n_L} p_j^{n_{L,p_j}-1}-1\right)$	4.36
(4)	$L=4\cdot\prod_{j=2}^{n_L} p_j^{n_{L,p_j}}$, $p_j>3$ and $n_{L,p_j}\geq 1, \forall j=\overline{2,n_L}$	$C_{L,4-PPs,ZF}=2\cdot\prod_{j=2}^{n_L}\left(p_j^{3\cdot(n_{L,p_j}-1)}\cdot(p_j-1)\right)\cdot\left(\prod_{j=2}^{n_L} p_j^{n_{L,p_j}-1}-1\right)$	4.37
(5)	$L=2\cdot 3^{n_{L,3}}\cdot\prod_{j=3}^{n_L} p_j^{n_{L,p_j}}$, $n_{L,3}\geq 1$, $p_j>3$ and $n_{L,p_j}\geq 1$, $\forall j=\overline{3,n_L}$	$C_{L,4-PPs,ZF}=2\cdot 3^{2\cdot(n_{L,3}-1)+\max\{0;n_{L,3}-2\}}\cdot\left(\prod_{j=3}^{n_L} p_j^{3\cdot(n_{L,p_j}-1)}\cdot(p_j-1)\right)\cdot\left(3^{\max\{0;n_{L,3}-2\}}\cdot\prod_{j=3}^{n_L} p_j^{n_{L,p_j}-1}-1\right)$	4.38

(continued)

Table 4.4 (continued)

Case	Decomposition of L	$C_{L,4-PPs,ZF}$	Theorems
(6)	$L = 2^{n_{L,2}} \cdot \prod_{j=2}^{n_L} p_j^{n_{L,p_j}}$, $n_{L,2} \geq 3$, $p_j > 3$ and $n_{L,p_j} \geq 1$, $\forall j = \overline{2, n_L}$	$C_{L,4-PPs,ZF} = 2^{3 \cdot n_{L,2}-5} \cdot \prod_{j=2}^{n_L} \left(p_j^{3 \cdot (n_{L,p_j}-1)} \cdot (p_j - 1) \right) \cdot \left(2^{n_{L,2}-3} \cdot \prod_{j=2}^{n_L} p_j^{n_{L,p_j}-1} - 1 \right)$	4.39
(7)	$L = 4 \cdot 3^{n_{L,3}} \cdot \prod_{j=3}^{n_L} p_j^{n_{L,p_j}}$, $n_{L,3} \geq 1$, $p_j > 3$ and $n_{L,p_j} \geq 1$, $\forall j = \overline{3, n_L}$	$C_{L,4-PPs,ZF} = 4 \cdot 3^{2 \cdot (n_{L,3}-1)+\max\{0; n_{L,3}-2\}} \cdot \prod_{j=3}^{n_L} \left(p_j^{3 \cdot (n_{L,p_j}-1)} \cdot (p_j - 1) \right) \cdot \left(3^{\max\{0; n_{L,3}-2\}} \cdot \prod_{j=3}^{n_L} p_j^{n_{L,p_j}-1} - 1 \right)$	4.40
(8)	$L = 2^{n_{L,2}} \cdot 3^{n_{L,3}} \cdot \prod_{j=3}^{n_L} p_j^{n_{L,p_j}}$, $n_{L,2} \geq 3, n_{L,3} \geq 1, p_j > 3$ and $n_{L,p_j} \geq 1, \forall j = \overline{3, n_L}$	$C_{L,4-PPs,ZF} = 2^{3 \cdot n_{L,2}-4} \cdot 3^{2 \cdot (n_{L,3}-1)+\max\{0; n_{L,3}-2\}} \cdot \prod_{j=3}^{n_L} \left(p_j^{3 \cdot (n_{L,p_j}-1)} \cdot (p_j - 1) \right) \cdot \left(2^{n_{L,2}-3} \cdot 3^{\max\{0; n_{L,3}-2\}} \cdot \prod_{j=3}^{n_L} p_j^{n_{L,p_j}-1} - 1 \right)$	4.41

Theorem 4.34 *If $2 \nmid L$ and $3 \nmid L$, i.e. $L = \prod_{j=1}^{n_L} p_j^{n_{L,p_j}}$, with $p_j > 3$ and $n_{L,p_j} \geq 1$, $\forall j = \overline{1, n_L}$, the number of ZF 4-PPs will be equal to:*

$$C_{L,4-PPs,ZF} = \left(\prod_{j=1}^{n_L} p_j^{3 \cdot (n_{L,p_j}-1)} \cdot (p_j - 1) \right) \cdot \left(\prod_{j=1}^{n_L} p_j^{n_{L,p_j}-1} - 1 \right) \tag{4.125}$$

Proof From case ZF2 at the beginning of Sect. 4.6, the number of possible combinations for coefficient q_1 is equal to $\prod_{j=1}^{n_L} \left(p_j^{n_{L,p_j}-1} \cdot (p_j - 1) \right)$ and the number of coefficients q_2, q_3 and q_4, respectively, is equal to $\prod_{j=1}^{n_L} p_j^{n_{L,p_j}-1}$. By removing the value $q_4 = 0$, the number of ZF 4-PPs will be equal to that in (4.125). ■

From (4.125) we see that the number of ZF 4-PPs is equal to 0 if L is a product of prime numbers greater than three, each of them to the power of 1.

Theorem 4.35 *If* $2 \mid L$, $3 \nmid L$ *and* $4 \nmid L$, *i.e.* $L = 2 \cdot \prod_{j=2}^{n_L} p_j^{n_{L,p_j}}$, *with* $p_j > 3$ *and* $n_{L,p_j} \geq 1, \forall j = \overline{2, n_L}$, *the number of ZF 4-PPs will be equal to:*

$$C_{L,4-PPs,ZF} = \left(\prod_{j=2}^{n_L} p_j^{3 \cdot (n_{L,p_j} - 1)} \cdot (p_j - 1) \right) \cdot \left(\prod_{j=2}^{n_L} p_j^{n_{L,p_j} - 1} - 1 \right) \tag{4.126}$$

Proof From the analysis for cases ZF1. (a) and ZF2 at the beginning of Sect. 4.6, the number of possible combinations for coefficient q_1 is equal to $1 \cdot \Phi(L/2) = \prod_{j=2}^{n_L} \left(p_j^{n_{L,p_j} - 1} \cdot (p_j - 1) \right)$, and the number of coefficients q_2, q_3 and q_4, respectively, is equal to $1 \cdot \prod_{j=2}^{n_L} p_j^{n_{L,p_j} - 1}$, out of which one value is zero. By removing the value $q_4 = 0$, the number of ZF 4-PPs will be equal to:

$$C_{L,4-PPs,ZF} = \left(\prod_{j=2}^{n_L} p_j^{n_{L,p_j} - 1} \cdot (p_j - 1) \right) \cdot \left(\prod_{j=2}^{n_L} p_j^{n_{L,p_j} - 1} \right) \cdot$$

$$\cdot \left(\prod_{j=2}^{n_L} p_j^{n_{L,p_j} - 1} \right) \cdot \left(\prod_{j=2}^{n_L} p_j^{n_{L,p_j} - 1} - 1 \right) =$$

$$= \left(\prod_{j=2}^{n_L} p_j^{3 \cdot (n_{L,p_j} - 1)} \cdot (p_j - 1) \right) \cdot \left(\prod_{j=2}^{n_L} p_j^{n_{L,p_j} - 1} - 1 \right) \tag{4.127}$$

■

From (4.126) we see that the number of ZF 4-PPs is equal to 0 if L is two times a product of prime numbers greater than three, each of them to the power of 1.

Theorem 4.36 *If* $2 \nmid L$ *and* $3 \mid L$, *i.e.* $L = 3^{n_{L,3}} \cdot \prod_{j=2}^{n_L} p_j^{n_{L,p_j}}$, *with* $n_{L,3} \geq 1$, $p_j > 3$ *and* $n_{L,p_j} \geq 1, \forall j = \overline{2, n_L}$, *the number of ZF 4-PPs will be equal to:*

$$C_{L,4-PPs,ZF} = 2 \cdot 3^{2 \cdot (n_{L,3} - 1) + \max\{0; n_{L,3} - 2\}} \cdot \prod_{j=2}^{n_L} \left(p_j^{3 \cdot (n_{L,p_j} - 1)} \cdot (p_j - 1) \right) \cdot$$

$$\cdot \left(3^{\max\{0; n_{L,3}-2\}} \cdot \prod_{j=2}^{n_L} p_j^{n_{L,p_j}-1} - 1 \right) \tag{4.128}$$

Proof The number of possible combinations for coefficient q_1 is equal to $2 \cdot 3^{n_{L,3}-1} \cdot \prod_{j=2}^{n_L} \left(p_j^{n_{L,p_j}-1} \cdot (p_j - 1) \right)$, the number of coefficients q_2 is equal to $3^{n_{L,3}-1} \cdot \prod_{j=2}^{n_L} p_j^{n_{L,p_j}-1}$, and the number of coefficients q_3 and q_4, respectively, is equal to $3^{\max\{0;n_{L,3}-2\}} \cdot \prod_{j=2}^{n_L} p_j^{n_{L,p_j}-1}$, out of which one value is zero. By removing the value $q_4 = 0$, the number of ZF 4-PPs will be equal to:

$$C_{L,4-PPs,ZF} = 2 \cdot 3^{n_{L,3}-1} \cdot \prod_{j=2}^{n_L} \left(p_j^{n_{L,p_j}-1} \cdot (p_j - 1) \right) \cdot$$

$$\cdot \left(3^{n_{L,3}-1} \cdot \prod_{j=2}^{n_L} p_j^{n_{L,p_j}-1} \right) \cdot 3^{\max\{0;n_{L,3}-2\}} \cdot \prod_{j=2}^{n_L} p_j^{n_{L,p_j}-1} \cdot$$

$$\cdot \left(3^{\max\{0;n_{L,3}-2\}} \cdot \prod_{j=2}^{n_L} p_j^{n_{L,p_j}-1} - 1 \right) =$$

$$= 2 \cdot 3^{2\cdot(n_{L,3}-1)+\max\{0;n_{L,3}-2\}} \cdot \prod_{j=2}^{n_L} \left(p_j^{3\cdot(n_{L,p_j}-1)} \cdot (p_j - 1) \right) \cdot$$

$$\cdot \left(3^{\max\{0;n_{L,3}-2\}} \cdot \prod_{j=2}^{n_L} p_j^{n_{L,p_j}-1} - 1 \right) \tag{4.129}$$

■

From (4.128) we see that the number of ZF 4-PPs is equal to 0 if L is three or nine times a product of prime numbers greater than three, each of them to the power of 1.

Theorem 4.37 *If* $3 \nmid L$, $4 \mid L$ *and* $8 \nmid L$, *i.e.* $L = 4 \cdot \prod_{j=2}^{n_L} p_j^{n_{L,p_j}}$, *with* $p_j > 3$ *and* $n_{L,p_j} \geq 1$, $\forall j = \overline{2, n_L}$, *the number of ZF 4-PPs will be equal to:*

$$C_{L,4-PPs,ZF} = 2 \cdot \prod_{j=2}^{n_L} \left(p_j^{3\cdot(n_{L,p_j}-1)} \cdot (p_j - 1) \right) \cdot \left(\prod_{j=2}^{n_L} p_j^{n_{L,p_j}-1} - 1 \right) \quad (4.130)$$

Proof From cases ZF1. (a) and ZF2 at the beginning of Sect. 4.6, it follows that the number of possible combinations for coefficient q_1 is equal to $\Phi(L) = 2 \cdot \prod_{j=2}^{n_L} \left(p_j^{n_{L,p_j}-1} \cdot (p_j - 1) \right)$, and the number of coefficients q_2, q_3 and q_4, respectively, is equal to $\prod_{j=2}^{n_L} p_j^{n_{L,p_j}-1}$, out of which one value is zero. By removing the value $q_4 = 0$, the number of ZF 4-PPs will be equal to:

$$C_{L,4-PPs,ZF} = 2 \cdot \prod_{j=2}^{n_L} \left(p_j^{n_{L,p_j}-1} \cdot (p_j - 1) \right) \cdot \left(\prod_{j=2}^{n_L} p_j^{n_{L,p_j}-1} \right) \cdot$$

$$\cdot \left(\prod_{j=2}^{n_L} p_j^{n_{L,p_j}-1} \right) \cdot \left(\prod_{j=2}^{n_L} p_j^{n_{L,p_j}-1} - 1 \right) =$$

$$= 2 \cdot \prod_{j=2}^{n_L} \left(p_j^{3\cdot(n_{L,p_j}-1)} \cdot (p_j - 1) \right) \cdot \left(\prod_{j=2}^{n_L} p_j^{n_{L,p_j}-1} - 1 \right) \quad (4.131)$$

■

From (4.130) we see that the number of ZF 4-PPs is equal to 0 if L is four times a product of prime numbers greater than three, each of them to the power of 1.

Theorem 4.38 *If* $6 \mid L$ *and* $4 \nmid L$, *i.e.* $L = 2 \cdot 3^{n_{L,3}} \cdot \prod_{j=3}^{n_L} p_j^{n_{L,p_j}}$, *with* $n_{L,3} \geq 1$, $p_j > 3$ *and* $n_{L,p_j} \geq 1$, $\forall j = \overline{3, n_L}$, *the number of ZF 4-PPs will be equal to:*

$$C_{L,4-PPs,ZF} = 2 \cdot 3^{2\cdot(n_{L,3}-1)+\max\{0;n_{L,3}-2\}} \cdot \left(\prod_{j=3}^{n_L} p_j^{3\cdot(n_{L,p_j}-1)} \cdot (p_j - 1) \right) \cdot$$

$$\cdot \left(3^{\max\{0;n_{L,3}-2\}} \cdot \prod_{j=3}^{n_L} p_j^{n_{L,p_j}-1} - 1 \right) \quad (4.132)$$

Proof The proof for Theorem 4.38 is similar to that of Theorem 4.36, by replacing L by $L/2$. Thus, the number of ZF 4-PPs will be equal to:

$$C_{L,4-PPs,ZF} = 2 \cdot 3^{n_{L,3}-1} \cdot \prod_{j=3}^{n_L} \left(p_j^{n_{L,p_j}-1} \cdot (p_j - 1) \right) \cdot$$

$$\cdot \left(3^{n_{L,3}-1} \cdot \prod_{j=3}^{n_L} p_j^{n_{L,p_j}-1} \right) \cdot 3^{\max\{0; n_{L,3}-2\}} \cdot \prod_{j=3}^{n_L} p_j^{n_{L,p_j}-1} \cdot$$

$$\cdot \left(3^{\max\{0; n_{L,3}-2\}} \cdot \prod_{j=3}^{n_L} p_j^{n_{L,p_j}-1} - 1 \right) =$$

$$= 2 \cdot 3^{2 \cdot (n_{L,3}-1) + \max\{0; n_{L,3}-2\}} \cdot \left(\prod_{j=3}^{n_L} p_j^{3 \cdot (n_{L,p_j}-1)} \cdot (p_j - 1) \right) \cdot$$

$$\cdot \left(3^{\max\{0; n_{L,3}-2\}} \cdot \prod_{j=3}^{n_L} p_j^{n_{L,p_j}-1} - 1 \right) \tag{4.133}$$

■

From (4.132) we see that the number of ZF 4-PPs is equal to 0 if L is 6 or 18 times a product of prime numbers greater than three, each of them to the power of 1.

Theorem 4.39 *If* $8 \mid L$ *and* $3 \nmid L$, *i.e.* $L = 2^{n_{L,2}} \cdot \prod_{j=2}^{n_L} p_j^{n_{L,p_j}}$, *with* $n_{L,2} \geq 3$, $p_j > 3$ *and* $n_{L,p_j} \geq 1$, $\forall j = \overline{2, n_L}$, *the number of ZF 4-PPs will be equal to:*

$$C_{L,4-PPs,ZF} = 2^{3 \cdot n_{L,2}-5} \cdot \prod_{j=2}^{n_L} \left(p_j^{3 \cdot (n_{L,p_j}-1)} \cdot (p_j - 1) \right) \cdot$$

$$\cdot \left(2^{n_{L,2}-3} \cdot \prod_{j=2}^{n_L} p_j^{n_{L,p_j}-1} - 1 \right) \tag{4.134}$$

Proof From cases ZF1. (b) and ZF2 at the beginning of Sect. 4.6, the number of possible combinations for coefficient q_1 is equal to $\Phi(L) = 2^{n_{L,2}-1} \cdot \prod_{j=2}^{n_L} \left(p_j^{n_{L,p_j}-1} \cdot (p_j - 1) \right)$, the number of coefficients q_3 is equal to $2^{n_{L,2}-2} \cdot \prod_{j=2}^{n_L} p_j^{n_{L,p_j}-1}$, the number of even or odd coefficients q_2 is equal to $2^{n_{L,2}-2} \cdot \prod_{j=2}^{n_L} p_j^{n_{L,p_j}-1}$, and the number of even or odd coefficients q_4 is equal to $2^{n_{L,2}-4} \cdot \prod_{j=2}^{n_L} p_j^{n_{L,p_j}-1}$ if $n_{L,2} \geq 4$, out of which one value is zero. If $n_{L,2} = 3$, there are $\prod_{j=2}^{n_L} p_j^{n_{L,p_j}-1}$ coefficients q_4 (all even), and $2 \cdot \prod_{j=2}^{n_L} p_j^{n_{L,p_j}-1}$ coefficients q_2 (all even). Therefore, if $n_{L,2} \geq 4$, the number of ZF 4-PPs will be equal to:

$$C_{L,4-PPs,ZF} = 2^{n_{L,2}-1} \cdot \prod_{j=2}^{n_L} \left(p_j^{n_{L,p_j}-1} \cdot (p_j - 1) \right) \cdot \left(2^{n_{L,2}-2} \cdot \prod_{j=2}^{n_L} p_j^{n_{L,p_j}-1}\right) \cdot \left(2^{n_{L,2}-2} \cdot \prod_{j=2}^{n_L} p_j^{n_{L,p_j}-1} \right) \cdot \left(2^{n_{L,2}-4} \cdot \prod_{j=2}^{n_L} p_j^{n_{L,p_j}-1} - 1 \right) +$$
$$+ 2^{n_{L,2}-1} \cdot \prod_{j=2}^{n_L} \left(p_j^{n_{L,p_j}-1} \cdot (p_j - 1) \right) \cdot \left(2^{n_{L,2}-2} \cdot \prod_{j=2}^{n_L} p_j^{n_{L,p_j}-1} \right) \cdot \left(2^{n_{L,2}-2} \cdot \prod_{j=2}^{n_L} p_j^{n_{L,p_j}-1} \right) \cdot \left(2^{n_{L,2}-4} \cdot \prod_{j=2}^{n_L} p_j^{n_{L,p_j}-1} \right) =$$
$$= 2^{3 \cdot n_{L,2}-5} \cdot \prod_{j=2}^{n_L} \left(p_j^{3 \cdot (n_{L,p_j}-1)} \cdot (p_j - 1) \right) \cdot \left(2^{n_{L,2}-3} \cdot \prod_{j=2}^{n_L} p_j^{n_{L,p_j}-1} - 1 \right) \quad (4.135)$$

We mention that if $n_{L,2} = 3$, the number of ZF 4-PPs will be equal to:

$$C_{L,4-PPs,ZF} = 2^{3-1} \cdot \prod_{j=2}^{n_L} \left(p_j^{n_{L,p_j}-1} \cdot (p_j - 1) \right) \cdot \left(2 \cdot \prod_{j=2}^{n_L} p_j^{n_{L,p_j}-1} \right) \cdot \left(2^{3-2} \cdot \prod_{j=2}^{n_L} p_j^{n_{L,p_j}-1} \right) \cdot \left(\prod_{j=2}^{n_L} p_j^{n_{L,p_j}-1} - 1 \right) =$$
$$= 16 \cdot \prod_{j=2}^{n_L} \left(p_j^{3 \cdot (n_{L,p_j}-1)} \cdot (p_j - 1) \right) \cdot \left(\prod_{j=2}^{n_L} p_j^{n_{L,p_j}-1} - 1 \right) \quad (4.136)$$

so that the relation (4.135) is also true. ■

From (4.134) we see that the number of ZF 4-PPs is equal to 0 if L is eight times a product of prime numbers greater than three, each of them to the power of 1.

Theorem 4.40 *If* $12 \mid L$ *and* $8 \nmid L$*, i.e.* $L = 4 \cdot 3^{n_{L,3}} \cdot \prod_{j=3}^{n_L} p_j^{n_{L,p_j}}$*, with* $n_{L,3} \geq 1$*,* $p_j > 3$ *and* $n_{L,p_j} \geq 1, \forall j = \overline{3, n_L}$*, the number of ZF 4-PPs will be equal to:*

$$C_{L,4-PPs,ZF} = 4 \cdot 3^{2\cdot(n_{L,3}-1)+\max\{0;n_{L,3}-2\}} \cdot \prod_{j=3}^{n_L} \left(p_j^{3\cdot(n_{L,p_j}-1)} \cdot (p_j - 1) \right) \cdot \left(3^{\max\{0;n_{L,3}-2\}} \cdot \prod_{j=3}^{n_L} p_j^{n_{L,p_j}-1} - 1 \right) \quad (4.137)$$

Proof From cases ZF1. (b) and ZF2 at the beginning of Sect. 4.6, the number of possible combinations for coefficients q_1 is equal to $\Phi(L) = 2 \cdot 2 \cdot 3^{n_{L,3}-1} \cdot \prod_{j=3}^{n_L} \left(p_j^{n_{L,p_j}-1} \cdot (p_j - 1) \right)$, the number of coefficients q_3 is equal to $3^{\max\{0;n_{L,3}-2\}} \cdot \prod_{j=3}^{n_L} p_j^{n_{L,p_j}-1}$, the number of coefficients q_2 is equal to $3^{n_{L,3}-1} \cdot \prod_{j=3}^{n_L} p_j^{n_{L,p_j}-1}$ and the number of coefficients q_4 is equal to $3^{\max\{0;n_{L,3}-2\}} \cdot \prod_{j=3}^{n_L} p_j^{n_{L,p_j}-1}$, out of which one is zero. Therefore, the number of ZF 4-PPs will be equal to:

$$C_{L,4-PPs,ZF} = 4 \cdot 3^{n_{L,3}-1} \cdot \prod_{j=3}^{n_L} \left(p_j^{n_{L,p_j}-1} \cdot (p_j - 1) \right) \cdot \left(3^{n_{L,3}-1} \cdot \prod_{j=3}^{n_L} p_j^{n_{L,p_j}-1} \right) \cdot \left(3^{\max\{0;n_{L,3}-2\}} \cdot \prod_{j=3}^{n_L} p_j^{n_{L,p_j}-1} \right) \cdot \left(3^{\max\{0;n_{L,3}-2\}} \cdot \prod_{j=3}^{n_L} p_j^{n_{L,p_j}-1} - 1 \right) =$$

$$= 4 \cdot 3^{2\cdot(n_{L,3}-1)+\max\{0;n_{L,3}-2\}} \cdot \prod_{j=3}^{n_L} \left(p_j^{3\cdot(n_{L,p_j}-1)} \cdot (p_j - 1) \right) \cdot$$

$$\cdot\left(3^{\max\{0;n_{L,3}-2\}}\cdot\prod_{j=3}^{n_L}p_j^{n_{L,p_j}-1}-1\right) \tag{4.138}$$

■

From (4.137) we see that the number of ZF 4-PPs is equal to 0 if L is 12 or 36 times a product of prime numbers, greater than three, each of them to the power of 1.

Theorem 4.41 *If* $24 \mid L$, *i.e.* $L = 2^{n_{L,2}}\cdot 3^{n_{L,3}}\cdot\prod_{j=3}^{n_L}p_j^{n_{L,p_j}}$, *with* $n_{L,2}\geq 3$, $n_{L,3}\geq 1$, $p_j > 3$ *and* $n_{L,p_j}\geq 1$, $\forall j=\overline{3,n_L}$, *the number of ZF 4-PPs will be equal to:*

$$C_{L,4-PPs,ZF} = 2^{3\cdot n_{L,2}-4}\cdot 3^{2\cdot(n_{L,3}-1)+\max\{0;n_{L,3}-2\}}\cdot$$

$$\cdot\prod_{j=3}^{n_L}\left(p_j^{3\cdot(n_{L,p_j}-1)}\cdot(p_j-1)\right)\cdot\left(2^{n_{L,2}-3}\cdot 3^{\max\{0;n_{L,3}-2\}}\cdot\prod_{j=3}^{n_L}p_j^{n_{L,p_j}-1}-1\right) \tag{4.139}$$

Proof From cases ZF1. (b) and ZF2 at the beginning of Sect. 4.6, the number of possible combinations for coefficient q_1 is equal to $\Phi(L) = 2^{n_{L,2}-1}\cdot 2\cdot 3^{n_{L,3}-1}\cdot\prod_{j=3}^{n_L}\left(p_j^{n_{L,p_j}-1}\cdot(p_j-1)\right)$, the number of coefficients q_3 is equal to $2^{n_{L,2}-2}\cdot 3^{\max\{0;n_{L,3}-2\}}\cdot\prod_{j=3}^{n_L}p_j^{n_{L,p_j}-1}$, the number of even or odd coefficients q_2 is equal to $2^{n_{L,2}-2}\cdot 3^{n_{L,3}-1}\cdot\prod_{j=3}^{n_L}p_j^{n_{L,p_j}-1}$, and the number of coefficients q_4 is equal to $2^{n_{L,2}-4}\cdot 3^{\max\{0;n_{L,3}-2\}}\cdot\prod_{j=3}^{n_L}p_j^{n_{L,p_j}-1}$ if $n_{L,2}\geq 4$, from which one is zero. If $n_{L,2}=3$, there are $2\cdot 3^{\max\{0;n_{L,3}-2\}}\cdot\prod_{j=3}^{n_L}p_j^{n_{L,p_j}-1}$ coefficients q_3, $3^{\max\{0;n_{L,3}-2\}}\cdot\prod_{j=3}^{n_L}p_j^{n_{L,p_j}-1}$ coefficients q_4 (all even), and $2\cdot 3^{n_{L,3}-1}\cdot\prod_{j=3}^{n_L}p_j^{n_{L,p_j}-1}$ coefficients q_2 (all even). Therefore, if $n_{L,2}\geq 4$, the number of ZF 4-PPs will be equal to:

$$C_{L,4-PPs,ZF} = 2^{n_{L,2}-1}\cdot 2\cdot 3^{n_{L,3}-1}\cdot\prod_{j=3}^{n_L}\left(p_j^{n_{L,p_j}-1}\cdot(p_j-1)\right)\cdot$$

$$\cdot\left(2^{n_{L,2}-2}\cdot 3^{n_{L,3}-1}\cdot\prod_{j=3}^{n_L} p_j^{n_{L,p_j}-1}\right)\cdot\left(2^{n_{L,2}-2}\cdot 3^{\max\{0;n_{L,3}-2\}}\cdot\prod_{j=3}^{n_L} p_j^{n_{L,p_j}-1}\right)\cdot$$

$$\cdot\left(2^{n_{L,2}-4}\cdot 3^{\max\{0;n_{L,3}-2\}}\cdot\prod_{j=3}^{n_L} p_j^{n_{L,p_j}-1}-1\right)+$$

$$+2^{n_{L,2}-1}\cdot 2\cdot 3^{n_{L,3}-1}\cdot\prod_{j=3}^{n_L}\left(p_j^{n_{L,p_j}-1}\cdot(p_j-1)\right)\cdot$$

$$\cdot\left(2^{n_{L,2}-2}\cdot 3^{n_{L,3}-1}\cdot\prod_{j=3}^{n_L} p_j^{n_{L,p_j}-1}\right)\cdot\left(2^{n_{L,2}-2}\cdot 3^{\max\{0;n_{L,3}-2\}}\cdot\prod_{j=3}^{n_L} p_j^{n_{L,p_j}-1}\right)\cdot$$

$$\cdot\left(2^{n_{L,2}-4}\cdot 3^{\max\{0;n_{L,3}-2\}}\cdot\prod_{j=3}^{n_L} p_j^{n_{L,p_j}-1}\right)=$$

$$=2^{3\cdot n_{L,2}-4}\cdot 3^{2\cdot(n_{L,3}-1)+\max\{0;n_{L,3}-2\}}\cdot\prod_{j=3}^{n_L}\left(p_j^{3\cdot(n_{L,p_j}-1)}\cdot(p_j-1)\right)\cdot$$

$$\cdot\left(2^{n_{L,2}-3}\cdot 3^{\max\{0;n_{L,3}-2\}}\cdot\prod_{j=3}^{n_L} p_j^{n_{L,p_j}-1}-1\right)\tag{4.140}$$

We mention that, if $n_{L,2}=3$, the number of ZF 4-PPs will be equal to:

$$C_{L,4-PPs,ZF}=2^{3-1}\cdot 2\cdot 3^{n_{L,3}-1}\cdot\prod_{j=3}^{n_L}\left(p_j^{n_{L,p_j}-1}\cdot(p_j-1)\right)\cdot$$

$$\cdot\left(2\cdot 3^{n_{L,3}-1}\cdot\prod_{j=3}^{n_L} p_j^{n_{L,p_j}-1}\right)\cdot\left(2\cdot 3^{\max\{0;n_{L,3}-2\}}\cdot\prod_{j=3}^{n_L} p_j^{n_{L,p_j}-1}\right)\cdot$$

$$\cdot\left(3^{\max\{0;n_{L,3}-2\}}\cdot\prod_{j=3}^{n_L} p_j^{n_{L,p_j}-1}-1\right)=$$

$$=32\cdot 3^{2\cdot(n_{L,3}-1)+\max\{0;n_{L,3}-2\}}\cdot\prod_{j=3}^{n_L}\left(p_j^{3\cdot(n_{L,p_j}-1)}\cdot(p_j-1)\right)\cdot$$

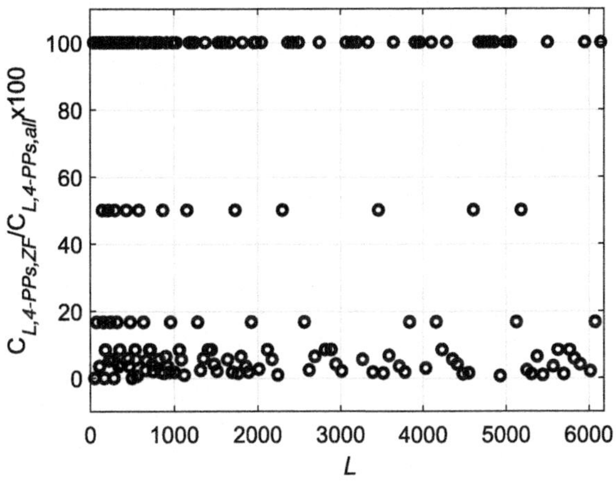

Fig. 4.2 The ratio (in percentage) between the number of true different ZF 4-PPs and the number of all true different 4-PPs, for all lengths used in LTE standard

$$\cdot\left(3^{\max\{0;n_{L,3}-2\}}\cdot\prod_{j=3}^{n_L}p_j^{n_{L,p_j}-1}-1\right) \tag{4.141}$$

so that relation (4.140) is also true. ■

From (4.139) we see that the number of ZF 4-PPs is equal to 0 if L is 24 or 72 times a product of prime numbers greater than three, each of them to the power of 1.

From Theorems 4.34–4.41 above, we conclude that the number of ZF 4-PPs is equal to zero, when the interleaver length L is of the form:

$$L=2^{n_{L,2}}\cdot 3^{n_{L,3}}\cdot\prod_{j=3}^{n_L}p_j,\ n_{L,2}=\overline{0,3},\ n_{L,3}=\overline{0,2},\ p_j>3,\ \forall j=\overline{3,n_L} \tag{4.142}$$

Such lengths have to be avoided in designing 4-PP-based interleavers, under Zhao and Fan sufficient conditions. We notice that the LTE standard (3GPP 2008) contains 27 lengths (multiples of 8, of the form (4.142) with $n_{L,2}=3$) for which there are no ZF 4-PPs, viz. the lengths 0, 56, 72, 88, 104, 120, 136, 152, 168, 184, 232, 248, 264, 280, 296, 312, 328, 344, 360, 376, 408, 424, 440, 456, 472, 488, and 504.

We compared the total number of true different 4-PPs, obtained with the algorithm from Sect. 4.7, for all lengths from the LTE standard with the number of true different ZF 4-PPs. Figure 4.2 shows the ratio (in percentage) between the number of true different ZF 4-PPs and the number of all true different 4-PPs for all the 188 lengths of the LTE standard. We note that for 23 from the 188 lengths the number of all true different 4-PPs is zero (i.e. there are no true 4-PPs for these lengths) and thus, the percentage is 100%. For 84 from the 188 lengths (from which 23 lengths are those mentioned above), the Zhao and Fan sufficient conditions are also necessary (i.e. this percentage is 100%). For other 4 lengths, this percentage is 0% (i.e. there are no true 4-PPs under Zhao and Fan sufficient conditions, but there are true different 4-PPs, for these lengths, namely 56, 168, 280, and 504), for 12 lengths the percentage is equal to 50% and for the other 88 lengths, the percentage is less than 20%.

In the following, we give an example with related comments for the most comprehensive case, i.e. for Theorem 4.41. The example provides information about all the values the coefficients of polynomials of degree 4 can take, so that they are 4-PPs modulo the considered length.

Example 4.3 (*Example for Theorem* 4.41) Let the interleaver length be $L = 1080 = 2^3 \cdot 3^3 \cdot 5$.

According to (4.139), the number of ZF 4-PPs will be equal to:

$$C_{1080,4-PPs,ZF} = 2^{3\cdot 3-4} \cdot 3^{2\cdot(3-1)+\max\{0;3-2\}} \cdot 5^{3\cdot(1-1)} \cdot (5-1)\cdot$$

$$\cdot\left(2^{3-3} \cdot 3^{\max\{0;3-2\}} \cdot 5^{1-1} - 1\right) = 32 \cdot 243 \cdot 1 \cdot 4 \cdot (3-1) = 62208.$$

Since $72 \mid L$, it is required that $q_4 < L/24 = 45$, $q_3 < L/6 = 180$ and $q_2 < L/2 = 540$. The coefficients q_4 and q_2 have to be multiples of $3 \cdot 5 = 15$, coefficient q_3 has to be multiple of $2 \cdot 3 \cdot 5 = 30$, and the coefficient q_1 has to be relatively prime with 1080. Moreover, the coefficients q_4 and q_2 have to meet the condition that the sum $q_4 + q_2$ is even. Therefore, coefficient q_3 can take 6 values: 0, 30, 60, 90, 120, 150, the pair of coefficients (q_4, q_2) can take the values $(15, 15), (15, 45), (15, 75), \ldots, (15, 525)$ (18 pairs with both coefficients odd) and $(30, 0), (30, 30), (30, 60), \ldots, (30, 510)$ (18 pairs with both coefficients even), i.e. a total of 36 pairs. Coefficient q_1 can take $\Phi(1080) = 1080 \cdot \frac{1}{2} \cdot \frac{2}{3} \cdot \frac{4}{5} = 288$ values. The total number of ZF 4-PPs is equal to $36 \cdot 6 \cdot 288 = 62208$, just the number found by the formula we determined. ■

4.6.3 Determining the Number of True Different Fifth Degree Permutation Polynomial-Based Interleavers Under Zhao and Fan Sufficient Conditions

As shown at the beginning Sect. 4.6, from the equivalence conditions for 5-PPs we must have:

- $q_2 < L/2, q_3 < L/2, q_4 < L/2$, and $q_5 < L/2$ when $2 \mid L, 3 \nmid L, 4 \nmid L$, and $5 \nmid L$;
- $q_2 < L/2, q_3 < L/2, q_4 < L/4$, and $q_5 < L/4$ when $3 \nmid L, 4 \mid L, 5 \nmid L$, and $8 \nmid L$;
- $q_2 < L/2, q_3 < L/2, q_4 < L/2$, and $q_5 < L/2$ when $4 \nmid L, 5 \nmid L, 6 \mid L$, and $9 \nmid L$;
- $q_2 < L/2$, $q_3 < L/2$, $q_4 < L/8$, and $q_5 < L/8$ when $3 \nmid L$, $5 \nmid L$, and $8 \mid L$;
- $q_3 < L/3$, $q_4 < L/3$, and $q_5 < L/3$ when $2 \nmid L$, $5 \nmid L$, and $9 \mid L$;
- $q_2 < L/2$, $q_3 < L/2$, $q_4 < L/2$, and $q_5 < L/2$ when $3 \nmid L$, $4 \nmid L$, $10 \mid L$, and $25 \nmid L$;
- $q_2 < L/2$, $q_3 < L/2$, $q_4 < L/4$, and $q_5 < L/4$ when $5 \nmid L, 8 \nmid L, 9 \nmid L$, and $12 \mid L$;

- $q_2 < L/2, q_3 < L/6, q_4 < L/6$, and $q_5 < L/6$ when $4 \nmid L, 5 \nmid L$, and $18 \mid L$;
- $q_2 < L/2, q_3 < L/2, q_4 < L/4$, and $q_5 < L/4$ when $3 \nmid L, 8 \nmid L, 20 \mid L$, and $25 \nmid L$;
- $q_2 < L/2, q_3 < L/2, q_4 < L/8$, and $q_5 < L/8$ when $5 \nmid L, 9 \nmid L$, and $24 \mid L$;
- $q_5 < L/5$ when $2 \nmid L, 3 \nmid L$, and $25 \mid L$;
- $q_2 < L/2, q_3 < L/2, q_4 < L/2$, and $q_5 < L/2$ when $4 \nmid L, 9 \nmid L, 25 \nmid L$, and $30 \mid L$;
- $q_2 < L/2, q_3 < L/6, q_4 < L/12$, and $q_5 < L/12$ when $5 \nmid L, 8 \nmid L$, and $36 \mid L$;
- $q_2 < L/2, q_3 < L/2, q_4 < L/8$, and $q_5 < L/8$ when $3 \nmid L, 25 \nmid L$, and $40 \mid L$;
- $q_3 < L/3, q_4 < L/3$, and $q_5 < L/3$ when $2 \nmid L, 25 \nmid L$, and $45 \mid L$;
- $q_2 < L/2, q_3 < L/2, q_4 < L/2$, and $q_5 < L/10$ when $3 \nmid L, 4 \nmid L$, and $50 \mid L$;
- $q_2 < L/2, q_3 < L/2, q_4 < L/4$, and $q_5 < L/4$ when $8 \nmid L, 9 \nmid L, 25 \nmid L$, and $60 \mid L$;
- $q_2 < L/2, q_3 < L/6, q_4 < L/24$, and $q_5 < L/24$ when $5 \nmid L$ and $72 \mid L$;
- $q_5 < L/5$ when $2 \nmid L, 9 \nmid L$, and $75 \mid L$;
- $q_2 < L/2, q_3 < L/6, q_4 < L/6$, and $q_5 < L/6$ when $4 \nmid L, 25 \nmid L$, and $90 \mid L$;
- $q_2 < L/2, q_3 < L/2, q_4 < L/4$, and $q_5 < L/20$ when $3 \nmid L, 8 \nmid L$, and $100 \mid L$;
- $q_2 < L/2, q_3 < L/2, q_4 < L/8$, and $q_5 < L/8$ when $9 \nmid L, 25 \nmid L$, and $120 \mid L$;
- $q_2 < L/2, q_3 < L/2, q_4 < L/2$, and $q_5 < L/10$ when $4 \nmid L, 9 \nmid L$, and $150 \mid L$;
- $q_2 < L/2, q_3 < L/6, q_4 < L/12$, and $q_5 < L/12$ when $8 \nmid L, 25 \nmid L$, and $180 \mid L$;
- $q_2 < L/2, q_3 < L/2, q_4 < L/8$, and $q_5 < L/40$ when $3 \nmid L$ and $200 \mid L$;
- $q_3 < L/3, q_4 < L/3$, and $q_5 < L/15$ when $2 \nmid L$ and $225 \mid L$;
- $q_2 < L/2, q_3 < L/2, q_4 < L/4$, and $q_5 < L/20$ when $8 \nmid L, 9 \nmid L$, and $300 \mid L$;
- $q_2 < L/2, q_3 < L/6, q_4 < L/24$, and $q_5 < L/24$ when $25 \nmid L$ and $360 \mid L$;
- $q_2 < L/2, q_3 < L/6, q_4 < L/6$, and $q_5 < L/30$ when $4 \nmid L$ and $450 \mid L$;
- $q_2 < L/2, q_3 < L/2, q_4 < L/8$, and $q_5 < L/40$ when $9 \nmid L$ and $600 \mid L$;
- $q_2 < L/2, q_3 < L/6, q_4 < L/12$, and $q_5 < L/60$ when $8 \nmid L$ and $900 \mid L$;
- $q_2 < L/2, q_3 < L/6, q_4 < L/24$, and $q_5 < L/120$ when $1800 \mid L$.

For ZF PPs of fifth degree, there are sixteen cases for the prime decomposition of L. The results regarding the number of true different ZF 5-PPs are given in Theorems 4.42–4.57 below and are summarized in Table 4.5 at the end of this subsection.

Theorem 4.42 *If* $2 \nmid L, 3 \nmid L$ *and* $5 \nmid L$, *i.e.* $L = \prod_{j=1}^{n_L} p_j^{n_{L,p_j}}$, *with* $p_j > 5$ *and* $n_{L,p_j} \geq 1, \forall j = \overline{1, n_L}$, *the number of ZF 5-PPs will be equal to:*

$$C_{L,5-PPs,ZF} = \prod_{j=1}^{n_L} \left(p_j^{4 \cdot (n_{L,p_j} - 1)} \cdot (p_j - 1) \right) \cdot \left(\prod_{j=1}^{n_L} p_j^{n_{L,p_j} - 1} - 1 \right) \tag{4.143}$$

Proof From case ZF2 at the beginning of Sect. 4.6, the number of possible combinations for coefficient q_1 is equal to $\prod_{j=1}^{n_L} \left(p_j^{n_{L,p_j} - 1} \cdot (p_j - 1) \right)$ and the number of coefficients q_2, q_3, q_4 and q_5, respectively, is equal to $\prod_{j=1}^{n_L} p_j^{n_{L,p_j}-1}$, out of which one is zero. By removing the value $q_5 = 0$, the number of ZF 5-PPs will be equal to that in (4.143). ■

From (4.143) we see that the number of ZF 5-PPs is equal to 0 if L is a product of prime numbers greater than five, each of them to the power of 1.

Theorem 4.43 *If* $2 \mid L$, $3 \nmid L$, $4 \nmid L$ *and* $5 \nmid L$, *i.e.* $L = 2 \cdot \prod_{j=2}^{n_L} p_j^{n_{L,p_j}}$, *with* $p_j > 5$ *and* $n_{L,p_j} \geq 1$, $\forall j = \overline{2, n_L}$, *the number of ZF 5-PPs will be equal to:*

$$C_{L,5-PPs,ZF} = \prod_{j=2}^{n_L} \left(p_j^{4 \cdot (n_{L,p_j}-1)} \cdot (p_j - 1) \right) \cdot \left(\prod_{j=2}^{n_L} p_j^{n_{L,p_j}-1} - 1 \right) \tag{4.144}$$

Proof From the analysis of cases ZF1. (a) and ZF2 at the beginning of Sect. 4.6, the number of possible combinations for coefficient q_1 is equal to $1 \cdot \Phi(L/2) = \prod_{j=2}^{n_L} \left(p_j^{n_{L,p_j}-1} \cdot (p_j - 1) \right)$, and the number of coefficients q_2, q_3, q_4 and q_5, respectively, is equal to $1 \cdot \prod_{j=2}^{n_L} p_j^{n_{L,p_j}-1}$, out of which one value is zero. By removing the value $q_5 = 0$, the number of ZF 5-PPs will be equal to:

$$C_{L,5-PPs,ZF} = \prod_{j=2}^{n_L} \left(p_j^{n_{L,p_j}-1} \cdot (p_j - 1) \right) \cdot \left(\prod_{j=2}^{n_L} p_j^{n_{L,p_j}-1} \right) \cdot$$

$$\cdot \left(\prod_{j=2}^{n_L} p_j^{n_{L,p_j}-1} \right) \cdot \left(\prod_{j=2}^{n_L} p_j^{n_{L,p_j}-1} \right) \cdot \left(\prod_{j=2}^{n_L} p_j^{n_{L,p_j}-1} - 1 \right) =$$

$$= \prod_{j=2}^{n_L} \left(p_j^{4 \cdot (n_{L,p_j}-1)} \cdot (p_j - 1) \right) \cdot \left(\prod_{j=2}^{n_L} p_j^{n_{L,p_j}-1} - 1 \right) \tag{4.145}$$

■

From (4.144) we see that the number of ZF 5-PPs is equal to 0 if L is two times a product of prime numbers greater than five, each of them to the power of 1.

Theorem 4.44 *If* $2 \nmid L$, $3 \mid L$ *and* $5 \nmid L$, *i.e.* $L = 3^{n_{L,3}} \cdot \prod_{j=2}^{n_L} p_j^{n_{L,p_j}}$, *with* $n_{L,3} \geq 1$, $p_j > 5$ *and* $n_{L,p_j} \geq 1$, $\forall j = \overline{2, n_L}$, *the number of ZF 5-PPs will be equal to:*

$$C_{L,5-PPs,ZF} = 2 \cdot 3^{2\cdot(n_{L,3}-1)+2\cdot\max\{0;n_{L,3}-2\}} \cdot \prod_{j=2}^{n_L}\left(p_j^{4\cdot(n_{L,p_j}-1)} \cdot (p_j-1)\right) \cdot$$

$$\cdot \left(3^{\max\{0;n_{L,3}-2\}} \cdot \prod_{j=2}^{n_L} p_j^{n_{L,p_j}-1} - 1\right) \tag{4.146}$$

Proof The number of possible combinations for coefficient q_1 is equal to $2 \cdot 3^{n_{L,3}-1} \cdot \prod_{j=2}^{n_L}\left(p_j^{n_{L,p_j}-1} \cdot (p_j-1)\right)$, the number of coefficients q_2 is equal to $3^{n_{L,3}-1} \cdot \prod_{j=2}^{n_L} p_j^{n_{L,p_j}-1}$, and the number of coefficients q_3, q_4 and q_5, respectively, is equal to $3^{\max\{0;n_{L,3}-2\}} \cdot \prod_{j=2}^{n_L} p_j^{n_{L,p_j}-1}$, out of which one value is zero. By removing the value $q_5 = 0$, the number of ZF 5-PPs will be equal to:

$$C_{L,5-PPs,ZF} = 2 \cdot 3^{n_{L,3}-1} \cdot \prod_{j=2}^{n_L}\left(p_j^{n_{L,p_j}-1} \cdot (p_j-1)\right) \cdot$$

$$\cdot \left(3^{n_{L,3}-1} \cdot \prod_{j=2}^{n_L} p_j^{n_{L,p_j}-1}\right) \cdot 3^{\max\{0;n_{L,3}-2\}} \cdot \prod_{j=2}^{n_L} p_j^{n_{L,p_j}-1} \cdot$$

$$\cdot 3^{\max\{0;n_{L,3}-2\}} \cdot \prod_{j=2}^{n_L} p_j^{n_{L,p_j}-1} \cdot \left(3^{\max\{0;n_{L,3}-2\}} \cdot \prod_{j=2}^{n_L} p_j^{n_{L,p_j}-1} - 1\right) =$$

$$= 2 \cdot 3^{2\cdot(n_{L,3}-1)+2\cdot\max\{0;n_{L,3}-2\}} \cdot \prod_{j=2}^{n_L}\left(p_j^{4\cdot(n_{L,p_j}-1)} \cdot (p_j-1)\right) \cdot$$

$$\cdot \left(3^{\max\{0;n_{L,3}-2\}} \cdot \prod_{j=2}^{n_L} p_j^{n_{L,p_j}-1} - 1\right) \tag{4.147}$$

■

From (4.146) we see that the number of ZF 5-PPs is equal to 0 if L is three or nine times a product of prime numbers greater than five, each of them to the power of 1.

Theorem 4.45 *If* $3 \nmid L$, $4 \mid L$, $5 \nmid L$ *and* $8 \nmid L$, *i.e.* $L = 4 \cdot \prod_{j=2}^{n_L} p_j^{n_{L,p_j}}$, *with* $p_j > 5$ *and* $n_{L,p_j} \geq 1$, $\forall j = \overline{2, n_L}$, *the number of ZF 5-PPs will be equal to:*

$$C_{L,5-PPs,ZF} = 2 \cdot \prod_{j=2}^{n_L} \left(p_j^{4 \cdot (n_{L,p_j} - 1)} \cdot (p_j - 1) \right) \cdot \left(\prod_{j=2}^{n_L} p_j^{n_{L,p_j} - 1} - 1 \right) \tag{4.148}$$

Proof From cases ZF1. (a) and ZF2 at the beginning of Sect. 4.6, the number of possible combinations for coefficient q_1 is equal to $\Phi(L) = 2 \cdot \prod_{j=2}^{n_L} \left(p_j^{n_{L,p_j} - 1} \cdot (p_j - 1) \right)$, and the number of coefficients q_2, q_3, q_4 and q_5, respectively, is equal to $\prod_{j=2}^{n_L} p_j^{n_{L,p_j} - 1}$, out of which one value is zero. By removing the value $q_5 = 0$, the number of ZF 5-PPs will be equal to:

$$C_{L,5-PPs,ZF} = 2 \cdot \prod_{j=2}^{n_L} \left(p_j^{n_{L,p_j} - 1} \cdot (p_j - 1) \right) \cdot \left(\prod_{j=2}^{n_L} p_j^{n_{L,p_j} - 1} \right) \cdot$$

$$\cdot \left(\prod_{j=2}^{n_L} p_j^{n_{L,p_j} - 1} \right) \cdot \left(\prod_{j=2}^{n_L} p_j^{n_{L,p_j} - 1} \right) \cdot \left(\prod_{j=2}^{n_L} p_j^{n_{L,p_j} - 1} - 1 \right) =$$

$$= 2 \cdot \prod_{j=2}^{n_L} \left(p_j^{4 \cdot (n_{L,p_j} - 1)} \cdot (p_j - 1) \right) \cdot \left(\prod_{j=2}^{n_L} p_j^{n_{L,p_j} - 1} - 1 \right) \tag{4.149}$$

■

From (4.148) we see that the number of ZF 5-PPs is equal to 0 if L is four times a product of prime numbers greater than five, each of them to the power of 1.

Theorem 4.46 *If* $2 \nmid L$, $3 \nmid L$ *and* $5 \mid L$, *i.e.* $L = 5^{n_{L,5}} \cdot \prod_{j=2}^{n_L} p_j^{n_{L,p_j}}$, *with* $n_{L,5} \geq 1$, $p_j > 5$ *and* $n_{L,p_j} \geq 1$, $\forall j = \overline{2, n_L}$, *the number of ZF 5-PPs will be equal to:*

$$C_{L,5-PPs,ZF} = 4 \cdot 5^{4 \cdot (n_{L,5} - 1)} \cdot \prod_{j=2}^{n_L} \left(p_j^{4 \cdot (n_{L,p_j} - 1)} \cdot (p_j - 1) \right) \cdot$$

$$\cdot \left(5^{\max\{0; n_{L,5} - 2\}} \cdot \prod_{j=2}^{n_L} p_j^{n_{L,p_j} - 1} - 1 \right) \tag{4.150}$$

Proof The number of possible combinations for coefficient q_1 is equal to $4 \cdot 5^{n_{L,5}-1} \cdot \prod_{j=2}^{n_L}\left(p_j^{n_{L,p_j}-1} \cdot (p_j-1)\right)$, the number of coefficients q_2, q_3, and q_4, respectively, is equal to $5^{n_{L,5}-1} \cdot \prod_{j=2}^{n_L} p_j^{n_{L,p_j}-1}$ and the number of coefficients q_5 is equal to $5^{\max\{0;n_{L,5}-2\}} \cdot \prod_{j=2}^{n_L} p_j^{n_{L,p_j}-1}$, out of which one value is zero. By removing the value $q_5 = 0$, the number of ZF 5-PPs will be equal to:

$$C_{L,5-PPs,ZF} = 4 \cdot 5^{n_{L,5}-1} \cdot \prod_{j=2}^{n_L}\left(p_j^{n_{L,p_j}-1} \cdot (p_j-1)\right) \cdot \left(5^{n_{L,5}-1} \cdot \prod_{j=2}^{n_L} p_j^{n_{L,p_j}-1}\right) \cdot \left(5^{n_{L,5}-1} \cdot \prod_{j=2}^{n_L} p_j^{n_{L,p_j}-1}\right) \cdot \left(5^{n_{L,5}-1} \cdot \prod_{j=2}^{n_L} p_j^{n_{L,p_j}-1}\right) \cdot \left(5^{\max\{0;n_{L,5}-2\}} \cdot \prod_{j=2}^{n_L} p_j^{n_{L,p_j}-1} - 1\right) =$$

$$= 4 \cdot 5^{4\cdot(n_{L,5}-1)} \cdot \prod_{j=2}^{n_L}\left(p_j^{4\cdot(n_{L,p_j}-1)} \cdot (p_j-1)\right) \cdot \left(5^{\max\{0;n_{L,5}-2\}} \cdot \prod_{j=2}^{n_L} p_j^{n_{L,p_j}-1} - 1\right) \tag{4.151}$$

■

From (4.150) we see that the number of ZF 5-PPs is equal to 0 if L is 5 or 25 times a product of prime numbers greater than five, each of them to the power of 1.

Theorem 4.47 *If* $4 \nmid L$, $5 \nmid L$ *and* $6 \mid L$, *i.e.* $L = 2 \cdot 3^{n_{L,3}} \cdot \prod_{j=3}^{n_L} p_j^{n_{L,p_j}}$, *with* $n_{L,3} \geq 1$, $p_j > 5$ *and* $n_{L,p_j} \geq 1$, $\forall j = \overline{3, n_L}$, *the number of ZF 5-PPs will be equal to:*

$$C_{L,5-PPs,ZF} = 2 \cdot 3^{2\cdot(n_{L,3}-1)+2\cdot\max\{0;n_{L,3}-2\}} \cdot \prod_{j=3}^{n_L}\left(p_j^{4\cdot(n_{L,p_j}-1)} \cdot (p_j-1)\right) \cdot \left(3^{\max\{0;n_{L,3}-2\}} \cdot \prod_{j=3}^{n_L} p_j^{n_{L,p_j}-1} - 1\right) \tag{4.152}$$

Proof The proof for Theorem 4.47 is similar to that of Theorem 4.44, by replacing L by $L/2$. Therefore, the number of ZF 5-PPs will be equal to:

$$C_{L,5-PPs,ZF} = 2 \cdot 3^{n_{L,3}-1} \cdot \prod_{j=3}^{n_L} \left(p_j^{n_{L,p_j}-1} \cdot (p_j - 1) \right) \cdot$$

$$\cdot \left(3^{n_{L,3}-1} \cdot \prod_{j=3}^{n_L} p_j^{n_{L,p_j}-1} \right) \cdot 3^{\max\{0; n_{L,3}-2\}} \cdot \prod_{j=3}^{n_L} p_j^{n_{L,p_j}-1} \cdot$$

$$\cdot 3^{\max\{0; n_{L,3}-2\}} \cdot \prod_{j=3}^{n_L} p_j^{n_{L,p_j}-1} \cdot \left(3^{\max\{0; n_{L,3}-2\}} \cdot \prod_{j=3}^{n_L} p_j^{n_{L,p_j}-1} - 1 \right) =$$

$$= 2 \cdot 3^{2\cdot(n_{L,3}-1)+2\cdot\max\{0; n_{L,3}-2\}} \cdot \prod_{j=3}^{n_L} \left(p_j^{4\cdot(n_{L,p_j}-1)} \cdot (p_j - 1) \right) \cdot$$

$$\cdot \left(3^{\max\{0; n_{L,3}-2\}} \cdot \prod_{j=3}^{n_L} p_j^{n_{L,p_j}-1} - 1 \right) \quad (4.153)$$

■

From (4.152) we see that the number of ZF 5-PPs is equal to 0 if L is 6 or 18 times a product of prime numbers greater than five, each of them to the power of 1.

Theorem 4.48 *If* $3 \nmid L$, $5 \nmid L$ *and* $8 \mid L$, *i.e.* $L = 2^{n_{L,2}} \cdot \prod_{j=2}^{n_L} p_j^{n_{L,p_j}}$, *with* $n_{L,2} \geq 3$, $p_j > 5$ *and* $n_{L,p_j} \geq 1$, $\forall j = \overline{2, n_L}$, *the number of ZF 5-PPs will be equal to:*

$$C_{L,5-PPs,ZF} = 2^{4\cdot(n_{L,2}-2)} \cdot \prod_{j=2}^{n_L} \left(p_j^{4\cdot(n_{L,p_j}-1)} \cdot (p_j - 1) \right) \cdot$$

$$\cdot \left(2^{n_{L,2}-3} \cdot \prod_{j=2}^{n_L} p_j^{n_{L,p_j}-1} - 1 \right) \quad (4.154)$$

Proof From cases ZF1. (b) and ZF2 at the beginning of Sect. 4.6, the number of possible combinations for coefficient q_1 is equal to $\Phi(L) = 2^{n_{L,2}-1} \cdot \prod_{j=2}^{n_L} \left(p_j^{n_{L,p_j}-1} \cdot (p_j - 1) \right)$, the number of even or odd coefficients q_2 and q_3, respectively, is equal to

$2^{n_{L,2}-2} \cdot \prod_{j=2}^{n_L} p_j^{n_{L,p_j}-1}$, the number of even or odd coefficients q_4 and q_5, respectively, is equal to $2^{n_{L,2}-4} \cdot \prod_{j=2}^{n_L} p_j^{n_{L,p_j}-1}$ if $n_{L,2} \geq 4$, out of which one value is zero. If $n_{L,2} = 3$, there are $\prod_{j=2}^{n_L} p_j^{n_{L,p_j}-1}$ coefficients q_4 and q_5, respectively, all of them even, out of which one is zero, and $2 \cdot \prod_{j=2}^{n_L} p_j^{n_{L,p_j}-1}$ coefficients q_2 and q_3, respectively, all of them even. Therefore, if $n_{L,2} \geq 4$, the number of ZF 5-PPs will be equal to:

$$
\begin{aligned}
C_{L,5-PPs,ZF} = {} & 2^{n_{L,2}-1} \cdot \prod_{j=2}^{n_L} \left(p_j^{n_{L,p_j}-1} \cdot (p_j - 1) \right) \cdot \\
& \cdot \left(2^{n_{L,2}-2} \cdot \prod_{j=2}^{n_L} p_j^{n_{L,p_j}-1} \right) \cdot \left(2^{n_{L,2}-4} \cdot \prod_{j=2}^{n_L} p_j^{n_{L,p_j}-1} \right) \cdot \\
& \cdot \left[\left(2^{n_{L,2}-2} \cdot \prod_{j=2}^{n_L} p_j^{n_{L,p_j}-1} \right) \cdot \left(2^{n_{L,2}-4} \cdot \prod_{j=2}^{n_L} p_j^{n_{L,p_j}-1} - 1 \right) + \right. \\
& \left. + \left(2^{n_{L,2}-2} \cdot \prod_{j=2}^{n_L} p_j^{n_{L,p_j}-1} \right) \cdot \left(2^{n_{L,2}-4} \cdot \prod_{j=2}^{n_L} p_j^{n_{L,p_j}-1} \right) \right] + \\
& + 2^{n_{L,2}-1} \cdot \prod_{j=2}^{n_L} \left(p_j^{n_{L,p_j}-1} \cdot (p_j - 1) \right) \cdot \left(2^{n_{L,2}-2} \cdot \prod_{j=2}^{n_L} p_j^{n_{L,p_j}-1} \right) \cdot \\
& \cdot \left(2^{n_{L,2}-4} \cdot \prod_{j=2}^{n_L} p_j^{n_{L,p_j}-1} \right) \cdot \\
& \cdot \left[\left(2^{n_{L,2}-2} \cdot \prod_{j=2}^{n_L} p_j^{n_{L,p_j}-1} \right) \cdot \left(2^{n_{L,2}-4} \cdot \prod_{j=2}^{n_L} p_j^{n_{L,p_j}-1} - 1 \right) + \right. \\
& \left. + \left(2^{n_{L,2}-2} \cdot \prod_{j=2}^{n_L} p_j^{n_{L,p_j}-1} \right) \cdot \left(2^{n_{L,2}-4} \cdot \prod_{j=2}^{n_L} p_j^{n_{L,p_j}-1} \right) \right] =
\end{aligned}
$$

$$= 2^{4\cdot(n_{L,2}-2)} \cdot \prod_{j=2}^{n_L} \left(p_j^{4\cdot(n_{L,p_j}-1)} \cdot (p_j - 1) \right) \cdot$$

$$\cdot \left(2^{n_{L,2}-3} \cdot \prod_{j=2}^{n_L} p_j^{n_{L,p_j}-1} - 1 \right) \tag{4.155}$$

and, if $n_{L,2} = 3$, this number will be:

$$C_{L,5-PPs,ZF} = 4 \cdot \prod_{j=2}^{n_L} \left(p_j^{n_{L,p_j}-1} \cdot (p_j - 1) \right) \cdot \left(2 \cdot \prod_{j=2}^{n_L} p_j^{n_{L,p_j}-1} \right) \cdot$$

$$\cdot \left(2 \cdot \prod_{j=2}^{n_L} p_j^{n_{L,p_j}-1} \right) \cdot \left(\prod_{j=2}^{n_L} p_j^{n_{L,p_j}-1} \right) \cdot \left(\prod_{j=2}^{n_L} p_j^{n_{L,p_j}-1} - 1 \right) =$$

$$= 16 \cdot \prod_{j=2}^{n_L} \left(p_j^{4\cdot(n_{L,p_j}-1)} \cdot (p_j - 1) \right) \cdot \left(\prod_{j=2}^{n_L} p_j^{n_{L,p_j}-1} - 1 \right) \tag{4.156}$$

so that relation (4.155) is also true. ∎

From (4.154) we see that the number of ZF 5-PPs is equal to 0 if L is eight times a product of prime numbers greater than five, each of them to the power of 1.

Theorem 4.49 *If* $2 \mid L$, $3 \nmid L$, $4 \nmid L$ *and* $5 \mid L$, *i.e.* $L = 2 \cdot 5^{n_{L,5}} \cdot \prod_{j=3}^{n_L} p_j^{n_{L,p_j}}$, *with* $n_{L,5} \geq 1$, $p_j > 5$ *and* $n_{L,p_j} \geq 1$, $\forall j = \overline{3, n_L}$, *the number of ZF 5-PPs will be equal to:*

$$C_{L,5-PPs,ZF} = 4 \cdot 5^{4\cdot(n_{L,5}-1)} \cdot \prod_{j=3}^{n_L} \left(p_j^{4\cdot(n_{L,p_j}-1)} \cdot (p_j - 1) \right) \cdot$$

$$\cdot \left(5^{\max\{0; n_{L,5}-2\}} \cdot \prod_{j=3}^{n_L} p_j^{n_{L,p_j}-1} - 1 \right) \tag{4.157}$$

Proof The proof for Theorem 4.49 is similar to that of Theorem 4.46, by replacing L by $L/2$. Therefore, the number of ZF 5-PPs will be equal to:

$$C_{L,5-PPs,ZF} = 4 \cdot 5^{n_{L,5}-1} \cdot \prod_{j=3}^{n_L} \left(p_j^{n_{L,p_j}-1} \cdot (p_j - 1) \right) \cdot$$

$$\cdot\left(5^{n_{L,5}-1}\cdot\prod_{j=3}^{n_L} p_j^{n_{L,p_j}-1}\right)\cdot\left(5^{n_{L,5}-1}\cdot\prod_{j=3}^{n_L} p_j^{n_{L,p_j}-1}\right)\cdot$$

$$\cdot\left(5^{n_{L,5}-1}\cdot\prod_{j=3}^{n_L} p_j^{n_{L,p_j}-1}\right)\cdot\left(5^{\max\{0;n_{L,5}-2\}}\cdot\prod_{j=3}^{n_L} p_j^{n_{L,p_j}-1}-1\right)=$$

$$=4\cdot 5^{4\cdot(n_{L,5}-1)}\cdot\prod_{j=3}^{n_L}\left(p_j^{4\cdot(n_{L,p_j}-1)}\cdot(p_j-1)\right)\cdot$$

$$\cdot\left(5^{\max\{0;n_{L,5}-2\}}\cdot\prod_{j=3}^{n_L} p_j^{n_{L,p_j}-1}-1\right) \quad (4.158)$$

■

From (4.157) we see that the number of ZF 5-PPs is equal to 0 if L is 10 or 50 times a product of prime numbers greater than five, each of them to the power of 1.

Theorem 4.50 *If* $5 \nmid L$, $8 \nmid L$ *and* $12 \mid L$, *i.e.* $L = 4\cdot 3^{n_{L,3}}\cdot\prod_{j=3}^{n_L} p_j^{n_{L,p_j}}$, *with* $n_{L,3}\geq 1$, $p_j > 5$ *and* $n_{L,p_j}\geq 1$, $\forall j=\overline{3,n_L}$, *the number of ZF 5-PPs will be equal to:*

$$C_{L,5-PPs,ZF}=4\cdot 3^{2\cdot(n_{L,3}-1)+2\cdot\max\{0;n_{L,3}-2\}}\cdot\prod_{j=3}^{n_L}\left(p_j^{4\cdot(n_{L,p_j}-1)}\cdot(p_j-1)\right)\cdot$$

$$\cdot\left(3^{\max\{0;n_{L,3}-2\}}\cdot\prod_{j=3}^{n_L} p_j^{n_{L,p_j}-1}-1\right) \quad (4.159)$$

Proof From cases ZF1. (b) and ZF2 at the beginning of Sect. 4.6, the number of possible combinations for coefficients q_1 is equal to $\Phi(L)=2\cdot 2\cdot 3^{n_{L,3}-1}\cdot\prod_{j=3}^{n_L}\left(p_j^{n_{L,p_j}-1}\cdot(p_j-1)\right)$, the number of coefficients q_2 is equal to $3^{n_{L,3}-1}\cdot\prod_{j=3}^{n_L} p_j^{n_{L,p_j}-1}$, the number of coefficients q_3, q_4 and q_5, respectively, is equal to $3^{\max\{0;n_{L,3}-2\}}\cdot\prod_{j=3}^{n_L} p_j^{n_{L,p_j}-1}$, out of which one is zero. Therefore, the number of ZF 5-PPs will be equal to:

$$C_{L,5-PPs,ZF}=4\cdot 3^{n_{L,3}-1}\cdot\prod_{j=3}^{n_L}\left(p_j^{n_{L,p_j}-1}\cdot(p_j-1)\right)\cdot$$

$$\cdot\left(3^{n_{L,3}-1}\cdot\prod_{j=3}^{n_L}p_j^{n_{L,p_j}-1}\right)\cdot 3^{\max\{0;n_{L,3}-2\}}\cdot\prod_{j=3}^{n_L}p_j^{n_{L,p_j}-1}\cdot$$

$$\cdot 3^{\max\{0;n_{L,3}-2\}}\cdot\prod_{j=3}^{n_L}p_j^{n_{L,p_j}-1}\cdot\left(3^{\max\{0;n_{L,3}-2\}}\cdot\prod_{j=3}^{n_L}p_j^{n_{L,p_j}-1}-1\right)=$$

$$=4\cdot 3^{2\cdot(n_{L,3}-1)+2\cdot\max\{0;n_{L,3}-2\}}\cdot\prod_{j=3}^{n_L}\left(p_j^{4\cdot(n_{L,p_j}-1)}\cdot(p_j-1)\right)\cdot$$

$$\cdot\left(3^{\max\{0;n_{L,3}-2\}}\cdot\prod_{j=3}^{n_L}p_j^{n_{L,p_j}-1}-1\right) \quad (4.160)$$

■

From (4.159) we see that the number of ZF 5-PPs is equal to 0 if L is 12 or 36 times a product of prime numbers greater than five, each of them to the power of 1.

Theorem 4.51 *If* $2\nmid L$, $3\mid L$ *and* $5\mid L$, *i.e.* $L=3^{n_{L,3}}\cdot 5^{n_{L,5}}\cdot\prod_{j=3}^{n_L}p_j^{n_{L,p_j}}$, *with* $n_{L,3}\geq 1$, $n_{L,5}\geq 1$, $p_j>5$ *and* $n_{L,p_j}\geq 1, \forall j=\overline{3,n_L}$, *the number of ZF 5-PPs will be equal to:*

$$C_{L,5-PPs,ZF}=8\cdot 3^{2\cdot(n_{L,3}-1)+2\cdot\max\{0;n_{L,3}-2\}}\cdot 5^{4\cdot(n_{L,5}-1)}\cdot$$

$$\cdot\prod_{j=3}^{n_L}\left(p_j^{4\cdot(n_{L,p_j}-1)}\cdot(p_j-1)\right)\cdot$$

$$\cdot\left(3^{\max\{0;n_{L,3}-2\}}\cdot 5^{\max\{0;n_{L,5}-2\}}\cdot\prod_{j=3}^{n_L}p_j^{n_{L,p_j}-1}-1\right) \quad (4.161)$$

Proof The number of possible combinations for coefficients q_1 is equal to $2\cdot 4\cdot 3^{n_{L,3}-1}\cdot 5^{n_{L,5}-1}\cdot\prod_{j=3}^{n_L}\left(p_j^{n_{L,p_j}-1}\cdot(p_j-1)\right)$, the number of coefficients q_2 is equal to $3^{n_{L,3}-1}\cdot 5^{n_{L,5}-1}\cdot\prod_{j=3}^{n_L}p_j^{n_{L,p_j}-1}$, the number of coefficients q_3 and q_4, respectively, is equal to $3^{\max\{0;n_{L,3}-2\}}\cdot 5^{n_{L,5}-1}\cdot\prod_{j=3}^{n_L}p_j^{n_{L,p_j}-1}$. The number of coefficients q_5 is equal to $3^{\max\{0;n_{L,3}-2\}} \cdot 5^{\max\{0;n_{L,5}-2\}} \cdot \prod_{j=3}^{n_L} p_j^{n_{L,p_j}-1}$, out of which one is zero. By removing the value $q_5 = 0$, the number of ZF 5-PPs will be equal to:

$$C_{L,5-PPs,ZF} = 8 \cdot 3^{n_{L,3}-1} \cdot 5^{n_{L,5}-1} \cdot \prod_{j=3}^{n_L} \left(p_j^{n_{L,p_j}-1} \cdot (p_j - 1) \right) \cdot$$

$$\cdot \left(3^{n_{L,3}-1} \cdot 5^{n_{L,5}-1} \cdot \prod_{j=3}^{n_L} p_j^{n_{L,p_j}-1} \right) \cdot \left(3^{\max\{0;n_{L,3}-2\}} \cdot 5^{n_{L,5}-1} \cdot \prod_{j=3}^{n_L} p_j^{n_{L,p_j}-1} \right) \cdot$$

$$\cdot \left(3^{\max\{0;n_{L,3}-2\}} \cdot 5^{n_{L,5}-1} \cdot \prod_{j=3}^{n_L} p_j^{n_{L,p_j}-1} \right) \cdot$$

$$\cdot \left(3^{\max\{0;n_{L,3}-2\}} \cdot 5^{\max\{0;n_{L,5}-2\}} \cdot \prod_{j=3}^{n_L} p_j^{n_{L,p_j}-1} - 1 \right) =$$

$$= 8 \cdot 3^{2\cdot(n_{L,3}-1)+2\cdot\max\{0;n_{L,3}-2\}} \cdot 5^{4\cdot(n_{L,5}-1)} \cdot \prod_{j=3}^{n_L} \left(p_j^{4\cdot(n_{L,p_j}-1)} \cdot (p_j - 1) \right) \cdot$$

$$\cdot \left(3^{\max\{0;n_{L,3}-2\}} \cdot 5^{\max\{0;n_{L,5}-2\}} \cdot \prod_{j=3}^{n_L} p_j^{n_{L,p_j}-1} - 1 \right) \quad (4.162)$$

■

From (4.161) we see that the number of ZF 5-PPs is 0 if L is 15, 45, 75, or 225 times a product of prime numbers greater than 5, each of them to the power of 1.

Theorem 4.52 *If* $3 \nmid L,\ 4 \mid L,\ 5 \mid L$ *and* $8 \nmid L$, *i.e.* $L = 4 \cdot 5^{n_{L,5}} \cdot \prod_{j=3}^{n_L} p_j^{n_{L,p_j}}$, *with* $n_{L,5} \geq 1,\ p_j > 5$ *and* $n_{L,p_j} \geq 1, \forall j = \overline{3, n_L}$, *the number of ZF 5-PPs will be equal to:*

$$C_{L,5-PPs,ZF} = 8 \cdot 5^{4\cdot(n_{L,5}-1)} \cdot \prod_{j=3}^{n_L} \left(p_j^{4\cdot(n_{L,p_j}-1)} \cdot (p_j - 1) \right) \cdot$$

$$\cdot \left(5^{\max\{0;n_{L,5}-2\}} \cdot \prod_{j=3}^{n_L} p_j^{n_{L,p_j}-1} - 1 \right) \quad (4.163)$$

Proof From cases ZF1. (a) and ZF2 at the beginning of Sect. 4.6, the number of possible combinations for coefficient q_1 is equal to $\Phi(L) = 2 \cdot 4 \cdot 5^{n_{L,5}-1} \cdot \prod_{j=3}^{n_L} \left(p_j^{n_{L,p_j}-1} \cdot (p_j - 1) \right)$, the number of coefficients q_2, q_3 and q_4, respectively, is equal to $5^{n_{L,5}-1} \cdot \prod_{j=3}^{n_L} p_j^{n_{L,p_j}-1}$ and the number of coefficients q_5 is equal to $5^{\max\{0;n_{L,5}-2\}} \cdot \prod_{j=3}^{n_L} p_j^{n_{L,p_j}-1}$, out of which one is zero. Therefore, the number of ZF 5-PPs will be equal to:

$$C_{L,5-PPs,ZF} = 8 \cdot 5^{n_{L,5}-1} \cdot \prod_{j=3}^{n_L} \left(p_j^{n_{L,p_j}-1} \cdot (p_j - 1) \right) \cdot$$

$$\cdot \left(5^{n_{L,5}-1} \cdot \prod_{j=3}^{n_L} p_j^{n_{L,p_j}-1} \right) \cdot \left(5^{n_{L,5}-1} \cdot \prod_{j=3}^{n_L} p_j^{n_{L,p_j}-1} \right) \cdot$$

$$\cdot \left(5^{n_{L,5}-1} \cdot \prod_{j=3}^{n_L} p_j^{n_{L,p_j}-1} \right) \cdot \left(5^{\max\{0;n_{L,5}-2\}} \cdot \prod_{j=3}^{n_L} p_j^{n_{L,p_j}-1} - 1 \right) =$$

$$= 8 \cdot 5^{4\cdot(n_{L,5}-1)} \cdot \prod_{j=3}^{n_L} \left(p_j^{4\cdot(n_{L,p_j}-1)} \cdot (p_j - 1) \right) \cdot$$

$$\cdot \left(5^{\max\{0;n_{L,5}-2\}} \cdot \prod_{j=3}^{n_L} p_j^{n_{L,p_j}-1} - 1 \right) \tag{4.164}$$

■

From (4.163) we see that the number of ZF 5-PPs is equal to 0 if L is 20 or 100 times a product of prime numbers greater than five, each of them to the power of 1.

Theorem 4.53 *If* $8 \mid L$, $3 \mid L$ *and* $5 \nmid L$, *i.e.* $L = 2^{n_{L,2}} \cdot 3^{n_{L,3}} \cdot \prod_{j=3}^{n_L} p_j^{n_{L,p_j}}$, *with* $n_{L,2} \geq 3$, $n_{L,3} \geq 1$, $p_j > 5$ *and* $n_{L,p_j} \geq 1$, $\forall j = \overline{3, n_L}$, *the number of ZF 5-PPs will be equal to:*

$$C_{L,5-PPs,ZF} = 2^{4\cdot n_{L,2}-7} \cdot 3^{2\cdot(n_{L,3}-1)+2\cdot\max\{0;n_{L,3}-2\}} \cdot$$

$$\cdot \prod_{j=3}^{n_L} \left(p_j^{4\cdot(n_{L,p_j}-1)} \cdot (p_j - 1) \right) \cdot \left(2^{n_{L,2}-3} \cdot 3^{\max\{0;n_{L,3}-2\}} \cdot \prod_{j=3}^{n_L} p_j^{n_{L,p_j}-1} - 1 \right) \tag{4.165}$$

Proof From cases ZF1. (b) and ZF2 at the beginning of Sect. 4.6, the number of possible combinations for coefficient q_1 is equal to $\Phi(L) = 2^{n_{L,2}-1} \cdot 2 \cdot 3^{n_{L,3}-1} \cdot \prod_{j=3}^{n_L} \left(p_j^{n_{L,p_j}-1} \cdot (p_j - 1) \right)$, the number of even or odd coefficients q_2 is equal to $2^{n_{L,2}-2} \cdot 3^{n_{L,3}-1} \cdot \prod_{j=3}^{n_L} p_j^{n_{L,p_j}-1}$, the number of even or odd coefficients q_3 is equal to $2^{n_{L,2}-2} \cdot 3^{\max\{0;n_{L,3}-2\}} \cdot \prod_{j=3}^{n_L} p_j^{n_{L,p_j}-1}$, and the number of even or odd coefficients q_4 and q_5, respectively, is equal to $2^{n_{L,2}-4} \cdot 3^{\max\{0;n_{L,3}-2\}} \cdot \prod_{j=3}^{n_L} p_j^{n_{L,p_j}-1}$ if $n_{L,2} \geq 4$, out of which one is zero. If $n_{L,2} = 3$, there are $2 \cdot 3^{n_{L,3}-1} \cdot \prod_{j=3}^{n_L} p_j^{n_{L,p_j}-1}$ coefficients q_2, all of them even, $2 \cdot 3^{\max\{0;n_{L,3}-2\}} \cdot \prod_{j=3}^{n_L} p_j^{n_{L,p_j}-1}$ coefficients q_3, all of them even, and $3^{\max\{0;n_{L,3}-2\}} \cdot \prod_{j=3}^{n_L} p_j^{n_{L,p_j}-1}$ coefficients q_4 and q_5, respectively, all of them even.
Therefore, if $n_{L,2} \geq 4$, the number of ZF 5-PPs will be equal to:

$$C_{L,5-PPs,ZF} = 2 \cdot 2^{n_{L,2}-1} \cdot 2 \cdot 3^{n_{L,3}-1} \cdot \prod_{j=3}^{n_L} \left(p_j^{n_{L,p_j}-1} \cdot (p_j - 1) \right) \cdot$$

$$\cdot \left(2^{n_{L,2}-2} \cdot 3^{n_{L,3}-1} \cdot \prod_{j=3}^{n_L} p_j^{n_{L,p_j}-1} \right) \cdot \left(2^{n_{L,2}-4} \cdot 3^{\max\{0;n_{L,3}-2\}} \cdot \prod_{j=3}^{n_L} p_j^{n_{L,p_j}-1} \right) \cdot$$

$$\cdot \left\{ \left(2^{n_{L,2}-2} \cdot 3^{\max\{0;n_{L,3}-2\}} \cdot \prod_{j=3}^{n_L} p_j^{n_{L,p_j}-1} \right) \cdot \right.$$

$$\cdot \left(2^{n_{L,2}-4} \cdot 3^{\max\{0;n_{L,3}-2\}} \cdot \prod_{j=3}^{n_L} p_j^{n_{L,p_j}-1} - 1 \right) +$$

$$+ \left(2^{n_{L,2}-2} \cdot 3^{\max\{0;n_{L,3}-2\}} \cdot \prod_{j=3}^{n_L} p_j^{n_{L,p_j}-1} \right) \cdot$$

$$\left.\left.\cdot\left(2^{n_{L,2}-4}\cdot 3^{\max\{0;n_{L,3}-2\}}\cdot\prod_{j=3}^{n_L}p_j^{n_{L,p_j}-1}\right)\right)\right\} =$$

$$= 2^{4\cdot n_{L,2}-7}\cdot 3^{2\cdot(n_{L,3}-1)+2\cdot\max\{0;n_{L,3}-2\}}\cdot\prod_{j=3}^{n_L}\left(p_j^{4\cdot(n_{L,p_j}-1)}\cdot(p_j-1)\right)\cdot$$

$$\cdot\left(2^{n_{L,2}-3}\cdot 3^{\max\{0;n_{L,3}-2\}}\cdot\prod_{j=3}^{n_L}p_j^{n_{L,p_j}-1}-1\right) \quad (4.166)$$

If $n_{L,2}=3$, the number of ZF 5-PPs will be equal to:

$$C_{L,5-PPs,ZF} = 4\cdot 2\cdot 3^{n_{L,3}-1}\cdot\prod_{j=3}^{n_L}\left(p_j^{n_{L,p_j}-1}\cdot(p_j-1)\right)\cdot$$

$$\cdot\left(2\cdot 3^{n_{L,3}-1}\cdot\prod_{j=3}^{n_L}p_j^{n_{L,p_j}-1}\right)\cdot\left(2\cdot 3^{\max\{0;n_{L,3}-2\}}\cdot\prod_{j=3}^{n_L}p_j^{n_{L,p_j}-1}\right)\cdot$$

$$\cdot\left(3^{\max\{0;n_{L,3}-2\}}\cdot\prod_{j=3}^{n_L}p_j^{n_{L,p_j}-1}\right)\cdot\left(3^{\max\{0;n_{L,3}-2\}}\cdot\prod_{j=3}^{n_L}p_j^{n_{L,p_j}-1}-1\right) =$$

$$= 32\cdot 3^{2\cdot(n_{L,3}-1)+2\cdot\max\{0;n_{L,3}-2\}}\cdot\prod_{j=3}^{n_L}\left(p_j^{4\cdot(n_{L,p_j}-1)}\cdot(p_j-1)\right)\cdot$$

$$\cdot\left(3^{\max\{0;n_{L,3}-2\}}\cdot\prod_{j=3}^{n_L}p_j^{n_{L,p_j}-1}-1\right), \quad (4.167)$$

so that relation (4.166) is also true. ∎

From (4.165) we see that the number of ZF 5-PPs is equal to 0 if L is 24 or 72 times a product of prime numbers greater than five, each of them to the power of 1.

Theorem 4.54 *If* $2\mid L$, $3\mid L$ *and* $4\nmid L$ *and* $5\mid L$, *i.e.* $L = 2\cdot 3^{n_{L,3}}\cdot 5^{n_{L,5}}\cdot\prod_{j=4}^{n_L}p_j^{n_{L,p_j}}$, *with* $n_{L,3}\geq 1$, $n_{L,5}\geq 1$, $p_j>5$ *and* $n_{L,p_j}\geq 1$, $\forall j=\overline{4,n_L}$, *the number of ZF 5-PPs will be equal to:*

$$C_{L,5-PPs,ZF} = 8\cdot 3^{2\cdot(n_{L,3}-1)+2\cdot\max\{0;n_{L,3}-2\}}\cdot 5^{4\cdot(n_{L,5}-1)}\cdot$$

$$\cdot \left(\prod_{j=4}^{n_L} p_j^{4\cdot(n_{L,p_j}-1)} \cdot (p_j - 1) \right) \cdot$$

$$\cdot \left(3^{\max\{0;n_{L,3}-2\}} \cdot 5^{\max\{0;n_{L,5}-2\}} \cdot \prod_{j=4}^{n_L} p_j^{n_{L,p_j}-1} - 1 \right) \tag{4.168}$$

Proof The proof for Theorem 4.54 is similar to that of Theorem 4.51, by replacing L by $L/2$. Thus, the number of ZF 5-PPs will be equal to:

$$C_{L,5-PPs,ZF} = 8 \cdot 3^{n_{L,3}-1} \cdot 5^{n_{L,5}-1} \cdot \prod_{j=4}^{n_L} \left(p_j^{n_{L,p_j}-1} \cdot (p_j - 1) \right) \cdot$$

$$\cdot \left(3^{n_{L,3}-1} \cdot 5^{n_{L,5}-1} \cdot \prod_{j=4}^{n_L} p_j^{n_{L,p_j}-1} \right) \cdot \left(3^{\max\{0;n_{L,3}-2\}} \cdot 5^{n_{L,5}-1} \cdot \prod_{j=4}^{n_L} p_j^{n_{L,p_j}-1} \right) \cdot$$

$$\cdot \left(3^{\max\{0;n_{L,3}-2\}} \cdot 5^{n_{L,5}-1} \cdot \prod_{j=4}^{n_L} p_j^{n_{L,p_j}-1} \right) \cdot$$

$$\cdot \left(3^{\max\{0;n_{L,3}-2\}} \cdot 5^{\max\{0;n_{L,5}-2\}} \cdot \prod_{j=4}^{n_L} p_j^{n_{L,p_j}-1} - 1 \right) =$$

$$= 8 \cdot 3^{2\cdot(n_{L,3}-1)+2\cdot\max\{0;n_{L,3}-2\}} \cdot 5^{4\cdot(n_{L,5}-1)} \cdot \left(\prod_{j=4}^{n_L} p_j^{4\cdot(n_{L,p_j}-1)} \cdot (p_j - 1) \right) \cdot$$

$$\cdot \left(3^{\max\{0;n_{L,3}-2\}} \cdot 5^{\max\{0;n_{L,5}-2\}} \cdot \prod_{j=4}^{n_L} p_j^{n_{L,p_j}-1} - 1 \right) \tag{4.169}$$

■

From (4.168) we see that the number of ZF 5-PPs is 0 if L is 30, 90, 150, or 450 times a product of prime numbers greater than 5, each of them to the power of 1.

Theorem 4.55 *If* $3 \nmid L$, $5 \mid L$ *and* $8 \mid L$, *i.e.* $L = 2^{n_{L,2}} \cdot 5^{n_{L,5}} \cdot \prod_{j=3}^{n_L} p_j^{n_{L,p_j}}$, *with* $n_{L,2} \geq 3$, $n_{L,5} \geq 1$, $p_j > 5$ *and* $n_{L,p_j} \geq 1$, $\forall j = \overline{3, n_L}$, *the number of ZF 5-PPs will be equal to:*

$$C_{L,5-PPs,ZF} = 2^{4\cdot n_{L,2}-6} \cdot 5^{4\cdot(n_{L,1}-1)} \cdot \prod_{j=3}^{n_L} \left(p_j^{4\cdot(n_{L,p_j}-1)} \cdot (p_j - 1) \right) \cdot$$

$$\cdot \left(2^{n_{L,2}-3} \cdot 5^{\max\{0; n_{L,5}-2\}} \cdot \prod_{j=3}^{n_L} p_j^{n_{L,p_j}-1} - 1 \right) \quad (4.170)$$

Proof From cases ZF1. (b) and ZF2 at the beginning of Sect. 4.6, the number of possible combinations for coefficient q_1 is equal to $\Phi(L) = 2^{n_{L,2}-1} \cdot 4 \cdot 5^{n_{L,5}-1} \cdot \prod_{j=3}^{n_L} \left(p_j^{n_{L,p_j}-1} \cdot (p_j - 1) \right)$, the number of even or odd coefficients q_2 and q_3, respectively, is equal to $2^{n_{L,2}-2} \cdot 5^{n_{L,5}-1} \cdot \prod_{j=3}^{n_L} p_j^{n_{L,p_j}-1}$, the number of even or odd coefficients q_4 is equal to $2^{n_{L,2}-4} \cdot 5^{n_{L,5}-1} \cdot \prod_{j=3}^{n_L} p_j^{n_{L,p_j}-1}$, and the number of even or odd coefficients q_5 is equal to $2^{n_{L,2}-4} \cdot 5^{\max\{0; n_{L,5}-2\}} \cdot \prod_{j=3}^{n_L} p_j^{n_{L,p_j}-1}$ if $n_{L,2} \geq 4$, out of which one value is zero. If $n_{L,2} = 3$, there are $2 \cdot 5^{n_{L,5}-1} \cdot \prod_{j=3}^{n_L} p_j^{n_{L,p_j}-1}$ coefficients q_2 and q_3, respectively, all of them even, $5^{n_{L,5}-1} \cdot \prod_{j=3}^{n_L} p_j^{n_{L,p_j}-1}$ coefficients q_4, all of them even, and $5^{\max\{0; n_{L,5}-2\}} \cdot \prod_{j=3}^{n_L} p_j^{n_{L,p_j}-1}$ coefficients q_5, all of them even, out of which one is zero. Therefore, if $n_{L,2} \geq 4$, the number of ZF 5-PPs will be equal to:

$$C_{L,5-PPs,ZF} = 2 \cdot 2^{n_{L,2}-1} \cdot 4 \cdot 5^{n_{L,5}-1} \cdot \prod_{j=3}^{n_L} \left(p_j^{n_{L,p_j}-1} \cdot (p_j - 1) \right) \cdot$$

$$\cdot \left(2^{n_{L,2}-2} \cdot 5^{n_{L,5}-1} \cdot \prod_{j=3}^{n_L} p_j^{n_{L,p_j}-1} \right) \cdot \left(2^{n_{L,2}-4} \cdot 5^{n_{L,5}-1} \cdot \prod_{j=3}^{n_L} p_j^{n_{L,p_j}-1} \right) \cdot$$

$$\cdot \left\{ \left(2^{n_{L,2}-2} \cdot 5^{n_{L,5}-1} \cdot \prod_{j=3}^{n_L} p_j^{n_{L,p_j}-1} \right) \cdot \right.$$

$$\cdot \left(2^{n_{L,2}-4} \cdot 5^{\max\{0; n_{L,5}-2\}} \cdot \prod_{j=3}^{n_L} p_j^{n_{L,p_j}-1} - 1 \right) +$$

$$+ \left(2^{n_{L,2}-2} \cdot 5^{n_{L,5}-1} \cdot \prod_{j=3}^{n_L} p_j^{n_{L,p_j}-1} \right) \cdot$$

$$\cdot \left(2^{n_{L,2}-4} \cdot 5^{\max\{0; n_{L,5}-2\}} \cdot \prod_{j=3}^{n_L} p_j^{n_{L,p_j}-1} \right) \Bigg\} =$$

$$= 2^{4 \cdot n_{L,2}-6} \cdot 5^{4 \cdot (n_{L,1}-1)} \cdot \prod_{j=3}^{n_L} \left(p_j^{4 \cdot (n_{L,p_j}-1)} \cdot (p_j - 1) \right) \cdot$$

$$\cdot \left(2^{n_{L,2}-3} \cdot 5^{\max\{0; n_{L,5}-2\}} \cdot \prod_{j=3}^{n_L} p_j^{n_{L,p_j}-1} - 1 \right) \tag{4.171}$$

If $n_{L,2} = 3$, the number of ZF 5-PPs will be equal to:

$$C_{L,5-PPs,ZF} = 4 \cdot 4 \cdot 5^{n_{L,5}-1} \cdot \prod_{j=3}^{n_L} \left(p_j^{n_{L,p_j}-1} \cdot (p_j - 1) \right) \cdot$$

$$\cdot \left(2 \cdot 5^{n_{L,5}-1} \cdot \prod_{j=3}^{n_L} p_j^{n_{L,p_j}-1} \right) \cdot \left(2 \cdot 5^{n_{L,5}-1} \cdot \prod_{j=3}^{n_L} p_j^{n_{L,p_j}-1} \right) \cdot$$

$$\cdot \left(5^{n_{L,5}-1} \cdot \prod_{j=3}^{n_L} p_j^{n_{L,p_j}-1} \right) \cdot \left(5^{\max\{0; n_{L,5}-2\}} \cdot \prod_{j=3}^{n_L} p_j^{n_{L,p_j}-1} - 1 \right) =$$

$$= 64 \cdot 5^{4 \cdot (n_{L,1}-1)} \cdot \prod_{j=3}^{n_L} \left(p_j^{4 \cdot (n_{L,p_j}-1)} \cdot (p_j - 1) \right) \cdot$$

$$\cdot \left(5^{\max\{0; n_{L,5}-2\}} \cdot \prod_{j=3}^{n_L} p_j^{n_{L,p_j}-1} - 1 \right), \tag{4.172}$$

so that relation (4.171) is also true. ■

From (4.170) we see that the number of ZF 5-PPs is equal to 0 if L is 40 or 200 times a product of prime numbers greater than five, each of them to the power of 1.

Theorem 4.56 *If* $3 \mid L,\ 4 \mid L,\ 5 \mid L$ *and* $8 \nmid L$, *i.e.* $L = 4 \cdot 3^{n_{L,3}} \cdot 5^{n_{L,5}} \cdot \prod_{j=4}^{n_L} p_j^{n_{L,p_j}}$, *with* $n_{L,3} \geq 1$, $n_{L,5} \geq 1$, $p_j > 5$ *and* $n_{L,p_j} \geq 1, \forall j = \overline{4, n_L}$, *the number of ZF 5-PPs will be equal to:*

$$C_{L,5-PPs,ZF} = 16 \cdot 3^{2\cdot(n_{L,3}-1)+2\cdot\max\{0;n_{L,3}-2\}} \cdot 5^{4\cdot(n_{L,5}-1)} \cdot \prod_{j=4}^{n_L} \left(p_j^{4\cdot(n_{L,p_j}-1)} \cdot (p_j-1)\right) \cdot \left(3^{\max\{0;n_{L,3}-2\}} \cdot 5^{\max\{0;n_{L,5}-2\}} \cdot \prod_{j=4}^{n_L} p_j^{n_{L,p_j}-1} - 1\right) \quad (4.173)$$

Proof The proof for Theorem 4.56 is similar to that of Theorem 4.51, by replacing L with $L/4$ and multiplying the number of coefficients q_1 by 2. The number of ZF 5-PPs will be equal to:

$$C_{L,5-PPs,ZF} = 16 \cdot 3^{n_{L,3}-1} \cdot 5^{n_{L,5}-1} \cdot \prod_{j=4}^{n_L} \left(p_j^{n_{L,p_j}-1} \cdot (p_j-1)\right) \cdot \left(3^{n_{L,3}-1} \cdot 5^{n_{L,5}-1} \cdot \prod_{j=4}^{n_L} p_j^{n_{L,p_j}-1}\right) \cdot \left(3^{\max\{0;n_{L,3}-2\}} \cdot 5^{n_{L,5}-1} \cdot \prod_{j=4}^{n_L} p_j^{n_{L,p_j}-1}\right) \cdot \left(3^{\max\{0;n_{L,3}-2\}} \cdot 5^{n_{L,5}-1} \cdot \prod_{j=4}^{n_L} p_j^{n_{L,p_j}-1}\right) \cdot \left(3^{\max\{0;n_{L,3}-2\}} \cdot 5^{\max\{0;n_{L,5}-2\}} \cdot \prod_{j=4}^{n_L} p_j^{n_{L,p_j}-1} - 1\right) =$$

$$= 16 \cdot 3^{2\cdot(n_{L,3}-1)+2\cdot\max\{0;n_{L,3}-2\}} \cdot 5^{4\cdot(n_{L,5}-1)} \cdot \prod_{j=4}^{n_L} \left(p_j^{4\cdot(n_{L,p_j}-1)} \cdot (p_j-1)\right) \cdot$$

$$\cdot\left(3^{\max\{0;n_{L,3}-2\}}\cdot 5^{\max\{0;n_{L,5}-2\}}\cdot\prod_{j=4}^{n_L} p_j^{n_{L,p_j}-1}-1\right) \quad (4.174)$$

■

From (4.173) we see that the number of ZF 5-PPs is equal to 0 if L is 60, 180, 300, or 900 times a product of prime numbers greater than 5, each of them to the power of 1.

Theorem 4.57 *If* $3 \mid L$, $5 \mid L$ *and* $8 \mid L$, *i.e.* $L = 2^{n_{L,2}}\cdot 3^{n_{L,3}}\cdot 5^{n_{L,5}}\cdot\prod_{j=4}^{n_L} p_j^{n_{L,p_j}}$, *with* $n_{L,2}\ge 3$, $n_{L,3}\ge 1$, $n_{L,5}\ge 1$, $p_j > 5$ *and* $n_{L,p_j}\ge 1$, $\forall j=\overline{4,n_L}$, *the number of ZF 5-PPs will be equal to:*

$$C_{L,5-PPs,ZF} = 2^{4\cdot n_{L,2}-5}\cdot 3^{2\cdot(n_{L,3}-1)+2\cdot\max\{0;n_{L,3}-2\}}\cdot 5^{4\cdot(n_{L,5}-1)}\cdot$$

$$\cdot\prod_{j=4}^{n_L}\left(p_j^{4\cdot(n_{L,p_j}-1)}\cdot(p_j-1)\right)\cdot$$

$$\cdot\left(2^{n_{L,2}-3}\cdot 3^{\max\{0;n_{L,3}-2\}}\cdot 5^{\max\{0;n_{L,5}-2\}}\cdot\prod_{j=4}^{n_L} p_j^{n_{L,p_j}-1}-1\right). \quad (4.175)$$

Proof From cases ZF1. (b) and ZF2 at the beginning of Sect. 4.6, the number of possible combinations for coefficient q_1 is equal to $2^{n_{L,2}-1}\cdot 2\cdot 3^{n_{L,3}-1}\cdot 4\cdot 5^{n_{L,5}-1}\cdot\prod_{j=4}^{n_L}\left(p_j^{n_{L,p_j}-1}\cdot(p_j-1)\right)$, the number of even or odd coefficients q_2 is equal to $2^{n_{L,2}-2}\cdot 3^{n_{L,3}-1}\cdot 5^{n_{L,5}-1}\cdot\prod_{j=4}^{n_L} p_j^{n_{L,p_j}-1}$ and the number of even or odd coefficients q_3 is equal to $2^{n_{L,2}-2}\cdot 3^{\max\{0;n_{L,3}-2\}}\cdot 5^{n_{L,5}-1}\cdot\prod_{j=4}^{n_L} p_j^{n_{L,p_j}-1}$. The number of even or odd coefficients q_4 is equal to $2^{n_{L,2}-4}\cdot 3^{\max\{0;n_{L,3}-2\}}\cdot 5^{n_{L,5}-1}\cdot\prod_{j=4}^{n_L} p_j^{n_{L,p_j}-1}$ if $n_{L,2}\ge 4$. If $n_{L,2}=3$, the number of coefficients q_4 is equal to $3^{\max\{0;n_{L,3}-2\}}\cdot 5^{n_{L,5}-1}\cdot\prod_{j=4}^{n_L} p_j^{n_{L,p_j}-1}$, all of them even. The number of even or odd coefficients q_5 is equal to $2^{n_{L,2}-4}\cdot 3^{\max\{0;n_{L,3}-2\}}\cdot 5^{\max\{0;n_{L,5}-2\}}\cdot\prod_{j=4}^{n_L} p_j^{n_{L,p_j}-1}$ if $n_{L,2}\ge 4$. If $n_{L,2}=3$, the number of coefficients q_5 is equal to $3^{\max\{0;n_{L,3}-2\}} \cdot 5^{\max\{0;n_{L,5}-2\}} \cdot \prod_{j=4}^{n_L} p_j^{n_{L,p_j}-1}$, all of them even, out of which one is zero. By removing the value $q_5 = 0$, if $n_{L,2} \geq 4$, the number of ZF 5-PPs will be equal to:

$$C_{L,5-PPs,ZF} = 2 \cdot 8 \cdot 2^{n_{L,2}-1} \cdot 3^{n_{L,3}-1} \cdot 5^{n_{L,5}-1} \cdot$$

$$\cdot \prod_{j=4}^{n_L} \left(p_j^{n_{L,p_j}-1} \cdot (p_j - 1) \right) \cdot \left(2^{n_{L,2}-2} \cdot 3^{n_{L,3}-1} \cdot 5^{n_{L,5}-1} \cdot \prod_{j=4}^{n_L} p_j^{n_{L,p_j}-1} \right) \cdot$$

$$\cdot \left(2^{n_{L,2}-4} \cdot 3^{\max\{0;n_{L,3}-2\}} \cdot 5^{n_{L,5}-1} \cdot \prod_{j=4}^{n_L} p_j^{n_{L,p_j}-1} \right) \cdot$$

$$\cdot \left\{ \left(2^{n_{L,2}-2} \cdot 3^{\max\{0;n_{L,3}-2\}} \cdot 5^{n_{L,5}-1} \cdot \prod_{j=4}^{n_L} p_j^{n_{L,p_j}-1} \right) \cdot \right.$$

$$\cdot \left(2^{n_{L,2}-4} \cdot 3^{\max\{0;n_{L,3}-2\}} \cdot 5^{\max\{0;n_{L,5}-2\}} \cdot \prod_{j=4}^{n_L} p_j^{n_{L,p_j}-1} - 1 \right) +$$

$$+ \left(2^{n_{L,2}-2} \cdot 3^{\max\{0;n_{L,3}-2\}} \cdot 5^{n_{L,5}-1} \cdot \prod_{j=4}^{n_L} p_j^{n_{L,p_j}-1} \right) \cdot$$

$$\left. \cdot \left(2^{n_{L,2}-4} \cdot 3^{\max\{0;n_{L,3}-2\}} \cdot 5^{\max\{0;n_{L,5}-2\}} \cdot \prod_{j=4}^{n_L} p_j^{n_{L,p_j}-1} \right) \right\} =$$

$$= 2^{4 \cdot n_{L,2}-5} \cdot 3^{2 \cdot (n_{L,3}-1)+2 \cdot \max\{0;n_{L,3}-2\}} \cdot 5^{4 \cdot (n_{L,5}-1)} \cdot$$

$$\cdot \prod_{j=4}^{n_L} \left(p_j^{4 \cdot (n_{L,p_j}-1)} \cdot (p_j - 1) \right) \cdot$$

$$\cdot \left(2^{n_{L,2}-3} \cdot 3^{\max\{0;n_{L,3}-2\}} \cdot 5^{\max\{0;n_{L,5}-2\}} \cdot \prod_{j=4}^{n_L} p_j^{n_{L,p_j}-1} - 1 \right) \tag{4.176}$$

If $n_{L,2} = 3$, the number of ZF 5-PPs will be equal to:

$$C_{L,5-PPs,ZF} = 8 \cdot 4 \cdot 3^{n_{L,3}-1} \cdot 5^{n_{L,5}-1} \cdot \prod_{j=4}^{n_L} \left(p_j^{n_{L,p_j}-1} \cdot (p_j - 1) \right) \cdot$$

$$\cdot \left(2 \cdot 3^{n_{L,3}-1} \cdot 5^{n_{L,5}-1} \cdot \prod_{j=4}^{n_L} p_j^{n_{L,p_j}-1} \right) \cdot$$

$$\cdot \left(2 \cdot 3^{\max\{0; n_{L,3}-2\}} \cdot 5^{n_{L,5}-1} \cdot \prod_{j=4}^{n_L} p_j^{n_{L,p_j}-1} \right) \cdot$$

$$\cdot \left(3^{\max\{0; n_{L,3}-2\}} \cdot 5^{n_{L,5}-1} \cdot \prod_{j=4}^{n_L} p_j^{n_{L,p_j}-1} \right) \cdot$$

$$\cdot \left(3^{\max\{0; n_{L,3}-2\}} \cdot 5^{\max\{0; n_{L,5}-2\}} \cdot \prod_{j=4}^{n_L} p_j^{n_{L,p_j}-1} - 1 \right) =$$

$$= 128 \cdot 3^{2 \cdot (n_{L,3}-1) + 2 \cdot \max\{0; n_{L,3}-2\}} \cdot 5^{4 \cdot (n_{L,5}-1)} \cdot$$

$$\cdot \prod_{j=4}^{n_L} \left(p_j^{4 \cdot (n_{L,p_j}-1)} \cdot (p_j - 1) \right) \cdot$$

$$\cdot \left(3^{\max\{0; n_{L,3}-2\}} \cdot 5^{\max\{0; n_{L,5}-2\}} \cdot \prod_{j=4}^{n_L} p_j^{n_{L,p_j}-1} - 1 \right), \tag{4.177}$$

so that relation (4.176) is also true. ■

From (4.175) we see that the number of ZF 5-PPs is equal to 0 if L is 120, 360, 600, or 1800 times a product of prime numbers greater than five, each of them to the power of 1.

From Theorems 4.42–4.57 above, we conclude that the number of ZF 5-PPs is equal to 0 if the interleaver length L is of the form:

$$L = 2^{n_{L,2}} \cdot 3^{n_{L,3}} \cdot 5^{n_{L,5}} \cdot \prod_{j=4}^{n_L} p_j, n_{L,2} = \overline{0,3}, n_{L,3} = \overline{0,2}, n_{L,5} = \overline{0,2},$$

$$p_j > 5, j = \overline{4, n_L} \tag{4.178}$$

Such lengths have to be avoided in designing 5-PP-based interleavers under Zhao and Fan sufficient conditions. We notice that the LTE standard (3GPP 2008) contains

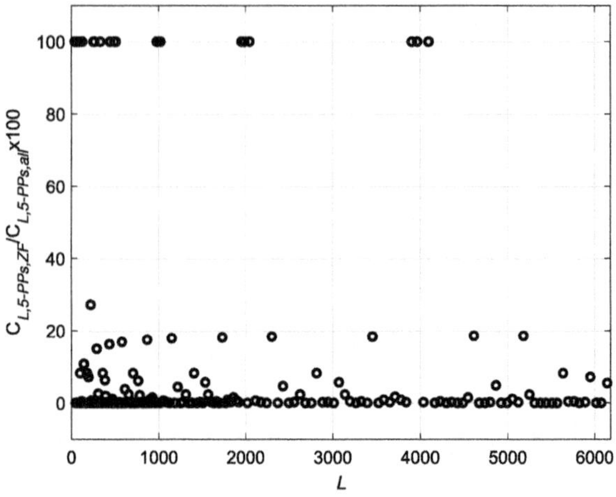

Fig. 4.3 The ratio (in percentage) between the number of true different ZF 5-PPs and the number of all true different 5-PPs, for all lengths used in LTE standard

28 lengths (of the form (4.178) with $n_{N,2} = 3$) for which there are no 5-PPs under Zhao and Fan sufficient conditions, viz. the lengths 40, 56, 72, 88, 104, 120, 136, 152, 168, 184, 200, 232, 248, 264, 280, 296, 312, 328, 344, 360, 376, 408, 424, 440, 456, 472, 488, and 504.

We compared the total number of true different 5-PPs obtained with the algoritm from Sect. 4.7 for all lengths from the LTE standard with the number of true different ZF 5-PPs. Figure 4.3 shows the ratio (in percentage) between the number of true different ZF 5-PPs and the number of all true different 5-PPs, for all the 188 lengths of the LTE standard. We note that for 8 of the 188 lengths the number of all true different 5-PPs is zero (i.e. there are no true different 5-PPs for these lengths, namely 40, 88, 120, 248, 264, 328, 440, and 488) and thus the percentage is 100%. For 31 from the 188 lengths (from which 8 lengths are those mentioned above), the Zhao and Fan sufficient conditions are also necessary (i.e. this percentage is 100%). For other 20 lengths, this percentage is 0% (i.e. there are no true 5-PPs under Zhao and Fan sufficient conditions, but there are true different 5-PPs, for these lengths) and for the other 137 lengths, the percentage is less than 20%.

In the following, we give two examples with related comments for the most comprehensive case, i.e. for Theorem 4.57. The examples provide information about all the values the coefficients of polynomials of degree 5 can take, so that they are 5-PPs modulo the considered length.

Example 4.4 (*Example 1 for Theorem* 4.57) Let the interleaver length be $L = 25200 = 2^4 \cdot 3^2 \cdot 5^2 \cdot 7$. According to (4.175), the number of ZF 5-PPs will be equal to:

$$C_{25200,5-PPs,ZF} = 2^{4\cdot 4-5} \cdot 3^{2\cdot(2-1)+2\cdot\max\{0;2-2\}} \cdot 5^{4\cdot(2-1)} \cdot 7^{4\cdot(1-1)} \cdot (7-1)\cdot$$

$$\cdot\left(2^{4-3} \cdot 3^{\max\{0;2-2\}} \cdot 5^{\max\{0;2-2\}} \cdot 7^{1-1} - 1\right) =$$

$$= (2048 \cdot 9 \cdot 625 \cdot 1 \cdot 6) \cdot (2 \cdot 1 \cdot 1 \cdot 1 - 1) = 69120000.$$

Since $1800 \mid L$, it is required that $q_5 < L/120 = 210$, $q_4 < L/24 = 1050$, $q_3 < L/6 = 4200$ and $q_2 < L/2 = 12600$. Coefficients q_5, q_4, q_3 and q_2 have to be multiples of $3 \cdot 5 \cdot 7 = 105$, and coefficient q_1 has to be relatively prime with 25200. Moreover, the coefficients q_5 and q_3 have to meet the condition that the sum $q_5 + q_3$ be even and the coefficients q_4 and q_2 have to meet the condition that the sum $q_4 + q_2$ be even. Therefore, the pair of coefficients (q_5, q_3) can take the values $(105, 105)$, $(105, 315)$, $(105, 525)$, ..., $(105, 4095)$ (20 pairs with both coefficients odd), and the pair of coefficients (q_4, q_2) can take the values $(105, 105)$, $(105, 315)$, $(105, 425)$, ..., $(105, 12495)$, $(315, 105)$, $(315, 315)$, $(315, 425)$, ..., $(315, 12495)$, $(525, 105)$, $(525, 315)$, $(525, 425)$, ..., $(525, 12495)$, $(735, 105)$, $(735, 315)$, $(735, 425)$, ..., $(735, 12495)$, $(945, 105)$, $(945, 315)$, $(945, 425)$, ..., $(945, 12495)$ ($5 \cdot 60 = 300$ pairs with both coefficients odd) and $(0, 0)$, $(0, 210)$, $(0, 420)$, ..., $(0, 12390)$, $(210, 0)$, $(210, 210)$, $(210, 420)$, ..., $(210, 12390)$, $(420, 0)$, $(420, 210)$, $(420, 420)$, ..., $(420, 12390)$, $(630, 0)$, $(630, 210)$, $(630, 420)$, ..., $(630, 12390)$, $(840, 0)$, $(840, 210)$, $(840, 420)$, ..., $(840, 12390)$, ($5 \cdot 60 = 300$ pairs with both coefficients even), i.e. a total of 600 pairs. Coefficient q_1 can take $\Phi(25200) = 25200 \cdot \frac{1}{2} \cdot \frac{2}{3} \cdot \frac{4}{5} \cdot \frac{6}{7} = 5760$ values. The total number of ZF 5-PPs is equal to $20 \cdot 600 \cdot 5760 = 69120000$, just the number we found by the formula we determined. ■

Example 4.5 (*Example 2 for Theorem* 4.57) Let the interleaver length be $L = 2400 = 2^5 \cdot 3 \cdot 5^2$. According to (4.175), the number of ZF 5-PPs will be equal to:

$$C_{2400,5-PPs,ZF} = 2^{4 \cdot 5-5} \cdot 3^{2 \cdot (1-1)+2 \cdot \max\{0;1-2\}} \cdot 5^{4 \cdot (2-1)} \cdot$$

$$\cdot \left(2^{5-3} \cdot 3^{\max\{0;1-2\}} \cdot 5^{\max\{0;2-2\}} - 1\right) =$$

$$= 32768 \cdot 1 \cdot 625 \cdot (4 \cdot 1 \cdot 1 - 1) = 61440000.$$

In this case, since $9 \nmid L$ and $600 \mid L$, it is required that $q_5 < L/40 = 60$, $q_4 < L/8 = 300$, $q_3 < L/2 = 1200$ and $q_2 < L/2 = 1200$. Coefficients q_5, q_4, q_3 and q_2 have to be multiples of $3 \cdot 5 = 15$. The other conditions coefficients q_5, q_4, q_3 and q_2 have to meet are the same as those for $L = 25200$ (in Example 4.4), and coefficient q_1 has to be relatively prime with 2400. Therefore, the pair of coefficients (q_5, q_3) can take the values $(15, 15)$, $(15, 45)$, $(15, 75)$, ..., $(15, 1185)$, $(45, 15)$, $(45, 45)$, $(45, 75)$, ..., $(45, 1185)$ ($2 \cdot 40 = 80$ pairs with both coefficients odd) and $(30, 0)$, $(30, 30)$, $(30, 60)$, ... $(30, 1170)$ (40 pairs with both coefficients even), i.e. a total of 120 pairs. The pair of coefficients (q_4, q_2) can take the values $(15, 15)$, $(15, 45)$, $(15, 75)$, ..., $(15, 1185)$, $(45, 15)$, $(45, 45)$, $(45, 75)$, ..., $(45, 1185)$, $(75, 15)$, $(75, 45)$, $(75, 75)$, ..., $(75, 1185)$, ..., $(285, 15)$, $(285, 45)$, $(285, 75)$, ..., $(285, 1185)$ ($10 \cdot 40 = 400$ pairs with both coefficients odd) and $(0, 0)$, $(0, 30)$, $(0, 60)$, ..., $(0, 1170)$, $(30, 0)$, $(30, 30)$, $(30, 60)$, ..., $(30, 1170)$,

(60, 0), (60, 30), (60, 60), ..., (60, 1170), (90, 0), (90, 30), (90, 60), ..., (90, 1170), ..., (270, 0), (270, 30), (270, 60), ..., (270, 1170) ($10 \cdot 40 = 400$ pairs with both coefficients even), i.e. a total of 800 pairs. Coefficient q_1 can take $\Phi(2400) = 2400 \cdot \frac{1}{2} \cdot \frac{2}{3} \cdot \frac{4}{5} = 640$ values. The total number of ZF 5-PPs is equal to $120 \cdot 800 \cdot 640 = 61440000$, just the number we found by the formula we determined.

We note that the equivalence conditions for 5-PPs in this case are: $q_5 < L/120 = 20$, $q_4 < L/24 = 100$, $q_3 < L/6 = 400$ and $q_2 < L/2 = 1200$. We note that every ZF 5-PP having coefficient $q_5 > 20$ (i.e. 30 or 45) and/or coefficient $q_4 > 100$ and/or coefficient $q_3 > 400$ and/or coefficient $q_2 > 1200$ has a equivalent 5-PP having coefficients $q_5 < 20$, $q_4 < 100$, $q_3 < 400$, and $q_2 < 1200$. For example, ZF 5-PP $1 \cdot x + 0 \cdot x^2 + 1185 \cdot x^3 + 270 \cdot x^4 + 45 \cdot x^5 \pmod{2400}$ is equivalent to 5-PP $1041 \cdot x + 200 \cdot x^2 + 185 \cdot x^3 + 70 \cdot x^4 + 5 \cdot x^5 \pmod{2400}$ only by NP $z(x) = 1040 \cdot x + 200 \cdot x^2 + 1400 \cdot x^3 + 2200 \cdot x^4 + 2360 \cdot x^5 \pmod{2400}$. This NP is not a ZF NP since $3 \nmid 2360$, $3 \nmid 2200$, $3 \nmid 1400$, and $3 \nmid 200$. ■

4.7 Determining the Number of True Different Permutation Polynomials of Degrees up to Five by Weng and Dong Algorithm

Section 3.11 presents an algorithm from Weng and Dong (2008) for getting all PPs of degrees up to five over $\mathbb{Z}_L$. It was derived on the base of normalized PPs given in Dickson (1896), Theorems 3.8 and 3.7, Propositions 3.12 and 3.13 and on the Chinese Remainder Theorem (Theorem 3.49).

We recall that if we consider $c = 0$ and $\bar{\pi}(x)$ any normalized PP from Table 3.7, from Proposition 3.12 all PPs $\pmod{p}$ of degree d ($d \leq 5$) are obtained with formula $a\bar{\pi}(x + b)$ for all $a \neq 0$, $b \in \mathbb{Z}_p$, when $p \nmid d$. When $p \mid d$, from Proposition 3.13 all PPs $\pmod{p}$ of degree d ($d \leq 5$) are obtained with formula $a\bar{\pi}(x)$ for all $a \in \mathbb{Z}_p^*$. We recall that for primes 2, 3 or 5, Table 3.4 provides all PPs of degree smaller than or equal to 5 modulo 2, 3 or 5. Further we consider the PPs obtained above where $\bar{\pi}(x)$ is any normalized PP given in Table 3.8. To obtain all PPs $\pmod{p^{n_{L,p}}}$ with $n_{L,p} > 1$, we have to add to the previously obtained PPs a polynomial $p \cdot \rho(x)$, where $\rho(x)$ is any polynomial over $\mathbb{Z}_{p^{n_{L,p}-1}}$. This means that each coefficient of $\rho(x)$ can take $p^{n_{L,p}-1}$ different values. Then, for $\rho(x)$ a polynomial of degree d, the number of all different polynomials $\rho(x)$ is $p^{d \cdot (n_{L,p}-1)}$.

The considerations above can be used to determine the number of all PPs of any degree from one to five over $\mathbb{Z}_L$ by means of the normalized PPs given in Tables 3.7 and 3.8, for $p > 5$. For $p = 2$, $p = 3$ and $p = 5$ we can use the coefficient conditions in Table 3.4. Tables 3.7 and 3.8 and the coefficient conditions in Table 3.4 are particularized for degrees up to five for every prime p. The numbers of PPs of degrees d from one to five over $\mathbb{Z}_p$ that permute $\mathbb{Z}_{p^{n_{L,p}}}$, denoted by $C_{p,d-PPs}$, with

Table 4.5 Number of true different 5-PP-based interleavers under Zhao and Fan sufficient conditions

Case	Decomposition of L	$C_{L,5-PPs,ZF}$	Theorems
(1)	$L=\prod_{j=1}^{n_L} p_j^{n_{L,p_j}}$, $p_j>5$ and $n_{L,p_j}\geq 1, \forall j=\overline{1,n_L}$	$C_{L,5-PPs,ZF}=\prod_{j=1}^{n_L}\left(p_j^{4\cdot(n_{L,p_j}-1)}\cdot(p_j-1)\right)\cdot\left(\prod_{j=1}^{n_L}p_j^{n_{L,p_j}-1}-1\right)$	4.42
(2)	$L=2\cdot\prod_{j=2}^{n_L} p_j^{n_{L,p_j}}$, $p_j>5$ and $n_{L,p_j}\geq 1, \forall j=\overline{2,n_L}$	$C_{L,5-PPs,ZF}=\prod_{j=2}^{n_L}\left(p_j^{4\cdot(n_{L,p_j}-1)}\cdot(p_j-1)\right)\cdot\left(\prod_{j=2}^{n_L}p_j^{n_{L,p_j}-1}-1\right)$	4.43
(3)	$L=3^{n_{L,3}}\cdot\prod_{j=2}^{n_L} p_j^{n_{L,p_j}}$, $n_{L,3}\geq 1$, $p_j>5$ and $n_{L,p_j}\geq 1, \forall j=\overline{2,n_L}$	$C_{L,5-PPs,ZF}=2\cdot 3^{2\cdot(n_{L,3}-1)+2\cdot\max\{0;n_{L,3}-2\}}\cdot\prod_{j=2}^{n_L}\left(p_j^{4\cdot(n_{L,p_j}-1)}\cdot(p_j-1)\right)\cdot\left(3^{\max\{0;n_{L,3}-2\}}\cdot\prod_{j=2}^{n_L}p_j^{n_{L,p_j}-1}-1\right)$	4.44
(4)	$L=4\cdot\prod_{j=2}^{n_L} p_j^{n_{L,p_j}}$, $p_j>5$ and $n_{L,p_j}\geq 1, \forall j=\overline{2,n_L}$	$C_{L,5-PPs,ZF}=2\cdot\prod_{j=2}^{n_L}\left(p_j^{4\cdot(n_{L,p_j}-1)}\cdot(p_j-1)\right)\cdot\left(\prod_{j=2}^{n_L}p_j^{n_{L,p_j}-1}-1\right)$	4.45
(5)	$L=5^{n_{L,5}}\cdot\prod_{j=2}^{n_L} p_j^{n_{L,p_j}}$, $n_{L,5}\geq 1$, $p_j>5$ and $n_{L,p_j}\geq 1, \forall j=\overline{2,n_L}$	$C_{L,5-PPs,ZF}=4\cdot 5^{4\cdot(n_{L,5}-1)}\cdot\prod_{j=2}^{n_L}\left(p_j^{4\cdot(n_{L,p_j}-1)}\cdot(p_j-1)\right)\cdot\left(5^{\max\{0;n_{L,5}-2\}}\cdot\prod_{j=2}^{n_L}p_j^{n_{L,p_j}-1}-1\right)$	4.46
(6)	$L=2\cdot 3^{n_{L,3}}\cdot\prod_{j=3}^{n_L} p_j^{n_{L,p_j}}$, $n_{L,3}\geq 1$, $p_j>5$ and $n_{L,p_j}\geq 1, \forall j=\overline{3,n_L}$	$C_{L,5-PPs,ZF}=2\cdot 3^{2\cdot(n_{L,3}-1)+2\cdot\max\{0;n_{L,3}-2\}}\cdot\prod_{j=3}^{n_L}\left(p_j^{4\cdot(n_{L,p_j}-1)}\cdot(p_j-1)\right)\cdot\left(3^{\max\{0;n_{L,3}-2\}}\cdot\prod_{j=3}^{n_L}p_j^{n_{L,p_j}-1}-1\right)$	4.47

(continued)

Table 4.5 (continued)

Case	Decomposition of L	$C_{L,5-PPs,ZF}$	Theorems
(7)	$L = 2^{n_{L,2}} \cdot \prod_{j=2}^{n_L} p_j^{n_{L,p_j}}$, $n_{L,2} \geq 3$, $p_j > 5$ and $n_{L,p_j} \geq 1, \forall j = \overline{2, n_L}$	$C_{L,5-PPs,ZF} = 2^{4\cdot(n_{L,2}-2)} \cdot \prod_{j=2}^{n_L} \left(p_j^{4\cdot(n_{L,p_j}-1)} \cdot (p_j - 1)\right) \cdot \left(2^{n_{L,2}-3} \cdot \prod_{j=2}^{n_L} p_j^{n_{L,p_j}-1} - 1\right)$	4.48
(8)	$L = 2 \cdot 5^{n_{L,5}} \cdot \prod_{j=3}^{n_L} p_j^{n_{L,p_j}}$, $n_{L,5} \geq 1$, $p_j > 5$ and $n_{L,p_j} \geq 1, \forall j = \overline{3, n_L}$	$C_{L,5-PPs,ZF} = 4 \cdot 5^{4\cdot(n_{L,5}-1)} \cdot \prod_{j=3}^{n_L} \left(p_j^{4\cdot(n_{L,p_j}-1)} \cdot (p_j - 1)\right) \cdot \left(5^{\max\{0;n_{L,53}-2\}} \cdot \prod_{j=3}^{n_L} p_j^{n_{L,p_j}-1} - 1\right)$	4.49
(9)	$L = 4 \cdot 3^{n_{L,3}} \cdot \prod_{j=3}^{n_L} p_j^{n_{L,p_j}}$, $n_{L,3} \geq 1$, $p_j > 5$ and $n_{L,p_j} \geq 1, \forall j = \overline{3, n_L}$	$C_{L,5-PPs,ZF} = 4 \cdot 3^{2\cdot(n_{L,3}-1)+2\cdot\max\{0;n_{L,3}-2\}} \cdot \prod_{j=3}^{n_L} \left(p_j^{4\cdot(n_{L,p_j}-1)} \cdot (p_j - 1)\right) \cdot \left(3^{\max\{0;n_{L,3}-2\}} \cdot \prod_{j=3}^{n_L} p_j^{n_{L,p_j}-1} - 1\right)$	4.50
(10)	$L = 3^{n_{L,3}} \cdot 5^{n_{L,5}} \cdot \prod_{j=3}^{n_L} p_j^{n_{L,p_j}}$, $n_{L,3} \geq 1$, $n_{L,5} \geq 1$, $p_j > 5$ and $n_{L,p_j} \geq 1, \forall j = \overline{3, n_L}$	$C_{L,5-PPs,ZF} = 8 \cdot 3^{2\cdot(n_{L,3}-1)+2\cdot\max\{0;n_{L,3}-2\}} \cdot 5^{4\cdot(n_{L,5}-1)} \cdot \prod_{j=3}^{n_L} \left(p_j^{4\cdot(n_{L,p_j}-1)} \cdot (p_j - 1)\right) \cdot \left(3^{n_{L,3}-2} \cdot 5^{n_{L,5}-2} \cdot \prod_{j=3}^{n_L} p_j^{n_{L,p_j}-1} - 1\right)$	4.51
(11)	$L = 4 \cdot 5^{n_{L,5}} \cdot \prod_{j=3}^{n_L} p_j^{n_{L,p_j}}$, $n_{L,5} \geq 1$, $p_j > 5$ and $n_{L,p_j} \geq 1, \forall j = \overline{3, n_L}$	$C_{L,5-PPs,ZF} = 8 \cdot 5^{4\cdot(n_{L,5}-1)} \cdot \prod_{j=3}^{n_L} \left(p_j^{4\cdot(n_{L,p_j}-1)} \cdot (p_j - 1)\right) \cdot \left(5^{\max\{0;n_{L,5}-2\}} \cdot \prod_{j=3}^{n_L} p_j^{n_{L,p_j}-1} - 1\right)$	4.52

(continued)

Table 4.5 (continued)

Case	Decomposition of L	$C_{L,5-PPs,ZF}$	Theorems
(12)	$L = 2^{n_{L,2}} \cdot 3^{n_{L,3}} \cdot \prod_{j=3}^{n_L} p_j^{n_{L,p_j}}$, $n_{L,2} \geq 3, n_{L,3} \geq 1, p_j > 5$ and $n_{L,p_j} \geq 1, \forall j = \overline{3, n_L}$	$C_{L,5-PPs,ZF} = 2^{4\cdot n_{L,2}-7} \cdot 3^{2\cdot(n_{L,3}-1)+2\cdot\max\{0;n_{L,3}-2\}} \cdot \prod_{j=3}^{n_L}\left(p_j^{4\cdot(n_{L,p_j}-1)} \cdot (p_j - 1)\right) \cdot \left(2^{n_{L,2}-3} \cdot 3^{\max\{0;n_{L,3}-2\}} \cdot \prod_{j=3}^{n_L} p_j^{n_{L,p_j}-1} - 1\right)$	
(13)	$L = 2 \cdot 3^{n_{L,3}} \cdot 5^{n_{L,5}} \cdot \prod_{j=4}^{n_L} p_j^{n_{L,p_j}}$, $n_{L,3} \geq 1, n_{L,5} \geq 1, p_j > 5$ and $n_{L,p_j} \geq 1, \forall j = \overline{4, n_L}$	$C_{L,5-PPs,ZF} = 8 \cdot 3^{2\cdot(n_{L,3}-1)+2\cdot\max\{0;n_{L,3}-2\}} \cdot 5^{4\cdot(n_{L,5}-1)} \cdot \left(\prod_{j=4}^{n_L} p_j^{4\cdot(n_{L,p_j}-1)} \cdot (p_j - 1)\right) \cdot \left(3^{\max\{0;n_{L,3}-2\}} \cdot 5^{\max\{0;n_{L,5}-2\}} \cdot \prod_{j=4}^{n_L} p_j^{n_{L,p_j}-1} - 1\right)$	4.54
(14)	$L = 2^{n_{L,2}} \cdot 5^{n_{L,5}} \cdot \prod_{j=3}^{n_L} p_j^{n_{L,p_j}}$, $n_{L,2} \geq 3, n_{L,5} \geq 1, p_j > 5$ and $n_{L,p_j} \geq 1, \forall j = \overline{3, n_L}$	$C_{L,5-PPs,ZF} = 2^{4\cdot n_{L,2}-6} \cdot 5^{4\cdot(n_{L,1}-1)} \cdot \prod_{j=3}^{n_L}\left(p_j^{4\cdot(n_{L,p_j}-1)} \cdot (p_j - 1)\right) \cdot \left(2^{n_{L,2}-3} \cdot 5^{\max\{0;n_{L,5}-2\}} \cdot \prod_{j=3}^{n_L} p_j^{n_{L,p_j}-1} - 1\right)$	4.55
(15)	$L = 4 \cdot 3^{n_{L,3}} \cdot 5^{n_{L,5}} \cdot \prod_{j=4}^{n_L} p_j^{n_{L,p_j}}$, $n_{L,3} \geq 1, n_{L,5} \geq 1, p_j > 5$ and $n_{L,p_j} \geq 1, \forall j = \overline{4, n_L}$	$C_{L,5-PPs,ZF} = 16 \cdot 3^{2\cdot(n_{L,3}-1)+2\cdot\max\{0;n_{L,3}-2\}} \cdot 5^{4\cdot(n_{L,5}-1)} \cdot \prod_{j=4}^{n_L}\left(p_j^{4\cdot(n_{L,p_j}-1)} \cdot (p_j - 1)\right) \cdot \left(3^{\max\{0;n_{L,3}-2\}} \cdot 5^{\max\{0;n_{L,5}-2\}} \cdot \prod_{j=4}^{n_L} p_j^{n_{L,p_j}-1} - 1\right)$	4.56
(16)	$L = 2^{n_{L,2}} \cdot 3^{n_{L,3}} \cdot 5^{n_{L,5}} \cdot \prod_{j=4}^{n_L} p_j^{n_{L,p_j}}$, $n_{L,2} \geq 3, n_{L,3} \geq 1, n_{L,5} \geq 1,$ $p_j > 5$ and $n_{L,p_j} \geq 1, \forall j = \overline{4, n_L}$	$C_{L,5-PPs,ZF} = 2^{4\cdot n_{L,2}-5} \cdot 3^{2\cdot(n_{L,3}-1)+2\cdot\max\{0;n_{L,3}-2\}} \cdot 5^{4\cdot(n_{L,5}-1)} \cdot \prod_{j=4}^{n_L}\left(p_j^{4\cdot(n_{L,p_j}-1)} \cdot (p_j - 1)\right) \cdot \left(2^{n_{L,2}-3} \cdot 3^{\max\{0;n_{L,3}-2\}} \cdot 5^{\max\{0;n_{L,5}-2\}} \cdot \prod_{j=4}^{n_L} p_j^{n_{L,p_j}-1} - 1\right)$	

Table 4.6 The number of all LPPs and QPPs over $\mathbb{Z}_p$ that permute $\mathbb{Z}_{p^{n_{L,p}}}$, with $n_{L,p} \geq 1$

p	$n_{L,p} = 1$		$n_{L,p} > 1$	
	$C_{p,LPPs}$	$C_{p,QPPs}$	$C_{p,LPPs}$	$C_{p,QPPs}$
2	1	2	1	1
$p > 2$	$p - 1$	$p - 1$	$p - 1$	$p - 1$

Table 4.7 The number of all CPPs and 4-PPs over $\mathbb{Z}_p$ that permute $\mathbb{Z}_{p^{n_{L,p}}}$, with $n_{L,p} \geq 1$

p	$n_{L,p} = 1$		$n_{L,p} > 1$	
	$C_{p,CPPs}$	$C_{p,4-PPs}$	$C_{p,CPPs}$	$C_{p,4-PPs}$
2	4	8	1	2
3	6	18	4	4
5	24	24	4	4
7	6	90	6	6
1 (mod 3), $p > 7$	$p - 1$	$p - 1$	$p - 1$	$p - 1$
2 (mod 3), $p > 5$	$p^2 - 1$	$p^2 - 1$	$p - 1$	$p - 1$

$n_{L,p} = 1$ or $n_{L,p} > 1$, are given in Tables 4.6, 4.7 and 4.8 (Trifina and Tarniceriu 2018a). In these tables and in the algorithm from this section, by 1-PPs, 2-PPs and 3-PPs, we mean LPPs, QPPs and CPPs, respectively. In the following, we give an example to show the computing of the value $C_{p,d-PPs}$, when $p = 7$ and $d = 5$.

Example 4.6 For $p = 7$, the normalized PPs of degree up to five are all normalized PPs from Table 3.7, except those for $p \neq 1 \pmod 3$ from the second line, and for $p = 13$ from the last line.

From the normalized PP $\bar{\pi}(x) = x$, six 5-PPs result, namely the PPs of the form $a \cdot x, \forall a \in \mathbb{Z}_7^*$.

As only 3, 5 and 6 are not square numbers in $\mathbb{Z}_7$, it results that 17 normalized PPs remain, viz. $x^4 + 3x^2$, $x^4 - 3x^2$, x^5, $x^5 + 2x^2$, $x^5 - 2x^2$, $x^5 + \alpha x^3 + x^2 + 3\alpha^2 x$, with $\alpha \in \{3, 5, 6\}$, $x^5 + \alpha x^3 - x^2 + 3\alpha^2 x$, with $\alpha \in \{3, 5, 6\}$, and $x^5 + \alpha x^3 + 5^{-1}\alpha^2 x$, with $\alpha \in \mathbb{Z}_7^*$. Each normalized PP $\bar{\pi}(x)$ from the previous 17, leads to $6 \cdot 7 = 42$ PPs, namely the PPs $a\bar{\pi}(x + b)$, $\forall a \neq 0, b \in \mathbb{Z}_7$. Totally, the 17 normalized PPs lead to $17 \cdot 6 \cdot 7 = 714$ 5-PPs over $\mathbb{Z}_7$.

Thus, overall, we have $714 + 6 = 720$ 5-PPs over $\mathbb{Z}_7$.

To obtain the 5-PPs over $\mathbb{Z}_7$ that permute $\mathbb{Z}_{7^{n_{L,7}}}$, with $n_{L,7} > 1$, we have to take into account only the normalized PPs from Table 3.8, i.e. one normalized PP of degree one ($\bar{\pi}(x) = x$) and six normalized PPs of degree five ($\bar{\pi}(x) = x^5 + \alpha x^3 + 5^{-1}\alpha^2 x$, with $\alpha \in \mathbb{Z}_7^*$). They lead to $6 + 6 \cdot 6 \cdot 7 = 258$ 5-PPs. ■

To determine the number of different PPs using the considerations above, we have to take into account the equivalence conditions between PPs. These equivalence

Table 4.8 The number of all 5-PPs over $\mathbb{Z}_p$ that permute $\mathbb{Z}_{p^{n_{L,p}}}$, with $n_{L,p} \geq 1$

p	$n_{L,p} = 1$	$n_{L,p} > 1$
	$C_{p,5-PPs}$	$C_{p,5-PPs}$
2	16	4
3	54	16
5	120	56
7	720	258
13	2976	1884
1 (mod 15)	$p-1$	$p-1$
11 (mod 15)	p^2-1	$p-1$
7 (mod 15) or 13 (mod 15)	p^3-p^2+p-1	p^3-2p^2+2p-1
2 (mod 15) or 8 (mod 15)	p^3-1	p^3-2p^2+2p-1
4 (mod 15)	p^2-1	$p-1$
14 (mod 15)	$(p-1)(2p+1)$	$p-1$

conditions were given in Theorem 4.8 in terms of NPs of any degree. Particular results for QNPs and CNPs were given in Theorems 4.4 and 4.6, respectively.

According to Sect. 4.6, for $\pi(x)$ given in (3.1), we can consider only those coefficients of different PPs for which $q_k < \dfrac{L}{\gcd(k!, L)}$, $\forall k = 2, \ldots, d$. Then, we can obtain the number of different PPs of any degree $d \leq 5$, by dividing the number of all PPs obtained with the above algorithm by the quantity $\prod_{k=2}^{d} \gcd(k!, L)$.

Taking into account the equivalence conditions for QPPs from Theorem 4.4, for CPPs from Theorem 4.6, and for 4-PPs and 5-PPs from Sect. 4.6, a PP of degree $d \leq 5$, with the coefficient of the maximum degree term $q_d = \dfrac{L}{\gcd(d!, L)} < L$, can be reduced to a PP of degree smaller than d. We remark that there are no NPs of degree one, so that LPPs cannot be further reduced and thus, the number of true different LPPs is equal to the number of different LPPs. The number of true different PPs of a certain degree d, with $2 \leq d \leq 5$, is obtained by subtracting the number of true different PPs of degrees smaller than d from the number of different PPs of degree d.

Below, we give the algorithm for determining the number of all true different PPs of any degree from one to five (Trifina and Tarniceriu 2018a).

1. Factor the interleaver length as

$$L = \prod_{k=1}^{n_{L1}} p_k \cdot \prod_{k=n_{L1}+1}^{n_{L1}+n_{L2}} p_k^{n_{L,p_k}}, \tag{4.179}$$

where $n_{L,p_k} > 1, \forall k = n_{L1}+1, \ldots, n_{L1}+n_{L2}, n_{L1} \geq 0$ is the number of primes at the power of one from the decomposition of L, $n_{L2} \geq 0$ is the number of primes at power greater than one from the decomposition of L, and $n_{L1}+n_{L2} \geq 1$.

2. According to the previous considerations from this section, compute the number of all PPs of degree d with formula:

$$C_{L,d-PPs,\text{all}} = \prod_{k=1}^{n_{L1}} C_{p_k,d-PPs} \cdot \prod_{k=n_{L1}+1}^{n_{L1}+n_{L2}} C_{p_k,d-PPs} \cdot (p_k)^{d \cdot (n_{L,p_k}-1)}, \tag{4.180}$$

where $C_{p_k,d-PPs}$ are given in Tables 4.6, 4.7 and 4.8 for every prime type at the power of one or greater than one and for any degree from one to five. $C_{p_k,d-PPs}$ in the first product from (4.180) is taken from columns with $n_{L,p} = 1$ in Tables 4.6, 4.7 and 4.8 and $C_{p_k,d-PPs}$ in the second product from (4.180) is taken from columns with $n_{L,p} > 1$ in Tables 4.6, 4.7 and 4.8.

3. Compute the number of different PPs of degree d, for $2 \leq d \leq 5$, with formula:

$$C_{L,d-PPs,\text{diff}} = \frac{C_{L,d-PPs,\text{all}}}{\prod_{k=2}^{d} \gcd(k!, L)} \tag{4.181}$$

4. Compute the number of true different PPs of degree d, for $2 \leq d \leq 5$, with recursive formula:

$$C_{L,d-PPs,\text{true diff}} = C_{L,d-PPs,\text{diff}} - \sum_{k=1}^{d-1} C_{L,d-PPs,\text{true diff}}, \tag{4.182}$$

where $C_{L,1-PPs,\text{true diff}} = C_{L,1-PPs,\text{all}}$.

Reference (Number of 1-5-PPs 2016) is a link to a file where we posted the number of true different LPPs, QPPs, CPPs, 4-PPs and 5-PPs, for any interleaver length $L \leq 100000$, using the above algorithm.

Using the algorithm from this section we can determine the number of true different ZF PPs for arbitrary degree d of PPs (Trifina and Tarniceriu 2018b). We denote by $C_{p_k,d-PPs,ZF}$ the values used in formula (4.180) in this case. To find the values for $C_{p_k,d-PPs,ZF}$, depending on the degree d of PPs, we consider the coefficient conditions from Table 3.6.

For $p = 2$ and $n_{L,p} = 1$ the condition is $(q_1 + q_2 + \cdots + q_d) \neq 0 \pmod 2$. It is fulfilled for $2^d/2 = 2^{d-1}$ combinations of coefficients $(q_1, q_2, \ldots, q_d)$.

For $p = 2$ and $n_{L,p} > 1$ the conditions are $q_1 \neq 0 \pmod 2$, $(q_2 + q_4 + q_6 + \cdots) = 0 \pmod 2$ and $(q_3 + q_5 + q_7 + \cdots) = 0 \pmod 2$.

The condition $q_1 \neq 0 \pmod 2$, with $q_1 \in \mathbb{Z}_2$, is fulfilled only for $q_1 = 1$.

For the other two conditions we consider the cases when the degree d is odd and even, respectively.

Table 4.9 The number of d-PPs ($d \geq 3$) under Zhao and Fan sufficient conditions over $\mathbb{Z}_p$ that permute $\mathbb{Z}_{p^{n_{L,p}}}$, with $n_{L,p} \geq 1$

p	$n_{L,p} = 1$	$n_{L,p} > 1$
	$C_{p,d-PPs,ZF}$	$C_{p,d-PPs,ZF}$
2	2^{d-1}	2^{d-3}
$p > 2$	$p - 1$	$p - 1$

For d an odd number ($d \geq 3$), i.e. $d = 2 \cdot k + 1$, with $k \in \mathbb{N}^*$, each of the sums $(q_2 + q_4 + q_6 + \cdots)$ and $(q_3 + q_5 + q_7 + \cdots)$ contains k coefficients. Thus, each of the sum is fulfilled for $2^k/2 = 2^{k-1}$ combinations of coefficients. It results that $C_{p_k,d-PPs,ZF} = 1 \cdot 2^{k-1} \cdot 2^{k-1} = 2^{2k-2} = 2^{d-3}$, for d odd.

For d an even number ($d \geq 4$), i.e. $d = 2 \cdot k$, with $k \in \mathbb{N}$, $k \geq 2$, the sum $(q_2 + q_4 + q_6 + \cdots)$ contains k coefficients and the sum $(q_3 + q_5 + q_7 + \cdots)$ contains $k - 1$ coefficients. Thus, the first sum is fulfilled for $2^k/2 = 2^{k-1}$ combinations of coefficients and the second sum is fulfilled for $2^{k-1}/2 = 2^{k-2}$ combinations of coefficients. It results that $C_{p_k,d-PPs,ZF} = 1 \cdot 2^{k-1} \cdot 2^{k-2} = 2^{2k-3} = 2^{d-3}$, for d even.

For $p > 2$ and $n_{L,p} \geq 1$ the conditions are $q_1 \neq 0 \pmod{p}$ and $q_2 = q_3 = \cdots = q_d = 0 \pmod{p}$. The condition $q_1 \neq 0 \pmod{p}$, with $q_1 \in \mathbb{Z}_p$, is fulfilled for $p - 1$ values. The condition $q_2 = q_3 = \cdots = q_d = 0 \pmod{p}$, with $q_i \in \mathbb{Z}_p$, $\forall i = \overline{2, d}$, is fulfilled only for $q_2 = q_3 = \cdots = q_d = 0$. Thus, in this case $C_{p_k,d-PPs,ZF} = (p - 1) \cdot 1 = p - 1$.

The values of $C_{p_k,d-PPs}$ used in formula (4.180) in the case of Zhao and Fan sufficient conditions are summarized in Table 4.9.

We mention that for LPPs and QPPs, the Zhao and Fan sufficient conditions become also necessary and, thus, we can use the same values of $C_{p_k,d-PPs}$ from Table 4.6.

As we pointed in the beggining of Sect. 4.6 the NPs under Zhao and Fan sufficient conditions have to fulfill the conditions from Eq. (4.109).

Therefore, the number of NPs of degrees up to d under Zhao and Fan sufficient conditions will not be equal to $\prod_{k=2}^{d} \gcd(k!, N)$ as used in formula (4.181). In the following we obtain the number of NPs of degrees up to d under Zhao and Fan sufficient conditions. We know that the general form of NPs of degrees up to d is that from (4.48).

We denote by g_k the quantity $\gcd(k!, L)$, $k \geq 3$. Let the prime factorization of g_k be of the form

$$g_k = \gcd(k!, L) = 2^{n_{g_k,2}} \cdot \prod_{j=2}^{n_{g_k}} p_{j,g_k}^{n_{g_k,p_j}}, \text{ with } n_{g_k} \geq 2, n_{g_k,2} \geq 1, n_{g_k,p_j} \geq 1,$$

$$\text{and } p_{j,g_k} > 2, \forall j = 2, 3, \ldots, n_{g_k}. \tag{4.183}$$

We define the *truth value function* $||x \bullet y||$, with $\bullet$ an operator between two positive integers x and y, as

$$||x \bullet y|| = \begin{cases} 1, & \text{if } x \bullet y \text{ is true;} \\ 0, & \text{if } x \bullet y \text{ is not true.} \end{cases} \tag{4.184}$$

In the following we will use the function in (4.184) with "*equality operator*" (==) and "*greater than or equal to*" operator ($\geq$).

Similarly as in the beggining of Section 4.6, if there exists a prime $p \leq d$ so that $n_{g_d,p} = n_{L,p}$, then for NPs under Zhao and Fan sufficient conditions we have to impose that $p \mid \tau_k$, $\forall k = k', k'+1, \ldots, d$, where k' is the least integer so that $n_{g_{k'},p} = n_{g_d,p}$. Therefore this prime will reduce the number of NPs of $p^{d-k'+1}$ times. If g_d is factorized as in (4.183) with $k = d$, then the number of NPs under Zhao and Fan sufficient conditions will be equal to

$$C_{NPs,ZF} = \frac{\prod_{k=2}^{d} \gcd(k!, L)}{\prod_{k=2}^{n_{g_d}} (p_{k,g_d})^{(d-k'_d+1)\cdot||n_{g_d},p_k == n_{L,p_k}||}}, \tag{4.185}$$

where k'_d is the least integer such that $n_{g_{k'_d},p_k} = n_{g_d,p_k}$.

The quantity from (4.185) must be used in formula (4.181) when we compute the number of true different ZF PPs using the algorithm from this section.

4.8 Determining the Lengths for Which the Number of True Different PPs of Degree up to Five is Equal to 0

In this section we determine the prime factorization for the lengths for which the number of true different PPs of degree up to five is equal to 0, using the algorithm from Sect. 4.7 (Trifina and Tarniceriu 2018b). Firstly, we obtain below a formula equivalent to (4.182) which is more convenient for this goal. For $d = 2$, using (4.181) in (4.182), we obtain:

$$C_{L,QPPs,\text{true diff}} = \frac{C_{L,2-PPs,\text{all}}}{\gcd(2, L)} - C_{L,1-PPs,\text{all}} \tag{4.186}$$

For $d = 3$, using (4.181) and (4.186) in (4.182), we obtain:

$$C_{L,CPPs,\text{true diff}} = \frac{C_{L,3-PPs,\text{all}}}{\gcd(2, L) \cdot \gcd(6, L)} - \frac{C_{L,2-PPs,\text{all}}}{\gcd(2, L)} \tag{4.187}$$

Similarly, for $d = 4$ and $d = 5$, we obtain:

$$C_{L,4-PPs,\text{true diff}} = \frac{C_{L,4-PPs,\text{all}}}{\gcd(2, L) \cdot \gcd(6, L) \cdot \gcd(24, L)} - \\ - \frac{C_{L,3-PPs,\text{all}}}{\gcd(2, L) \cdot \gcd(6, L)} \tag{4.188}$$

and

$$C_{L,5-PPs,\text{true diff}} = \frac{C_{L,5-PPs,\text{all}}}{\gcd(2, L) \cdot \gcd(6, L) \cdot \gcd(24, L) \cdot \gcd(120, L)} - \\ - \frac{C_{L,4-PPs,\text{all}}}{\gcd(2, L) \cdot \gcd(6, L) \cdot \gcd(24, L)} \tag{4.189}$$

We note that formula

$$C_{L,d-PPs,\text{true diff}} = \frac{C_{L,d-PPs,\text{all}}}{\prod_{k=2}^{d} \gcd(k!, L)} - \frac{C_{L,(d-1)-PPs,\text{all}}}{\prod_{k=2}^{d-1} \gcd(k!, L)} \tag{4.190}$$

is valid for each degree $d \geq 2$, but we don't know the quantities $C_{p_{1,k},d-PPs}$ and $C_{p_{2,k},d-PPs}$ in (4.180), as in Tables 4.6, 4.7 and 4.8, for each degree d.

Now, using equations (4.180) and (4.186)–(4.189) and Tables 4.6, 4.7 and 4.8 we can obtain the lengths for which the number of true different PPs of degree up to five is equal to 0.

Before to proceed, we give in the following a lemma which states a necessary condition to obtain $C_{L,d-PPs,\text{true diff}} = 0$.

Lemma 4.58 *The number of true different PPs of degree d is equal to 0 only if* $n_{L,p_k} - 1$ *and only if* $C_{p_k,d-PPs} = C_{p_k,(d-1)-PPs}$ *or at most* $\frac{C_{p_k,d-PPs}}{C_{p_k,(d-1)-PPs}} \mid \gcd(d!, L)$ *for each* $p_k \mid L$ *so that* $p_k \nmid \gcd(d!, L)$.

Proof Condition $C_{L,d-PPs,\text{true diff}} = 0$ in (4.182) is equivalent to

$$\frac{C_{L,d-PPs,\text{all}}}{C_{L,(d-1)-PPs,\text{all}}} = \gcd(d!, L), \tag{4.191}$$

or taking into account (4.180),

$$\prod_{k=1}^{n_{L_1}} \frac{C_{p_{1,k},d-PPs}}{C_{p_{1,k},(d-1)-PPs}} \cdot \prod_{k=n_{L_1}+1}^{n_{L_1}+n_{L_2}} \frac{C_{p_{2,k},d-PPs}}{C_{p_{2,k},(d-1)-PPs}} \cdot (p_{2,k})^{(n_{L,p_{2,k}}-1)} = \gcd(d!, L), \tag{4.192}$$

Table 4.10 The values $\frac{C_{p,d-PPs}}{C_{p,(d-1)-PPs}}$ for $d = 2, 3$

p	$n_{L,p} = 1$	$n_{L,p} > 1$	$n_{L,p} = 1$	$n_{L,p} > 1$
	$\frac{C_{p,QPPs}}{C_{p,LPPs}}$	$\frac{C_{p,QPPs}}{C_{p,LPPs}}$	$\frac{C_{p,CPPs}}{C_{p,QPPs}}$	$\frac{C_{p,CPPs}}{C_{p,QPPs}}$
2	2	1	2	1
3	1	1	3	2
5	1	1	6	1
1 (mod 3), $p > 5$	1	1	1	1
2 (mod 3), $p > 5$	1	1	$p + 1$	1

Table 4.11 The values $\frac{C_{p,4-PPs}}{C_{p,CPPs}}$

p	$n_{L,p} = 1$	$n_{L,p} > 1$
	$\frac{C_{p,4-PPs}}{C_{p,CPPs}}$	$\frac{C_{p,4-PPs}}{C_{p,CPPs}}$
2	2	2
3	3	1
5	1	1
7	15	1
$p > 7$	1	1

It is clear that $C_{p_{1,k},d-PPs} \geq C_{p_{1,k},(d-1)-PPs}$ $\forall k = \overline{1, n_{L_1}}$, and $C_{p_{2,k},d-PPs} \geq C_{p_{2,k},(d-1)-PPs}$, and $(p_{2,k})^{(n_{L,p_{2,k}}-1)} > 1$ for $n_{L,p_{2,k}} > 1$ $\forall k = \overline{n_{L_1}+1, n_{L_1}+n_{L_2}}$. Then, if $p_{2,k} \nmid \gcd(d!, L)$ for some $k \in \{n_{L_1}+1, \ldots, n_{L_1}+n_{L_2}\}$, Eq. (4.192) can be fulfilled only if $n_{L,p_{2,k}} = 1$, and $C_{p_{2,k},d-PPs} = C_{p_{2,k},(d-1)-PPs}$ or $\frac{C_{p_{2,k},d-PPs}}{C_{p_{2,k},(d-1)-PPs}} \mid \gcd(d!, L)$, and if $p_{1,k} \nmid \gcd(d!, L)$ for some $k \in \{1, \ldots, n_{L_1}\}$, Eq. (4.192) can be fulfilled only if $C_{p_{1,k},d-PPs} = C_{p_{1,k},(d-1)-PPs}$ or $\frac{C_{p_{1,k},d-PPs}}{C_{p_{1,k},(d-1)-PPs}} \mid \gcd(d!, L)$. ■

We approach the cases when $p_k \mid L$ and $p_k \mid \gcd(d!, L)$, for degrees of 2 up to 5, separately in the next subsections. To help in this goal we give in Tables 4.10, 4.11 and 4.12 the values of $\frac{C_{p_{1,k},d-PPs}}{C_{p_{1,k},(d-1)-PPs}}$ and $\frac{C_{p_{2,k},d-PPs}}{C_{p_{2,k},(d-1)-PPs}}$, for $d = 2, 3, 4, 5$.

4.8.1 *Determining the Lengths for Which the Number of True Different QPPs is Equal to 0*

From Table 4.10 we see that the conditions from Lemma 4.58 are fulfilled for each prime $p > 2$.

Table 4.12 The values $\frac{C_{p,5-PPs}}{C_{p,4-PPs}}$

p	$n_{L,p} = 1$ $\frac{C_{p,5-PPs}}{C_{p,4-PPs}}$	$n_{L,p} > 1$ $\frac{C_{p,5-PPs}}{C_{p,4-PPs}}$
2	2	2
3	3	4
5	5	14
7	8	43
13	248	157
1 (mod 15)	1	1
11 (mod 15)	1	1
7 (mod 15) or 13 (mod 15), $p > 13$	$p^2 + 1$	$p^2 - p + 1$
2 (mod 15) or 8 (mod 15), $p > 2$	$(p^2 + p + 1)/(p + 1)$	$p^2 - p + 1$
4 (mod 15)	$p + 1$	1
14 (mod 15)	$(2p + 1)/(p + 1)$	1

If $\gcd(2!, L) = 2$ it means that $2 \mid L$. In this case we have two subcases.

If $p = 2$ and $n_{L,2} = 1$, the condition $C_{L,QPPs,\text{true diff}} = 0$ is fulfilled if $\frac{C_{2,QPPs}}{C_{2,LPPs}} = 2$, which, as we see in Table 4.10, is true.

If $p = 2$ and $n_{L,2} > 1$, the condition $C_{L,QPPs,\text{true diff}} = 0$ is fulfilled if $\frac{C_{2,QPPs}}{C_{2,LPPs}} \cdot 2^{n_{L,2}-1} = 2$, or, equivalently, $1 \cdot 2^{n_{L,2}-1} = 2$, which is true for $n_{L,2} = 2$.

Thus, we conclude that the number of true different QPPs is 0, for the lengths of the form:

$$L_{C_{L,QPPs,\text{true diff}=0}} = 2^{n_{L,2}} \cdot \prod_{k=2}^{n_L} p_k, \text{ with } n_{L,2} = \overline{0, 2}, \ p_k > 2, \forall k = \overline{2, n_L}, \tag{4.193}$$

as it was previously obtained in Eq. (4.56).

4.8.2 *Determining the Lengths for Which the Number of True Different CPPs is Equal to 0*

From Table 4.10 we see that the conditions from Lemma 4.58 are fulfilled for primes p of type $p = 1 \pmod 3$. Also, if $\gcd(3!, L) = 6$, the condition $\frac{C_{p,CPPs}}{C_{p,QPPs}} \mid \gcd(3!, L)$ is fulfilled for $p = 5$.

If $\gcd(3!, L) > 1$, we have the following cases.

(1) $\gcd(3!, L) = 2$, i.e. $2 \mid L$ and $3 \nmid L$.
If $p = 2$ and $n_{L,2} = 1$, the condition $C_{L,CPPs,\text{true diff}} = 0$ is fulfilled if $\frac{C_{2,CPPs}}{C_{2,QPPs}} = 2$, which, as we see in Table 4.10, is true.
If $p = 2$ and $n_{L,2} > 1$, the condition $C_{L,CPPs,\text{true diff}} = 0$ is fulfilled if $\frac{C_{2,CPPs}}{C_{2,QPPs}} \cdot 2^{n_{L,2}-1} = 2$, or, equivalently, $1 \cdot 2^{n_{L,2}-1} = 2$, which is true for $n_{L,2} = 2$.

(2) $\gcd(3!, L) = 3$, i.e. $2 \nmid L$ and $3 \mid L$.
If $p = 3$ and $n_{L,3} = 1$, the condition $C_{L,CPPs,\text{true diff}} = 0$ is fulfilled if $\frac{C_{3,CPPs}}{C_{3,QPPs}} = 3$, which, as we see in Table 4.10, is true.
If $p = 3$ and $n_{L,3} > 1$, the condition $C_{L,CPPs,\text{true diff}} = 0$ is fulfilled if $\frac{C_{3,CPPs}}{C_{3,QPPs}} \cdot 3^{n_{L,3}-1} = 3$, or, equivalently, $2 \cdot 3^{n_{L,3}-1} = 3$, which is not true $\forall\, n_{L,3} > 1$.

(3) $\gcd(3!, L) = 6$, i.e. $2 \mid L$ and $3 \mid L$.
Since each of the cases when $p = 2$ and $n_{L,2} \in \{1, 2\}$, and when $p = 3$ and $n_{L,3} = 1$ have one solution, we have to consider only the case when $n_{L,2} > 2$ and $n_{L,3} > 1$.
If $n_{L,2} > 2$ and $n_{L,3} > 1$, the condition $C_{L,CPPs,\text{true diff}} = 0$ is fulfilled if $\frac{C_{2,CPPs}}{C_{2,QPPs}} \cdot 2^{n_{L,2}-1} \cdot \frac{C_{3,CPPs}}{C_{3,QPPs}} \cdot 3^{n_{L,3}-1} = 2 \cdot 3$, or, equivalently, $1 \cdot 2^{n_{L,2}-1} \cdot 2 \cdot 3^{n_{L,3}-1} = 2 \cdot 3$, which is not true $\forall\, n_{L,2} > 2$ and $\forall\, n_{L,3} > 1$. If $5 \mid L$, $n_{L,5} = 1$, $n_{L,2} > 2$ and $n_{L,3} > 1$, the condition $C_{L,CPPs,\text{true diff}} = 0$ is fulfilled if $\frac{C_{2,CPPs}}{C_{2,QPPs}} \cdot 2^{n_{L,2}-1} \cdot \frac{C_{3,CPPs}}{C_{3,QPPs}} \cdot 3^{n_{L,3}-1} \cdot \frac{C_{5,CPPs}}{C_{5,QPPs}} = 2 \cdot 3$, or, equivalently, $1 \cdot 2^{n_{L,2}-1} \cdot 2 \cdot 3^{n_{L,3}-1} \cdot 6 = 2 \cdot 3$, which, is also not true $\forall\, n_{L,2} > 2$ and $\forall\, n_{L,3} > 1$.

Thus, we conclude that the number of true different CPPs is equal to 0 for the lengths of the form:

$$L_{C_{L,CPPs,\text{true diff}=0}} = 2^{n_{L,2}} \cdot 3^{n_{L,3}} \cdot \prod_{k=3}^{n_L} p_k,$$

$$\text{with } n_{L,2} = \overline{0, 2},\, n_{L,3} = \overline{0, 1},\, p_k > 3, \text{ with } p_k = 1 \pmod 3,\, k = \overline{3, n_L}, \quad (4.194)$$

as it was previously obtained in Eq. (4.106).

4.8.3 Determining the Lengths for Which the Number of True Different 4-PPs is Equal to 0

From Table 4.11 we see that the conditions from Lemma 4.58 are fulfilled for each prime $p > 7$, and for prime $p = 5$.

If $\gcd(4!, L) > 1$, we have the following cases.

(1) $\gcd(4!, L) = 2^{n_{L,2}}$, with $n_{L,2} \in \{1, 2, 3\}$, i.e. $2^{n_{L,2}} \mid L$.
If $p = 2$ and $n_{L,2} = 1$, the condition $C_{L,4-PPs,\text{true diff}} = 0$ is fulfilled if $\frac{C_{2,4-PPs}}{C_{2,CPPs}} = 2$, which, as we see in Table 4.11, is true.
If $p = 2$ and $n_{L,2} \in \{2, 3\}$, the condition $C_{L,4-PPs,\text{true diff}} = 0$ is fulfilled if $\frac{C_{2,4-PPs}}{C_{2,CPPs}} \cdot 2^{n_{L,2}-1} = 2^{n_{L,2}}$, or, equivalently, $2 \cdot 2^{n_{L,2}-1} = 2^{n_{L,2}}$, which is true both for $n_{L,2} = 2$ and $n_{L,2} = 3$.
If $p = 2$ and $n_{L,2} > 3$, the condition $C_{L,4-PPs,\text{true diff}} = 0$ is fulfilled if $\frac{C_{2,4-PPs}}{C_{2,CPPs}} \cdot 2^{n_{L,2}-1} = 2^3$, or, equivalently, $2 \cdot 2^{n_{L,2}-1} = 2^3$, which is not true $\forall\, n_{L,2} > 3$.

(2) $\gcd(4!, L) = 3$, i.e. $2 \nmid L$ and $3 \mid L$.
If $p = 3$ and $n_{L,3} = 1$, the condition $C_{L,4-PPs,\text{true diff}} = 0$ is fulfilled if $\frac{C_{3,4-PPs}}{C_{3,CPPs}} = 3$, which, as we see in Table 4.11, is true.
If $p = 3$ and $n_{L,3} > 1$, the condition $C_{L,4-PPs,\text{true diff}} = 0$ is fulfilled if $\frac{C_{3,4-PPs}}{C_{3,CPPs}} \cdot 3^{n_{L,3}-1} = 3$, or, equivalently, $1 \cdot 3^{n_{L,3}-1} = 3$, which is true for $n_{L,3} = 2$.

(3) $\gcd(4!, L) = 2^{n_{L,2}} \cdot 3$, with $n_{L,2} \in \{1, 2, 3\}$, i.e. $2^{n_{L,2}} \mid L$ and $3 \mid L$.
Since each of the above cases have solutions, we have not to consider this case because it will lead to the same solutions.

Thus, we conclude that the number of true different 4-PPs is equal to 0 for the lengths of the form:

$$L_{C_{L,4-PPs,\text{true diff}=0}} = 2^{n_{L,2}} \cdot 3^{n_{L,3}} \cdot 5^{n_{L,5}} \cdot \prod_{k=4}^{n_L} p_k,$$

$$\text{with } n_{L,2} = \overline{0,3},\, n_{L,3} = \overline{0,2},\, n_{L,5} = \overline{0,1},\, p_k > 7,\, k = \overline{4, n_L}. \tag{4.195}$$

4.8.4 Determining the Lengths for Which the Number of True Different 5-PPs is Equal to 0

From Table 4.12 we see that the conditions from Lemma 4.58 are fulfilled for primes p of types $p = 1 \pmod{15}$ and $p = 11 \pmod{15}$. Also, if $8 \mid \gcd(5!, L)$, the condition

$\frac{C_{p,5-PPs}}{C_{p,4-PPs}} \mid \gcd(5!, L)$ is fullfilled for $p = 7$ and if $20 \mid \gcd(5!, L)$, the condition $\frac{C_{p,5-PPs}}{C_{p,4-PPs}} \mid \gcd(5!, L)$ is fullfilled for $p = 19$.

If $\gcd(5!, L) > 1$, we have the following cases.

(1) $\gcd(5!, L) = 2^{n_{L,2}}$, with $n_{L,2} \in \{1, 2, 3\}$, i.e. $2^{n_{L,2}} \mid L$, $3 \nmid L$, and $5 \nmid L$.
If $p = 2$ and $n_{L,2} = 1$, the condition $C_{L,5-PPs,\text{true diff}} = 0$ is fulfilled if $\frac{C_{2,5-PPs}}{C_{2,4-PPs}} = 2$, which, as we see in Table 4.12, is true.
If $p = 2$ and $n_{L,2} \in \{2, 3\}$, the condition $C_{L,5-PPs,\text{true diff}} = 0$ is fulfilled if $\frac{C_{2,5-PPs}}{C_{2,4-PPs}} \cdot 2^{n_{L,2}-1} = 2^{n_{L,2}}$, or, equivalently, $2 \cdot 2^{n_{L,2}-1} = 2^{n_{L,2}}$, which is true both for $n_{L,2} = 2$ and $n_{L,2} = 3$.
If $p = 2$ and $n_{L,2} > 3$, the condition $C_{L,5-PPs,\text{true diff}} = 0$ is fulfilled if $\frac{C_{2,5-PPs}}{C_{2,4-PPs}} \cdot 2^{n_{L,2}-1} = 2^3$, or, equivalently, $2 \cdot 2^{n_{L,2}-1} = 2^3$, which is is not true $\forall\, n_{L,2} > 3$.
If $7 \mid L$, $n_{L,7} = 1$, and $n_{L,2} > 3$, the condition $C_{L,5-PPs,\text{true diff}} = 0$ is fulfilled if $\frac{C_{2,5-PPs}}{C_{2,4-PPs}} \cdot 2^{n_{L,2}-1} \cdot \frac{C_{7,5-PPs}}{C_{5,4-PPs}} = 2^3$, or, equivalently, $2 \cdot 2^{n_{L,2}-1} \cdot 8 = 2^3$, which, is also not true $\forall\, n_{L,2} > 3$.

(2) $\gcd(3!, L) = 3$, i.e. $2 \nmid L$, $3 \mid L$, and $5 \nmid L$.
If $p = 3$ and $n_{L,3} = 1$, the condition $C_{L,5-PPs,\text{true diff}} = 0$ is fulfilled if $\frac{C_{3,5-PPs}}{C_{3,4-PPs}} = 3$, which, as we see in Table 4.12, is true.
If $p = 3$ and $n_{L,3} > 1$, the condition $C_{L,5-PPs,\text{true diff}} = 0$ is fulfilled if $\frac{C_{3,5-PPs}}{C_{3,4-PPs}} \cdot 3^{n_{L,3}-1} = 3$, or, equivalently, $4 \cdot 3^{n_{L,3}-1} = 3$, which is not true $\forall\, n_{L,3} > 1$.

(3) $\gcd(5!, L) = 2^{n_{L,2}} \cdot 3$, with $n_{L,2} \in \{1, 2, 3\}$, i.e. $2^{n_{L,2}} \mid L$, $3 \mid L$, and $5 \nmid L$.
Since each of the cases when $p = 2$ and $n_{L,2} \in \{1, 2, 3\}$, and when $p = 3$ and $n_{L,3} = 1$ have one solution, we have to consider only the case when $n_{L,2} > 3$ and $n_{L,3} > 1$.
If $n_{L,2} > 3$ and $n_{L,3} > 1$, the condition $C_{L,5-PPs,\text{true diff}} = 0$ is fulfilled if $\frac{C_{2,5-PPs}}{C_{2,4-PPs}} \cdot 2^{n_{L,2}-1} \cdot \frac{C_{3,5-PPs}}{C_{3,4-PPs}} \cdot 3^{n_{L,3}-1} = 2^3 \cdot 3$, or, equivalently, $2 \cdot 2^{n_{L,2}-1} \cdot 4 \cdot 3^{n_{L,3}-1} = 2^3 \cdot 3$, which is not true $\forall\, n_{L,2} > 3$ and $\forall\, n_{L,3} > 1$.
If $7 \mid L$, $n_{L,7} = 1$, $n_{L,2} > 3$ and $n_{L,3} > 1$ the condition $C_{L,5-PPs,\text{true diff}} = 0$ is fulfilled if $\frac{C_{2,5-PPs}}{C_{2,4-PPs}} \cdot 2^{n_{L,2}-1} \cdot \frac{C_{3,5-PPs}}{C_{3,4-PPs}} \cdot 3^{n_{L,3}-1} \cdot \frac{C_{7,5-PPs}}{C_{5,4-PPs}} = 2^3 \cdot 3$, or, equivalently, $2 \cdot 2^{n_{L,2}-1} \cdot 4 \cdot 3^{n_{L,3}-1} \cdot 8 = 2^3 \cdot 3$, which, is also not true $\forall\, n_{L,2} > 3$ and $\forall\, n_{L,3} > 1$.

(4) $\gcd(5!, L) = 5$, i.e. $2 \nmid L$, $3 \nmid L$, and $5 \mid L$.
If $p = 5$ and $n_{L,5} = 1$, the condition $C_{L,5-PPs,\text{true diff}} = 0$ is fulfilled if $\frac{C_{5,5-PPs}}{C_{5,4-PPs}} = 5$, which, as we see in Table 4.12, is true.

If $p = 5$ and $n_{L,5} > 1$, the condition $C_{L,5-PPs,\text{true diff}} = 0$ is fulfilled if $\frac{C_{5,5-PPs}}{C_{5,4-PPs}} \cdot 5^{n_{L,5}-1} = 5$, or, equivalently, $14 \cdot 5^{n_{L,5}-1} = 5$, which is not true $\forall\, n_{L,5} > 1$.

We do not have to consider the cases of other combinations of prime factors 2, 3, and 5, as they will lead to the same solutions.

Thus, we conclude that the number of true different 5-PPs is equal to 0 for the lengths of the form:

$$L_{C_{L,5-PPs,\text{true diff}=0}} = 2^{n_{L,2}} \cdot 3^{n_{L,3}} \cdot 5^{n_{L,5}} \cdot \prod_{k=4}^{n_L} p_k, \text{ with } n_{L,2} = \overline{0,3},$$

$$n_{L,3} = \overline{0,1}, n_{L,5} = \overline{0,1}, \ p_k > 5, \text{ with } p_k = 1 \pmod{15} \text{ or}$$

$$p_k = 11 \pmod{15}, k = \overline{4, n_L}. \tag{4.196}$$

4.9 Determining the Lengths for Which the Number of True Different d-PPs Under Zhao and Fan Sufficient Conditions is Equal to 0

In this section we determine the prime factorization for the lengths for which the number of true different PPs of any degree under Zhao and Fan sufficient conditions is equal to 0, using the algorithm from Sect. 4.7 (Trifina and Tarniceriu 2018b). The formula analog to (4.190) for the number of true different $d-PPs$ under Zhao and Fan sufficient conditions is

$$C_{L,d-PPs,ZF,\text{true diff}} = C_{L,d-PPs,ZF,\text{all}} \cdot \frac{\prod_{k=2}^{n_{g_d}} (p_{k,g_d})^{(d-k'_d+1)\cdot||n_{g_d,p_k}==n_{L,p_k}||}}{\prod_{k=2}^{d} \gcd(k!, L)} -$$

$$-C_{L,(d-1)-PPs,\text{all}} \cdot \frac{\prod_{k=2}^{n_{g_{d-1}}} (p_{k,g_{d-1}})^{(d-k'_{d-1})\cdot||n_{g_{d-1},p_k}==n_{L,p_k}||}}{\prod_{k=2}^{d-1} \gcd(k!, L)} \tag{4.197}$$

The values $\frac{C_{p_{1,k},d-PPs,ZF}}{C_{p_{1,k},(d-1)-PPs,ZF}}$ and $\frac{C_{p_{2,k},d-PPs,ZF}}{C_{p_{2,k},(d-1)-PPs},ZF}$, for $d \geq 3$, are given in Table 4.13 .

Table 4.13 The values $\frac{C_{p,d-PPs}}{C_{p,(d-1)-PPs}}$ for d-PPs ($d \geq 3$) under Zhao and Fan sufficient conditions

p	$n_{N,p} = 1$	$n_{N,p} > 1$
	$\frac{C_{p,d-PPs}}{C_{p,(d-1)-PPs}}$	$\frac{C_{p,d-PPs}}{C_{p,(d-1)-PPs}}$
2	2	$\begin{cases} 2, & \text{if } d > 3, \\ 1, & \text{if } d = 3. \end{cases}$
$p > 2$	1	1

Theorem 4.59 (Lengths for which $C_{L,d-PPs,ZF,true\ diff} = 0$) *Let the prime factorization of $d!$ be*

$$d! = 2^{n_{d!,2}} \cdot \prod_{k=2}^{n_{d!}} p_{k,d!}^{n_{d!,p_k}},$$

with $n_{d!} \geq 2, n_{d!,2} \geq 1, n_{d!,p_k} \geq 1,$ and $2 < p_{k,d!} \leq d, \forall k = 2, 3, \ldots, n_{d!}$. (4.198)

Then the number of true different PPs of degree d under Zhao and Fan sufficient conditions is equal to zero ($C_{L,d-PPs,ZF,true\ diff} = 0$) if L is of the form

$$L_{C_{L,d-PPs,ZF,true\ diff=0}} = 2^{n_{L,2}} \cdot \prod_{k=2}^{n_{d!}} p_{k,d!}^{n_{L,p_k}} \cdot \prod_{k=n_{d!}+1}^{n_L} p_k,$$

with $0 \leq n_{L,2} \leq n_{d!,2}$ for $d > 3$ and $0 \leq n_{L,2} \leq 2$ for $d = 3$,

$$0 \leq n_{L,p_k} \leq n_{d!,p_k} + 1, 2 < p_{k,d!} < d, \forall k = 2, 3, \ldots, n_{d!}, \text{ and } p_k > d,$$

$$\forall k = n_{d!} + 1, \ldots, n_L. \tag{4.199}$$

Proof Imposing that $C_{L,d-PPs,ZF,\text{true diff}} = 0$ in (4.197), we obtain

$$\frac{C_{L,d-PPs,ZF,\text{all}}}{C_{L,(d-1)-PPs,\text{all}}} = \gcd(d!, L) \cdot \frac{\prod_{k=2}^{n_{g_{d-1}}} (p_{k,g_{d-1}})^{(d-k'_{d-1}) \cdot ||n_{g_{d-1},p_k} == n_{L,p_k}||}}{\prod_{k=2}^{n_{g_d}} (p_{k,g_d})^{(d-k'_d+1) \cdot ||n_{g_d,p_k} == n_{L,p_k}||}} \tag{4.200}$$

In the following we will analyze the cases when d is a prime number and d is not a prime number.

(1) d - a prime number
If d is a prime number then the prime factorizations of g_{d-1} and $g_d/d^{(||n_{L,d} \geq 1||)}$ are the same. Further $n_{g_d,d} = 1$ and $k'_d = d$. Thus we have

$$\frac{\prod_{k=2}^{n_{gd-1}} (p_k)^{(d-k'_{d-1})\cdot||n_{gd-1},p_k==n_{L,p_k}||}}{\prod_{k=2}^{n_{gd}} (p_k)^{(d-k'_d+1)\cdot||n_{gd},p_k==n_{L,p_k}||}} = \frac{1}{d^{(||n_{L,d}==1||)}} \tag{4.201}$$

In this case condition (4.200) becomes

$$\frac{C_{L,d-PPs,ZF,\text{all}}}{C_{L,(d-1)-PPs,\text{all}}} = \frac{\gcd(d!, L)}{d^{(||n_{L,d}==1||)}} \tag{4.202}$$

Taking into account Eq. (4.180) and the values from Table 4.13, we obtain for $d = 3$ $C_{L,CPPs,ZF,\text{true diff}} = 0$ if L is of the form

$$L_{C_{L,CPPs,ZF,\text{true diff}=0}} = 2^{n_{L,2}} \cdot 3^{n_{L,3}} \cdot \prod_{k=3}^{n_L} p_k, \text{ with } n_{L,2} = \overline{0,2},$$

$$n_{L,3} = \overline{0,2},\ p_k > 3, \forall k = \overline{3, n_L}, \tag{4.203}$$

as in (4.199) for $d = 3$. The same result was obtained in Eq. (4.124).
Then, if d is a prime number, $d > 3$, from (4.180) and the values from Table 4.13, (4.202) is equivalent to

$$2 \cdot 2^{n_{L,2}-1} \cdot \prod_{k=n_{L_1}+1}^{n_{L_1}+n_{L_2}} 1 \cdot (p_{2,k})^{(n_{L,p_{2,k}}-1)} = 2^{n_{gd,2}} \cdot \prod_{k=2}^{n_{d!}-1} p_{k,d!}^{n_{gd},p_k} \cdot$$

$$\cdot d^{(||n_{L,d}\geq 1||)-(||n_{L,d}==1||)}, \text{ with } 0 \leq n_{gd,2} \leq n_{d!,2},\ 0 \leq n_{gd,p_k} \leq n_{d!,p_k},$$

$$\text{and } 2 < p_{k,d!} < d, \forall k = 2, 3, \ldots, n_{d!} - 1. \tag{4.204}$$

From (4.204) it results that for d a prime number, $d > 3$, $C_{L,d-PPs,ZF,\text{true diff}} = 0$ if L is of the form

$$L_{C_{L,d-PPs,ZF,\text{true diff}=0}} = 2^{n_{L,2}} \cdot \prod_{k=2}^{n_{d!}-1} p_{k,d!}^{n_{L,p_k}} \cdot d^{n_{L,d}} \cdot \prod_{k=n_{d!}+1}^{n_L} p_k,$$

$$\text{with } 0 \leq n_{L,2} \leq n_{d!,2},\ 0 \leq n_{L,p_k} \leq n_{d!,p_k} + 1, \text{ and}$$

$$2 < p_{k,d!} < d, \forall k = 2, 3, \ldots, n_{d!} - 1, 0 \leq n_{L,d} \leq 2,\ p_k > d,$$

$$\forall k = n_{d!} + 1, \ldots, n_L. \tag{4.205}$$

Equation (4.205) is the same as (4.199) for d a prime number.

(2) d - not a prime number
If d is not a prime number then the prime factors from the prime factorization of g_d are the same as those from the prime factorization of g_{d-1}, possibly with greater powers of some factors. The maximum powers of the primes $p_{k,d!}$ in the factorization of g_d are $n_{d!,p_k}$, $\forall k = 2, 3, \ldots, n_{d!}$.
If $p_{k,d!} \mid g_d$, $p_{k,d!} \nmid d$, and $n_{d!,p_k} \geq n_{L,p_k}$, then $n_{g_{d-1},p_k} = n_{g_d,p_k} = n_{L,p_k}$ and $k'_{d-1} = k'_d$. Thus the term corresponding to factor $p_{k,d!}$ in the ratio from the right hand side of (4.200) is $\dfrac{1}{p_{k,d!}}$. The same observation is valid if $p_{k,d!} \mid g_d$, $p_{k,d!} \mid d$, and $n_{d!,p_k} - n_{d,p_k} \geq n_{L,p_k}$.
If $p_{k,d!} \mid g_d$, $p_{k,d!} \mid d$, and $n_{d!,p_k} - n_{d,p_k} < n_{L,p_k} \leq n_{d!,p_k}$, then $n_{g_{d-1},p_k} < n_{g_d,p_k} = n_{L,p_k}$, $||n_{g_{d-1},p_k} == n_{L,p_k}|| = 0$, $||n_{g_d,p_k} == n_{L,p_k}|| = 1$, and $k'_d = d$. Thus the term corresponding to factor $p_{k,d!}$ in the ratio from the right hand side of (4.200) is also $\dfrac{1}{p_{k,d!}}$.
If $p_{k,d!} \mid g_d$ and $n_{L,p_k} > n_{d!,p_k}$, then $n_{g_{d-1},p_k} < n_{L,p_k}$, $n_{g_d,p_k} < n_{L,p_k}$, $||n_{g_{d-1},p_k} == n_{L,p_k}|| = 0$, and $||n_{g_d,p_k} == n_{L,p_k}|| = 0$. Thus the term corresponding to factor $p_{k,d!}$ in the ratio from the right hand side of (4.200) is equal to 1.
From those above it results that if d is not a prime number, then (4.200) is equivalent to

$$\frac{C_{L,d-PPs,ZF,\text{all}}}{C_{L,(d-1)-PPs,\text{all}}} = \gcd(d!, L) \cdot \frac{1}{\prod_{k=2}^{n_{d!}} (p_{k,d!})^{||n_{d!,p_k} \geq n_{L,p_k}||}} \tag{4.206}$$

Similarly to (4.204), (4.206) is equivalent to

$$2 \cdot 2^{n_{L,2}-1} \cdot \prod_{k=n_{L_1}+1}^{n_{L_1}+n_{L_2}} 1 \cdot (p_{2,k})^{(n_{L,p_{2,k}}-1)} = 2^{n_{g_d,2}} \cdot \frac{\prod_{k=2}^{n_{d!}} p_{k,d!}^{n_{g_d,p_k}}}{\prod_{k=2}^{n_{d!}} (p_{k,d!})^{||n_{d!,p_k} \geq n_{L,p_k}||}},$$

$$\text{with } 0 \leq n_{g_d,2} \leq n_{d!,2},\ 0 \leq n_{g_d,p_k} \leq n_{d!,p_k},\ \text{and } 2 < p_{k,d!} < d,$$

$$\forall k = 2, 3, \ldots, n_{d!}. \tag{4.207}$$

From (4.207) it results that if d is not a prime number, $C_{L,d-PPs,ZF,\text{true diff}} = 0$ if L is of the form

$$L_{C_{L,d-PPs,ZF,\text{true diff}=0}} = 2^{n_{L,2}} \cdot \prod_{k=2}^{n_{d!}} p_{k,d!}^{n_{L,p_k}} \cdot \prod_{k=n_{d!}+1}^{n_L} p_k,$$

$$\text{with } 0 \le n_{L,2} \le n_{d!,2},\, 0 \le n_{L,p_k} \le n_{d!,p_k} + 1, \text{ and } 2 < p_{k,d!} < d,$$
$$\forall k = 2, 3, \ldots, n_{d!},\, p_k > d, \forall k = n_{d!} + 1, \ldots, n_L. \tag{4.208}$$

We note that formula (4.208) is also valid if d is a prime number. Thus the theorem is proved. ■

In the following we will give two examples for the form of L when d is a prime number and when d is not a prime number.

Example 4.7 (*Example of L for which $C_{L,11-PPs,ZF,\text{true diff}} = 0$*) For $d = 11$ we have

$$11! = 2^8 \cdot 3^4 \cdot 5^2 \cdot 7^1 \cdot 11^1, \tag{4.209}$$

and $C_{L,11-PPs,ZF,\text{true diff}} = 0$ if L is of the form

$$L_{C_{L,11-PPs,ZF,\text{true diff}=0}} = 2^{n_{L,2}} \cdot 3^{n_{L,3}} \cdot 5^{n_{L,5}} \cdot 7^{n_{L,7}} \cdot 11^{n_{L,11}} \cdot \prod_{k=6}^{n_L} p_k,$$

$$\text{with } 0 \le n_{L,2} \le 8,\, 0 \le n_{L,3} \le 5,\, 0 \le n_{L,5} \le 3,\, 0 \le n_{L,7} \le 2,$$
$$0 \le n_{L,11} \le 2, \text{ and } p_k > 11, \forall k = 6, \ldots, n_L. \tag{4.210}$$

■

Example 4.8 (*Example of L for which $C_{L,12-PPs,ZF,\text{true diff}} = 0$*) For $d = 12$ we have

$$12! = 2^{10} \cdot 3^5 \cdot 5^2 \cdot 7^1 \cdot 11^1, \tag{4.211}$$

and $C_{L,12-PPs,ZF,\text{true diff}} = 0$ if L is of the form

$$L_{C_{L,12-PPs,ZF,\text{true diff}=0}} = 2^{n_{L,2}} \cdot 3^{n_{L,3}} \cdot 5^{n_{L,5}} \cdot 7^{n_{L,7}} \cdot 11^{n_{L,11}} \cdot \prod_{k=6}^{n_L} p_k,$$

$$\text{with } 0 \le n_{L,2} \le 10,\, 0 \le n_{L,3} \le 6,\, 0 \le n_{L,5} \le 3,\, 0 \le n_{L,7} \le 2,$$
$$0 \le n_{L,11} \le 2, \text{ and } p_k > 11, \forall k = 6, \ldots, n_L. \tag{4.212}$$

■

We note that for $d = 4$ and $d = 5$ we obtain the same results as in Eqs. (4.142) and (4.178), respectively, i.e.

$$L_{C_{L,4-PPs,ZF,\text{true diff}=0}} = 2^{n_{L,2}} \cdot 3^{n_{L,3}} \cdot \prod_{k=3}^{n_L} p_k,$$

$$\text{with } 0 \le n_{L,2} \le 3, 0 \le n_{L,3} \le 2,$$

$$\text{and } p_k > 3, \forall k = 3, \ldots, n_L. \tag{4.213}$$

and

$$L_{C_{L,5-PPs,ZF,\text{true diff}=0}} = 2^{n_{L,2}} \cdot 3^{n_{L,3}} \cdot 5^{n_{L,5}} \cdot \prod_{k=4}^{n_L} p_k,$$

$$\text{with } 0 \le n_{L,2} \le 3, 0 \le n_{L,3} \le 2, 0 \le n_{L,5} \le 2,$$

$$\text{and } p_k > 5, \forall k = 4, \ldots, n_L. \tag{4.214}$$

References

2015, http://telecom.etti.tuiasi.ro/tti/papers/Text_files/Number_of_true_diff_LPPs_QPPs_CPPs_N_000002_100000.txt

2016, http://telecom.etti.tuiasi.ro/tti/papers/Text_files/Number_of_true_diff_LPPs_QPPs_CPPs_4PPs_5PPs_N_000002_100000.txt

3GPP TS 36.212 V8.3.0, 3rd generation partnership project, Multiplexing and channel coding (Release 8) (2008), http://www.etsi.org

L.E. Dickson, The analytic representation of substitutions on a power of a prime number of letters with a discussion of the liner group. Ann. Math. **11**(1–6), 65–120 (1896)

G.H. Hardy, E.M. Wright, *An Introduction to the Theory of Numbers*, 4th edn. (Oxford University Press, Oxford, 1975)

S. Li, Null polynomials modulo m (2005), arXiv:math/0510217v2

J. Ryu, Permutation polynomial based interleavers for turbo codes over integer rings. Ph.D. thesis. Ohio State University (2007), https://etd.ohiolink.edu/rws_etd/document/get/osu1181139404/inline

J. Ryu, O.Y. Takeshita, On inverses for quadratic permutation polynomials over integers rings (2011), arXiv:1102.2223

D. Tarniceriu, L. Trifina, V. Munteanu, About minimum distance for QPP interleavers. Ann. Telecommun. **64**(11–12), 745–751 (2009)

L. Trifina, D. Tarniceriu, Analysis of cubic permutation polynomials for turbo codes. Wirel. Pers. Commun. **69**(1), 1–22 (2013)

L. Trifina, D. Tarniceriu, Improved method for searching interleavers from a certain set using Garello's method with applications for the LTE standard. Ann. Telecommun. **69**(5–6), 251–272 (2014)

L. Trifina, D. Tarniceriu, When the number of true different permutation polynomials is equal to 0? Submitted for possible publication (under review) (2018a)

L. Trifina, D. Tarniceriu, Determining the number of true different permutation polynomials of degrees up to five by Weng and Dong algorithm. Telecommun. Syst. **67**(2), 211–215 (2018b)

L. Trifina, D. Tarniceriu, V. Munteanu, Improved QPP interleavers for LTE standard, in *IEEE International Symposium on Signals, Circuits and Systems (ISSCS), Iasi, Romania*, June 30–July 1 2011, pp. 403–406

G. Weng, C. Dong, A note on permutation polynomial over Zn. IEEE Trans. Inf. Theory **54**(9), 4388–4390 (2008)

H. Zhao, P. Fan, Simple method for generating mth-order permutation polynomials over integer rings. Electron. Lett. **43**(8), 449–451 (2007)

H. Zhao, P. Fan, V. Tarokh, On the equivalence of interleavers for turbo codes using quadratic permutation polynomials over integer rings. IEEE Commun. Lett. **14**(3), 236–238 (2010)

Chapter 5
Minimum Distance of Turbo Codes with Permutation Polynomial-Based Interleavers

5.1 Inverse Permutation Polynomials of a QPP

First, in this section, the necessary and sufficient condition for a QPP to admit at least one quadratic inverse is given (Ryu and Takeshita 2006). The necessary lemmas and the main theorems along with their proofs follow the same pattern as in Ryu and Takeshita (2006).

In the following lemma it is shown that for any QPP there exists at least one quadratic polynomial that inverts it at three points $x = 0, 1, 2$. The reason for this partially inverting polynomial is because it becomes the basis for the quadratic inverse polynomial if it exists.

Lemma 5.1 *Let* L *be a positive integer and let* $\pi(x) = q_1 \cdot x + q_2 \cdot x^2 \pmod{L}$ *be a QPP. Then there exists at least one quadratic polynomial* $\rho(x) = r_1 \cdot x + r_2 \cdot x^2 \pmod{L}$ *that inverts* $\pi(x)$ *at these three points:* $x = 0, 1, 2$. *If* L *is odd, there is exactly one quadratic polynomial* $\rho(x) = r_1 \cdot x + r_2 \cdot x^2 \pmod{L}$ *and its coefficients can be obtained by solving the linear congruences:*

$$r_2(q_1 + q_2)(q_1 + 2q_2)(q_1 + 3q_2) \equiv (-q_2) \pmod{L} \tag{5.1}$$

$$r_1(q_1 + q_2) + r_2(q_1 + q_2)^2 \equiv 1 \pmod{L} \tag{5.2}$$

If L *is even, there are exactly two quadratic polynomials* $\rho_1(x) = r_{1,1} \cdot x + r_{1,2} \cdot x^2 \pmod{L}$, $\rho_2(x) = r_{2,1} \cdot x + r_{2,2} \cdot x^2 \pmod{L}$ *and the coefficients of the polynomial* $\rho_1(x) = r_{1,1} \cdot x + r_{1,2} \cdot x^2 \pmod{L}$ *can be obtained by solving the linear congruences:*

$$r_{1,2}(q_1 + q_2)(q_1 + 2q_2)(q_1 + 3q_2) \equiv (-q_2)\left(\bmod \frac{L}{2}\right) \tag{5.3}$$

$$r_{1,1}(q_1 + q_2) + r_{1,2}(q_1 + q_2)^2 \equiv 1 \pmod{L} \tag{5.4}$$

L. Trifina and D. Tarniceriu, *Permutation Polynomial Interleavers for Turbo Codes*, Signals and Communication Technology,
https://doi.org/10.1007/978-981-13-2625-7_5

After computing coefficients $(r_{1,1}, r_{1,2})$ *of the polynomial* $\rho_1(x)$, *coefficients* $(r_{2,1}, r_{2,2})$ *of the polynomial* $\rho_2(x)$ *can be obtained by* $r_{2,1} \equiv (r_{1,1} + L/2) \pmod{L}$ *and* $r_{2,2} \equiv (r_{1,2} + L/2) \pmod{L}$.

Proof Let $\rho \circ \pi$ the operation of composing functions ρ and π. Then $\rho(x) = r_1 \cdot x + r_2 \cdot x^2 \pmod{L}$ inverts $\pi(x)$ at the points: $x = 1$ and $x = 2$ if and only if the following two congruences have at least one solution set (r_1, r_2):

$$\begin{aligned}(\rho \circ \pi)(1) = \rho(\pi(1)) = \rho(q_1 + q_2) = \\ = r_1 \cdot (q_1 + q_2) + r_2 \cdot (q_1 + q_2)^2 \equiv 1 \pmod{L}\end{aligned} \tag{5.5}$$

$$\begin{aligned}(\rho \circ \pi)(2) = \rho(\pi(2)) = \rho(2q_1 + 4q_2) = \\ = r_1 \cdot (2q_1 + 4q_2) + r_2 \cdot (2q_1 + 4q_2)^2 \equiv 2 \pmod{L}\end{aligned} \tag{5.6}$$

It is trivial that $\rho(x)$ inverts $\pi(x)$ at the third point $x = 0$, because $(\rho \circ \pi)(0) = \rho(\pi(0)) = \rho(0) = 0 \pmod{L}$.

By multiplying (5.5) by $(2q_1 + 4q_2)$ and (5.6) by $(q_1 + q_2)$, we get

$$\begin{aligned}r_1 \cdot (2q_1 + 4q_2) \cdot (q_1 + q_2) + r_2 \cdot (2q_1 + 4q_2) \cdot (q_1 + q_2)^2 \equiv \\ \equiv (2q_1 + 4q_2) \pmod{L}\end{aligned} \tag{5.7}$$

$$\begin{aligned}r_1 \cdot (2q_1 + 4q_2) \cdot (q_1 + q_2) + r_2 \cdot (2q_1 + 4q_2)^2 \cdot (q_1 + q_2) \equiv \\ \equiv 2 \cdot (q_1 + q_2) \pmod{L}\end{aligned} \tag{5.8}$$

By subtracting (5.7) from (5.8), we obtain

$$\begin{aligned}r_2 \cdot (2q_1 + 4q_2) \cdot (q_1 + q_2) \cdot (2q_1 + 4q_2 - q_1 - q_2) \equiv \\ \equiv (2q_1 + 2q_2 - 2q_1 - 4q_2) \pmod{L} \Leftrightarrow\end{aligned}$$

$$\Leftrightarrow 2r_2 \cdot (q_1 + q_2) \cdot (q_1 + 2q_2) \cdot (q_1 + 3q_2) \equiv (-2q_2) \pmod{L} \tag{5.9}$$

It can be shown that there exists at least one r_2 that satisfies (5.9), as follows.

If either $2 \nmid L$ or $4 \mid L$, from Table 3.1, cases 1(b) and (2), it results that for all p's so that $p \mid L$, we have $p \nmid q_1$ and $p \mid q_2$. Thus $p \nmid (q_1 + q_2)$. Therefore $\gcd(q_1 + q_2, L) = 1$. Similarly, since $p \mid (2q_2)$ and $p \mid (3q_2)$, we have $\gcd(q_1 + 2q_2, L) = 1$ and $\gcd(q_1 + 3q_2, L) = 1$. Thus

$$\gcd\big((q_1 + q_2)(q_1 + 2q_2)(q_1 + 3q_2), L\big) = 1 \tag{5.10}$$

Consequently, if $2 \nmid L$,

$$\gcd\big(2(q_1 + q_2)(q_1 + 2q_2)(q_1 + 3q_2), L\big) = 1 \tag{5.11}$$

and if $4 \mid L$,

$$\gcd\big(2(q_1 + q_2)(q_1 + 2q_2)(q_1 + 3q_2), L\big) = 2 \tag{5.12}$$

If $2 \mid L$ or $4 \nmid L$, $\gcd(2, L) = 2$ and, as above

$$\gcd\left((q_1 + q_2)(q_1 + 2q_2)(q_1 + 3q_2), p\right) = 1 \tag{5.13}$$

for every p's, with $p \neq 2$, so that $p \mid L$. Thus

$$\gcd\left(2(q_1 + q_2)(q_1 + 2q_2)(q_1 + 3q_2), L\right) = 2 \tag{5.14}$$

To summarize, from Theorem 57 in Hardy and Wright (1975), if L is an even number, we have exactly two solution sets, and if L is an odd number we have exactly one solution set.

When L is an even number, let $(r_{1,1}, r_{1,2})$ and $(r_{2,1}, r_{2,2})$ be the two solution sets. Then,

$$r_{1,2}(q_1 + q_2)(q_1 + 2q_2)(q_1 + 3q_2) \equiv (-q_2)\left(\text{mod } \frac{L}{2}\right) \tag{5.15}$$

and $r_{2,2} \equiv (r_{1,2} + L/2) \pmod{L}$ according to the same theorem.

When L is an odd number, let (r_1, r_2) be the solution set. Then, according to Theorem 55 from Hardy and Wright (1975), the congruence (5.9) can be rewritten as

$$r_2(q_1 + q_2)(q_1 + 2q_2)(q_1 + 3q_2) \equiv (-q_2) \pmod{L} \tag{5.16}$$

since $\gcd(2, L) = 1$.

After computing r_2 using (5.16), or $(r_{1,2}, r_{2,2})$ using (5.15) and Theorem 57 from Hardy and Wright (1975), we can compute the corresponding r_1 or $(r_{1,1}, r_{2,1})$ using (5.5). Specifically, it can be verified that $r_{2,1} \equiv (r_{1,1} + L/2) \pmod{L}$. Thus, for a given QPP $\pi(x)$, we can find at least one quadratic polynomial that inverts the polynomial $\pi(x)$ at three points: $x = 0, 1, 2$.

It is well known that linear congruences can be efficiently solved by using the extended Euclidean algorithm. We denote by s^* the arithmetic inverse of s modulo L, i.e. a number s^* so that $s \cdot s^* \equiv 1 \pmod{L}$. The Algorithm 2 provided at the end of this section can be used to calculate such an inverse. Thus, given q_1 and q_2, from (5.1) we compute r_2, as follows

$$r_2 \equiv \left\{(-q_2) \cdot [(q_1 + q_2)(q_1 + 2q_2)(q_1 + 3q_2)]^*\right\} \pmod{L}, \tag{5.17}$$

and then r_1, as

$$r_1 \equiv \left\{[1 - r_2(q_1 + q_2)^2] \cdot (q_1 + q_2)^*\right\} \pmod{L}. \tag{5.18}$$

The congruences (5.3) and (5.4) are similarly solved. ■

In the following lemma, it is shown that the polynomials $\rho(x)$, $\rho_1(x)$, and $\rho_2(x)$ obtained by solving the congruences (5.1), (5.2), (5.3) and (5.4) are PPs.

Lemma 5.2 *The polynomials $\rho(x)$, $\rho_1(x)$, and $\rho_2(x)$ obtained in Lemma 5.1 are permutation polynomials.*

Proof In the proof we will use the obvious fact that, from $u \equiv v \pmod{L}$ and $K \mid L$ it results $u \equiv v \pmod{K}$.

Case (a) $2 \nmid L$

If $2 \nmid L$, for p a prime number so that $p \mid L$, we can reduce (5.9) to

$$2r_2 \cdot q_1 \cdot q_1 \cdot q_1 \equiv 0 \pmod{p}, \tag{5.19}$$

since, from Table 3.1, $q_2 = 0 \pmod{p}$. Thus $r_2 = 0 \pmod{p}$, since $q_1 \neq 0 \pmod{p}$ and then $\gcd(2 \cdot q_1 \cdot q_1 \cdot q_1, p) = 1$, $\forall p$ such that $p \mid L$.

By multiplying (5.5) by $(2q_1 + 4q_2)^2$ and (5.6) by $(q_1 + q_2)^2$ we have

$$\begin{aligned} r_1 \cdot (2q_1 + 4q_2)^2 \cdot (q_1 + q_2) + r_2 \cdot (2q_1 + 4q_2)^2 \cdot (q_1 + q_2)^2 \equiv \\ \equiv (2q_1 + 4q_2)^2 \pmod{L} \end{aligned} \tag{5.20}$$

$$\begin{aligned} r_1 \cdot (2q_1 + 4q_2) \cdot (q_1 + q_2)^2 + r_2 \cdot (2q_1 + 4q_2)^2 \cdot (q_1 + q_2)^2 \equiv \\ \equiv 2(q_1 + q_2)^2 \pmod{L} \end{aligned} \tag{5.21}$$

By subtracting (5.21) from (5.20), we obtain

$$\begin{aligned} r_1 \cdot (q_1 + q_2) \cdot (2q_1 + 4q_2) \cdot (q_1 + 3q_2) \equiv \\ \equiv (4q_1^2 + 16q_2^2 + 16q_1q_2 - 2q_1^2 - 2q_2^2 - 4q_1q_2) \pmod{L} \Leftrightarrow \\ 2r_1 \cdot (q_1 + q_2) \cdot (q_1 + 2q_2) \cdot (q_1 + 3q_2) \equiv (2q_1^2 + 14q_2^2 + 12q_1q_2) \pmod{L} \end{aligned} \tag{5.22}$$

Considering Table 3.1 we can reduce (5.22) to

$$2r_1 \cdot q_1 \cdot q_1 \cdot q_1 \equiv (2q_1^2) \pmod{p}, \tag{5.23}$$

Since $\gcd(2 \cdot q_1 \cdot q_1 \cdot q_1, p) = 1$, if $p \mid r_1$, from (5.23) we have that $p \mid (2q_1^2)$, which contradicts the condition for q_1 from Table 3.1 when $p \neq 2$. Therefore $p \nmid r_1$ or $r_1 \neq 0 \pmod{p}$. Since $r_1 \neq 0 \pmod{p}$ and $r_2 = 0 \pmod{p}$, from Table 3.1 it results that $\rho(x)$ is a QPP.

Case (b) $2 \mid L$ and $4 \nmid L$

For p a prime number so that $p \mid L$ and $p \neq 2$, similarly to the previous case, it can be shown that $\rho_1(x)$ and, thus, also $\rho_2(x)$, are QPPs modulo $L/2$.

Now we show that $\rho_1(x)$ and $\rho_2(x)$ are QPPs modulo 2. Since $\pi(x)$ is a QPP, from Table 3.1 it results that $q_1 + q_2$ is an odd number. Thus $(q_1 + q_2)^2$ is also odd. Let us assume that $r_{1,1} + r_{1,2}$ is even, i.e., both $r_{1,1}$ and $r_{1,2}$ are even or odd numbers. Then the quantity $r_{1,1} \cdot (q_1 + q_2) + r_{1,2} \cdot (q_1 + q_2)^2$ from (5.5) becomes an even number, being the sum of two numbers, both of them even or odd. But from (5.5) we have a contradiction, since L is even and an even number modulo an even number is also even, but $1 \pmod{L}$ is odd. Therefore, $r_{1,1} + r_{1,2}$ must be an odd number. Thus, from Table 3.1 it results that $\rho_1(x)$ is a QPP modulo 2. Because $r_{1,1} + r_{1,2}$ is odd, and $r_{2,1} \equiv (r_{1,1} + L/2) \pmod{L}$ and $r_{2,2} \equiv (r_{1,2} + L/2) \pmod{L}$, we have that $r_{2,1} + r_{2,2}$ is also odd, being the sum of an odd number $(r_{1,1} + r_{1,2})$ and an even number $(L/2 + L/2 = L)$. As a result, $\rho_2(x)$ is a QPP modulo 2. Finally, since both $\rho_1(x)$ and $\rho_2(x)$ are QPPs modulo 2 and modulo $L/2$, from Theorem 3.8 it results that they are QPPs modulo L.

Case (c) $4 \mid L$

This case is approached similarly to case (a). For p a prime number so that $p \mid L$ and $p \neq 2$ it can be shown that $\rho_1(x)$, and thus also $\rho_2(x)$, are QPPs modulo $L/2^{n_{L,2}}$, where $n_{L,2} \geq 2$ is the power of 2 from the factorization of L.

Now we show that $\rho_1(x)$ and $\rho_2(x)$ are QPPs modulo $2^{n_{L,2}}$. Since $\pi(x)$ is a QPP, from Table 3.1 it results that q_1 is an odd number and q_2 is an even number. Then $q_1 + q_2$, $q_1 + 2q_2$ and $q_1 + 3q_2$ are all odd numbers and the product $(q_1 + q_2)(q_1 + 2q_2)(q_1 + 3q_2)$ is also an odd number. Consequently, from (5.3) $r_{1,2}$ must be an even number, since $(-q_2)\left(\bmod \frac{L}{2}\right)$ is an even number, and an even number modulo an even number must be even. From (5.5) we have that $r_{1,1} \cdot (q_1 + q_2) + r_{1,2} \cdot (q_1 + q_2)^2$ is odd. Since $r_{1,2} \cdot (q_1 + q_2)^2$ is even and $q_1 + q_2$ is odd, $r_{1,1}$ must be an odd number. Since $r_{1,1}$ is odd and $r_{1,2}$ is even, from Table 3.1 it results that $\rho_1(x)$ is a QPP modulo $2^{n_{L,2}}$. Since $L/2$ is even, it results that $r_{2,1} \equiv (r_{1,1} + L/2) \pmod{L}$ is odd and $r_{2,2} \equiv (r_{1,2} + L/2) \pmod{L}$ is even. Thus $\rho_2(x)$ is a QPP modulo $2^{n_{L,2}}$. Finally, from Theorem 3.8, $\rho_1(x)$ and $\rho_2(x)$ are QPPs modulo L. ■

Considering Lemmas 5.1 and 5.2, at least one QPP $\rho(x)$ exists that inverts any QPP $\pi(x)$ at three points $x = 0, 1, 2$. It does not necessarily mean that $\rho(x)$ is an inverse polynomial of $\pi(x)$.

Some exponents $n_{\rho,p}$s of r_2 obtained in Lemma 5.1 are determined by exponents $n_{\pi,p}$, as the next lemma shows.

Lemma 5.3 *Let the interleaver length be* $L = \prod_{p \in \mathcal{P}} p^{n_{L,p}}$. *Let* $\pi(x) = q_1 \cdot x + q_2 \cdot x^2 \pmod{L}$ *be a QPP and let* $\rho(x) = r_1 \cdot x + r_2 \cdot x^2 \pmod{L}$ *be a QPP as in Lemmas 5.1 and 5.2. Then,* $q_2 = \prod_{p \in \mathcal{P}} p^{n_{\pi,p}}$ *and* $r_2 = \prod_{p \in \mathcal{P}} p^{n_{\rho,p}}$ *satisfy the conditions from Table 3.1. Moreover, the following holds.*

Case (a) $2 \nmid L$ *(i.e.* $n_{L,2} = 0$*)*

If p is a factor of L (i.e. $n_{L,p} \geq 1$*), we have*

$$\begin{cases} n_{\rho,p} = n_{\pi,p}, & \text{if } 1 \leq n_{\pi,p} < n_{L,p} \\ n_{\rho,p} \geq n_{L,p}, & \text{if } n_{\pi,p} \geq n_{L,p} \end{cases} \tag{5.24}$$

Case (b) $2 \mid L$ *and* $4 \nmid L$ *(i.e.* $n_{L,2} = 1$*)*

$p = 2$ *is a factor of L, but for the moment we make no considerations how* $n_{\rho,2}$ *is determined by* $n_{\pi,2}$. *This will be explained in the proof of Theorem 5.6.*

If $p \neq 2$ *and L contains p as a factor (i.e.* $n_{L,p} \geq 1$*), we have*

$$\begin{cases} n_{\rho,p} = n_{\pi,p}, & \textit{if}\ 1 \le n_{\pi,p} < n_{L,p} \\ n_{\rho,p} \ge n_{L,p}, & \textit{if}\ n_{\pi,p} \ge n_{L,p} \end{cases} \tag{5.25}$$

Case (c) 4 | L *(i.e.* $n_{L,2} \ge 2$*)*
2^2 *is a factor of* L *(i.e.* $n_{L,2} \ge 2$*).*
If $p = 2$,

$$\begin{cases} n_{\rho,2} = n_{\pi,2}, & \textit{if}\ 1 \le n_{\pi,2} < n_{L,2} - 1 \\ n_{\rho,2} \ge n_{L,2} - 1, & \textit{if}\ n_{\pi,2} \ge n_{L,2} - 1 \end{cases} \tag{5.26}$$

If $p \neq 2$ *and* p *is a factor of* L *(i.e.* $n_{L,p} \ge 1$*), we have*

$$\begin{cases} n_{\rho,p} = n_{\pi,p}, & \textit{if}\ 1 \le n_{\pi,p} < n_{L,p} \\ n_{\rho,p} \ge n_{L,p}, & \textit{if}\ n_{\pi,p} \ge n_{L,p} \end{cases} \tag{5.27}$$

Proof Case (a) $2 \nmid L$

We consider that $n_{\pi,p} < n_{L,p}$ and $n_{\pi,p} < n_{\rho,p}$, where p is a prime number so that $p \mid L$. Reducing (5.16) modulo $p^{\min(n_{\rho,p},n_{L,p})}$, since $r_2 \equiv 0 \pmod{p^{\min(n_{\rho,p},n_{L,p})}}$, we have $0 \equiv (-q_2) \pmod{p^{\min(n_{\rho,p},n_{L,p})}}$. This is a contradiction since $n_{\pi,p} < \min(n_{\rho,p}, n_{L,p})$.

Now, we consider that $n_{\pi,p} < n_{L,p}$ and $n_{\rho,p} < n_{\pi,p}$. Reducing (5.16) modulo $p^{n_{\pi,p}}$, we have $r_2 \cdot q_1 \cdot q_1 \cdot q_1 \equiv 0 \pmod{p^{n_{\pi,p}}}$. But, from Table 3.1 $\gcd(q_1 \cdot q_1 \cdot q_1, p^{n_{\pi,p}}) = 1$, and then it would result that $r_2 = 0$, which is a contradiction. Therefore, $n_{\rho,p} = n_{\pi,p}$.

If $n_{\pi,p} \ge n_{L,p}$, reducing (5.16) modulo $p^{n_{L,p}}$, we have $r_2 \cdot q_1 \cdot q_1 \cdot q_1 \equiv 0 \pmod{p^{n_{L,p}}}$, which forces $n_{\rho,p} \ge n_{L,p}$, for $r_2 \neq 0$, since $\gcd(q_1 \cdot q_1 \cdot q_1, p^{n_{L,p}}) = 1$.

Case (b) $2 \mid L$ and $4 \nmid L$

The proof is similar to that in previous case, taking into account (5.15) instead of (5.16).

Case (c) $4 \mid L$

The proof for $p \neq 2$ is similar to that in case (a), taking into account (5.15) instead of (5.16).

Now we give the proof for $p = 2$.

Suppose that $n_{\pi,2} < n_{L,2} - 1$ and $n_{\pi,2} < n_{\rho,2}$. Reducing (5.15) modulo $2^{\min(n_{\rho,2},n_{L,2}-1)}$, since $r_2 \equiv 0 \left(\bmod\ 2^{\min(n_{\rho,2},n_{L,2}-1)}\right)$, we have $0 \equiv (-q_2)$ $\left(\bmod\ 2^{\min(n_{\rho,2},n_{L,2}-1)}\right)$ (at exponent appears $n_{L,2} - 1$, and not $n_{L,2}$, since (5.15) is evaluated modulo $L/2$). This is a contradiction since $n_{\pi,2} < \min(n_{\rho,2}, n_{L,2} - 1)$.

Now we suppose that $n_{\pi,2} < n_{L,2} - 1$ and $n_{\rho,2} < n_{\pi,2}$. Reducing (5.15) modulo $2^{n_{\pi,2}}$, we have $r_2 \cdot q_1 \cdot q_1 \cdot q_1 \equiv 0 \pmod{2^{n_{\pi,2}}}$. But, from Table 3.1, $\gcd(q_1 \cdot q_1 \cdot q_1, 2^{n_{\pi,2}}) = 1$, and then it would result that $r_2 = 0$, which is a contradiction. Thus $n_{\rho,2} = n_{\pi,2}$.

If $n_{\pi,2} \ge n_{L,2} - 1$, reducing (5.15) modulo $2^{n_{L,2}-1}$, we have $r_2 \cdot q_1 \cdot q_1 \cdot q_1 \equiv 0 \pmod{2^{n_{L,2}-1}}$, which forces $n_{\rho,2} \ge n_{L,2} - 1$, for $r_2 \neq 0$, since $\gcd(q_1 \cdot q_1 \cdot q_1, 2^{n_{L,2}-1}) = 1$. ■

Before proceeding further, we need the following lemma.

Lemma 5.4 *Let there be* $F(x) = f_1x + f_2x^2 + f_3x^3 + f_4x^4 \pmod{L}$ *and* $F(0) \equiv F(1) \equiv F(2) \equiv 0 \pmod{L}$. *Then* $F(x) \equiv 0 \pmod{L}$, $\forall x \in \mathbb{Z}_L$ *if and only if* $24f_4 \equiv 0 \pmod{L}$ *and* $6f_3 + 36f_4 \equiv 0 \pmod{L}$.

Proof "$\Rightarrow$" Define

$$F_0(x) = F(x) = f_1x + f_2x^2 + f_3x^3 + f_4x^4 \pmod{L} \tag{5.28}$$

and

$$F_n(x) = F_{n-1}(x+1) - F_{n-1}(x), \forall n \geq 1 \tag{5.29}$$

If $F(x) = F_0(x) \equiv 0 \pmod{L}$, $\forall x \in \mathbb{Z}_L$, then $F_n(x) \equiv 0 \pmod{L}$, $\forall x \in \mathbb{Z}_L$, $\forall n \geq 0$. We have

$$\begin{gathered}F_1(x) = F_0(x+1) - F_0(x) = f_1(x+1) + f_2(x+1)^2 + f_3(x+1)^3 + \\ + f_4(x+1)^4 - f_1x - f_2x^2 - f_3x^3 - f_4x^4 = \\ = f_1(x+1) + f_2(x^2+2x+1) + f_3(x^3+3x^2+3x+1) + \\ \\ + f_4(x^4+4x^3+6x^2+4x+1) - f_1x - f_2x^2 - f_3x^3 - f_4x^4 = \\ = (f_1+f_2+f_3+f_4) + (2f_2+3f_3+4f_4)x + \\ \\ +(3f_3+6f_4)x^2 + 4f_4x^3 \equiv 0 \pmod{L}\end{gathered} \tag{5.30}$$

$$\begin{gathered}F_2(x) = F_1(x+1) - F_1(x) = (f_1+f_2+f_3+f_4) + \\ +(2f_2+3f_3+4f_4)(x+1) + (3f_3+6f_4)(x+1)^2 + 4f_4(x+1)^3 - \\ -(f_1+f_2+f_3+f_4) - (2f_2+3f_3+4f_4)x - (3f_3+6f_4)x^2 - 4f_4x^3 = \\ = (2f_2+6f_3+14f_4) + (6f_3+24f_4)x + 12f_4x^2 \equiv 0 \pmod{L}\end{gathered} \tag{5.31}$$

$$\begin{gathered}F_3(x) = F_2(x+1) - F_2(x) = (2f_2+6f_3+14f_4) + \\ +(6f_3+24f_4)(x+1) + 12f_4(x+1)^2 - (2f_2+6f_3+14f_4) - \\ -(6f_3+24f_4)x - 12f_4x^2 = \\ \\ = (6f_3+36f_4) + 24f_4x \equiv 0 \pmod{L}\end{gathered} \tag{5.32}$$

In order to demonstrate $F_3(x) \equiv 0 \pmod{L}$, $\forall x \in \mathbb{Z}_L$, we must have

$$24f_4 \equiv 0 \pmod{L} \tag{5.33}$$

and

$$6f_3 + 36f_4 \equiv 0 \pmod{L} \tag{5.34}$$

"$\Leftarrow$" We define $F_0(x)$, $F_1(x)$, $F_2(x)$ and $F_3(x)$ as above. Then, we have $F_3(x) \equiv 0 \pmod{L}$, $\forall x \in \mathbb{Z}_L$, and $F_2(0) = F_1(1) - F_1(0) = F_0(2) - F_0(1) - F_0(1) +$

$F_0(0) = F(2) - 2F(1) + F(0) \equiv 0 \pmod{L}$. Using the mathematical induction method, we have $F_2(x) \equiv 0 \pmod{L}$, $\forall x \in \mathbb{Z}_L$, since $F_2(x+1) = F_2(x) + F_3(x)$. Similarly, $F_1(0) = F_0(1) - F_0(0) = F(1) - F(0) \equiv 0 \pmod{L}$ and $F_1(x) \equiv 0 \pmod{L}$, $\forall x \in \mathbb{Z}_L$, since $F_1(x+1) = F_1(x) + F_2(x)$. Finally $F(x) = F_0(x) \equiv 0 \pmod{L}$ as above. ■

The following theorem results as a combination of Lemmas 5.1 and 5.4.

Theorem 5.5 *Let $\pi(x)$ be a QPP and let $\rho(x)$ be a quadratic polynomial given in Lemma 5.1. $\rho(x)$ is a quadratic inverse polynomial of $\pi(x)$ if and only if $12q_2r_2 \equiv 0 \pmod{L}$.*

Proof We have $(\rho \circ \pi)(x) \equiv x \pmod{L}$ if and only if $\rho(x)$ is the quadratic inverse polynomial of $\pi(x)$.

$$\begin{aligned}(\rho \circ \pi)(x) = \rho(\pi(x)) = r_1 \cdot \pi(x) + r_2 \cdot (\pi(x))^2 \equiv \\ \equiv r_1 \cdot (q_1x + q_2x^2) + r_2 \cdot (q_1x + q_2x^2)^2 \equiv \\ \equiv r_1 \cdot (q_1x + q_2x^2) + r_2 \cdot (q_1^2x^2 + 2q_1q_2x^3 + q_2^2x^4) \equiv\end{aligned}$$

$$\equiv q_1r_1x + (q_2r_1 + q_1^2r_2)x^2 + 2q_1q_2r_2x^3 + q_2^2r_2x^4 \equiv x \pmod{L} \tag{5.35}$$

$\rho(x)$ results the quadratic inverse polynomial of $\pi(x)$ if and only if the following condition is satisfied:

$$(q_1r_1 - 1)x + (q_2r_1 + q_1^2r_2)x^2 + 2q_1q_2r_2x^3 + q_2^2r_2x^4 \equiv 0 \pmod{L} \tag{5.36}$$

Let there be

$$F(x) = (\rho \circ \pi)(x) - x =$$

$$= (q_1r_1 - 1)x + (q_2r_1 + q_1^2r_2)x^2 + 2q_1q_2r_2x^3 + q_2^2r_2x^4 \tag{5.37}$$

According to Lemma 5.1, we have

$$F(0) = \rho(\pi(0)) - 0 \equiv 0 \pmod{L} \tag{5.38}$$

$$F(1) = \rho(\pi(1)) - 1 \equiv 0 \pmod{L} \tag{5.39}$$

$$F(2) = \rho(\pi(2)) - 2 \equiv 0 \pmod{L} \tag{5.40}$$

According to Lemma 5.4, taking into account that $f_3 = 2q_1q_2r_2$ and $f_4 = q_2^2r_2$, $F(x) = 0 \pmod{L}$, $\forall x \in \mathbb{Z}_L$, if and only if

$$24q_2^2r_2 \equiv 0 \pmod{L} \tag{5.41}$$

and

$$12q_1q_2r_2 + 36q_2^2r_2 \equiv 0 \pmod{L} \tag{5.42}$$

The congruence (5.42) is equivalent to

$$12q_2r_2(q_1 + 3q_2) \equiv 0 \pmod{L} \tag{5.43}$$

which, considering Theorem 57 in Hardy and Wright (1975), is true only if $12q_2r_2 \equiv 0 \pmod{L}$, since $\gcd(q_1 + 3q_2, L) = 1$, from the conditions in Table 3.1. Since for $12q_2r_2 \equiv 0 \pmod{L}$, (5.41) trivially holds, the congruences (5.41) and (5.42) can be reduced to

$$12q_2r_2 \equiv 0 \pmod{L} \tag{5.44}$$

■

The next theorem states the necessary and sufficient condition for the existence of a quadratic inverse polynomial for a QPP $\pi(x)$, by simply checking the inequalities involving the exponents for the prime factors of L and the second degree coefficient of $\pi(x)$.

Theorem 5.6 *Let the interleaver length be* $L = \prod_{p\in\mathcal{P}} p^{n_{L,p}}$. *Let* $\pi(x) = q_1 \cdot x + q_2 \cdot x^2 \pmod{L}$ *be a QPP and let* $q_2 = \prod_{p\in\mathcal{P}} p^{n_{\pi,p}}$ *be the second degree coefficient of* $\pi(x)$. *Then,* $\pi(x)$ *has at least one quadratic inverse polynomial if and only if*

$$n_{\pi,2} \geq \begin{cases} \max\left(\left\lceil \frac{n_{L,2}-2}{2} \right\rceil, 1\right), & \text{if } n_{L,2} > 1 \\ 0, & \text{if } n_{L,2} = 0, 1 \end{cases} \tag{5.45}$$

$$n_{\pi,3} \geq \begin{cases} \max\left(\left\lceil \frac{n_{L,3}-1}{2} \right\rceil, 1\right), & \text{if } n_{L,3} > 0 \\ 0, & \text{if } n_{L,3} = 0 \end{cases} \tag{5.46}$$

$$n_{\pi,p} \geq \left\lceil \frac{n_{L,p}}{2} \right\rceil, \text{ if } p \neq 2, 3 \tag{5.47}$$

Proof "$\Rightarrow$" Considering Lemmas 5.1 and 5.2, there exists at least one QPP $\rho(x) = r_1 \cdot x + r_2 \cdot x^2 \pmod{L}$ that inverts a QPP $\pi(x)$ at three points $x = 0, 1, 2$. As $\rho(x)$ must invert $\pi(x)$ at these points, we only need to check whether $\rho(x)$ is a quadratic inverse polynomial or not.

Now we show that if $\rho(x)$ is a quadratic inverse polynomial, then the condition on $n_{\pi,p}$ holds for $p = 2$.

If $n_{L,2} = 0, 1$, irrespective of whether $\rho(x)$ is or not a quadratic inverse, $n_{\pi,2} \geq 0$ trivially holds and this is why we do not need to determine $n_{\rho,2}$ in Lemma 5.3, case (b), when $n_{L,2} = 1$.

If $n_{L,2} = 2, 3, 4$ we have $\max\left(\left\lceil \frac{n_{L,2}-2}{2} \right\rceil, 1\right) = 1$. Since $\pi(x)$ is a QPP, from Table 3.1 it results that $n_{\pi,2} \geq 1$ and thus the condition in Theorem 5.6 is satisfied.

Assume that $\rho(x)$ is a quadratic inverse polynomial, but

$$n_{\pi,2} < \left\lceil \frac{n_{L,2}-2}{2} \right\rceil, \text{ for } n_{L,2} \geq 5 \tag{5.48}$$

Because $\rho(x)$ is a quadratic inverse polynomial, according to Theorem 5.5, $12q_2r_2 \equiv 0 \pmod{L}$ holds, i.e.

$$\left(\prod_{p\in\mathcal{P}} p^{n_{L,p}}\right) \mid \left(2^2 \cdot 3 \cdot \prod_{p\in\mathcal{P}} p^{n_{\pi,p}} \cdot \prod_{p\in\mathcal{P}} p^{n_{\rho,p}}\right) \tag{5.49}$$

We consider two cases

(1) $n_{L,2}$ is odd

$\left\lceil \frac{n_{L,2}-2}{2} \right\rceil = \frac{n_{L,2}-1}{2}$ and consequently, $n_{\pi,2} \leq \frac{n_{L,2}-1}{2} - 1 = \frac{n_{L,2}-3}{2} < n_{L,2} - 1$.

Considering Lemma 5.3, $n_{\rho,2} = n_{\pi,2}$, thus $2 + n_{\pi,2} + n_{\rho,2} \leq n_{L,2} - 1 < n_{L,2}$, which is a contradiction, since $L \mid (12q_2r_2)$ implies $n_{L,2} \leq 2 + n_{\pi,2} + n_{\rho,2}$.

(2) $n_{L,2}$ is even

$\left\lceil \frac{n_{L,2}-2}{2} \right\rceil = \frac{n_{L,2}-2}{2}$ and consequently, $n_{\pi,2} \leq \frac{n_{L,2}-2}{2} - 1 = \frac{n_{L,2}-4}{2} < n_{L,2} - 1$.

Considering Lemma 5.3, $n_{\rho,2} = n_{\pi,2}$, thus $2 + n_{\pi,2} + n_{\rho,2} \leq n_{L,2} - 2 < n_{L,2}$, which is a contradiction, since $L \mid (12q_2r_2)$ implies $n_{L,2} \leq 2 + n_{\pi,2} + n_{\rho,2}$.

Now we show that if $\rho(x)$ is a quadratic inverse polynomial, then the condition on $n_{\pi,p}$ for $p = 3$ holds.

If $n_{L,3} = 0$, irrespective of $\rho(x)$ is a quadratic inverse or not, $n_{\pi,3} \geq 0$ trivially holds.

If $n_{L,3} = 1, 2, 3$ we have $\max\left(\left\lceil \frac{n_{L,3}-1}{2} \right\rceil, 1\right) = 1$. Since $\pi(x)$ is a QPP, from Table 3.1 it results that $n_{\pi,3} \geq 1$ and thus the condition in Theorem 5.6 is satisfied.

Suppose that $\rho(x)$ is a quadratic inverse polynomial, but

$$n_{\pi,3} < \left\lceil \frac{n_{L,3}-1}{2} \right\rceil, \text{ for } n_{L,2} \geq 4 \tag{5.50}$$

Since $\rho(x)$ is a quadratic inverse polynomial, according to Theorem 5.5 $12q_2r_2 \equiv 0 \pmod{L}$ holds, i.e. (5.49) is true.

Then, when

(1) $n_{L,3}$ is odd

$\left\lceil \frac{n_{L,3}-1}{2} \right\rceil = \frac{n_{L,3}-1}{2}$, thus, $n_{\pi,3} \leq \frac{n_{L,3}-1}{2} - 1 = \frac{n_{L,3}-3}{2} < n_{L,3}$.

Taking into account Lemma 5.3, $n_{\rho,3} = n_{\pi,3}$, thus $1 + n_{\pi,3} + n_{\rho,3} \leq n_{L,3} - 2 < n_{L,3}$, which is a contradiction, since $L \mid (12q_2r_2)$ implies $n_{L,3} \leq 1 + n_{\pi,3} + n_{\rho,3}$.

(2) $n_{L,3}$ is even

$\left\lceil \frac{n_{L,3}-1}{2} \right\rceil = \frac{n_{L,3}}{2}$, thus, $n_{\pi,3} \leq \frac{n_{L,3}}{2} - 1 = \frac{n_{L,3}-2}{2} < n_{L,3}$.

Taking into account Lemma 5.3, $n_{\rho,3} = n_{\pi,3}$, thus $1 + n_{\pi,3} + n_{\rho,3} \leq n_{L,3} - 1 < n_{L,3}$, which is a contradiction, since $L \mid (12q_2r_2)$ implies $n_{L,3} \leq 1 + n_{\pi,3} + n_{\rho,3}$.

Finally, we show that if $\rho(x)$ is a quadratic inverse polynomial, then the condition on $n_{\pi,p}$ for $p \neq 2, 3$ holds.

In this case, we have $n_{L,p} \geq 1$.

If $n_{L,p} = 1, 2$, we have $\left\lceil \frac{n_{L,p}}{2} \right\rceil = 1$. Since $\pi(x)$ is a QPP from Table 3.1, it results that $n_{\pi,p} \geq 1$ and thus the condition in Theorem 5.6 is satisfied.

Assume that $\rho(x)$ is a quadratic inverse polynomial, but

$$n_{\pi,p} < \left\lceil \frac{n_{L,p}}{2} \right\rceil, \text{ for } n_{L,p} \geq 3 \tag{5.51}$$

Since $\rho(x)$ is a quadratic inverse polynomial, considering Theorem 5.5 $12q_2r_2 \equiv 0 \pmod{L}$ holds, i.e. (5.49) is true.

Then when

(1) $n_{L,p}$ is odd

$\left\lceil \frac{n_{L,p}}{2} \right\rceil = \frac{n_{L,p}+1}{2}$, thus, $n_{\pi,p} \leq \frac{n_{L,p}+1}{2} - 1 = \frac{n_{L,p}-1}{2} < n_{L,p}$.

Taking into account Lemma 5.3, $n_{\rho,p} = n_{\pi,p}$, thus $n_{\pi,p} + n_{\rho,p} \leq n_{L,p} - 1 < n_{L,p}$, which is a contradiction, since $L \mid (12q_2r_2)$ implies $n_{L,p} \leq n_{\pi,p} + n_{\rho,p}$.

(2) $n_{L,p}$ is even

$\left\lceil \frac{n_{L,p}}{2} \right\rceil = \frac{n_{L,p}}{2}$, thus, $n_{\pi,p} \leq \frac{n_{L,p}}{2} - 1 = \frac{n_{L,p}-2}{2} < n_{L,p}$.

Taking into account Lemma 5.3, $n_{\rho,p} = n_{\pi,p}$, thus $n_{\pi,p} + n_{\rho,p} \leq n_{L,p} - 2 < n_{L,p}$, which is a contradiction, since $L \mid (12q_2r_2)$ implies $n_{L,p} \leq n_{\pi,p} + n_{\rho,p}$.

"$\Leftarrow$" We show that if the conditions on $n_{\pi,p}$ in theorem hold, then $12q_2r_2 \equiv 0 \pmod{L}$, i.e. (5.49) is true. We will separately show that (5.49) holds for $p = 2$, for $p = 3$ and for $p \neq 2, 3$ so that $p \mid L$.

For $p = 2$ we have to show that $2^{n_{L,2}} \mid \left(2^2 \cdot 2^{n_{\pi,2}} \cdot 2^{n_{\rho,2}}\right)$.

We consider three cases.

(1) $n_{L,2} = 0, 1$

If $n_{\pi,2} \geq 0$, then $n_{L,2} \leq 2 + n_{\pi,2}$. Thus, $2^{n_{L,2}} \mid \left(2^2 \cdot 2^{n_{\pi,2}} \cdot 2^{n_{\rho,2}}\right)$.

(2) $n_{L,2} = 2, 3, 4$

If $n_{\pi,2} \geq \max\left(\left\lceil \frac{n_{L,2}-2}{2} \right\rceil, 1\right) = 1$, then as it is required by Lemma 5.3, $n_{\rho,2} \geq 1$. Thus, $n_{L,2} \leq 2 + n_{\pi,2} + n_{\rho,2}$ holds and, consequently, $2^{n_{L,2}} \mid \left(2^2 \cdot 2^{n_{\pi,2}} \cdot 2^{n_{\rho,2}}\right)$.

(3) $n_{L,2} \geq 5$

In this case, $\left\lceil \frac{n_{L,2}-2}{2} \right\rceil > 1$.

Considering Lemma 5.3, if $n_{L,2} - 1 > n_{\pi,2} \geq \left\lceil \frac{n_{L,2} - 2}{2} \right\rceil$, then $n_{\rho,2} = n_{\pi,2}$. Consequently, if $n_{L,2}$ is even, $\left\lceil \frac{n_{L,2} - 2}{2} \right\rceil = \frac{n_{L,2} - 2}{2}$, and then $2 + n_{\pi,2} + n_{\rho,2} = 2 + 2 \cdot n_{\pi,2} \geq 2 + n_{L,2} - 2 = n_{L,2}$. If $n_{L,2}$ is odd, $\left\lceil \frac{n_{L,2} - 2}{2} \right\rceil = \frac{n_{L,2} - 1}{2}$, and then $2 + n_{\pi,2} + n_{\rho,2} = 2 + 2 \cdot n_{\pi,2} \geq 2 + n_{L,2} - 1 = n_{L,2} + 1 > n_{L,2}$. Therefore, $2^{n_{L,2}} \mid \left(2^2 \cdot 2^{n_{\pi,2}} \cdot 2^{n_{\rho,2}}\right)$.

If $n_{\pi,2} \geq n_{L,2} - 1$, considering Lemma 5.3, $n_{\rho,2} \geq n_{L,2} - 1$. Thus $2 + n_{\pi,2} + n_{\rho,2} \geq 2 \cdot n_{L,2} > n_{L,2}$, and, consequently, $2^{n_{L,2}} \mid \left(2^2 \cdot 2^{n_{\pi,2}} \cdot 2^{n_{\rho,2}}\right)$.

For $p = 3$ we have to show that $3^{n_{L,3}} \mid \left(3 \cdot 3^{n_{\pi,3}} \cdot 3^{n_{\rho,3}}\right)$.

We consider three cases.

(1) $n_{L,3} = 0$

If $n_{\pi,3} \geq 0$, then $n_{L,3} \leq 1 + n_{\pi,3}$. Thus, $3^{n_{L,3}} \mid \left(3 \cdot 3^{n_{\pi,3}} \cdot 3^{n_{\rho,3}}\right)$.

(2) $n_{L,3} = 1, 2, 3$

If $n_{\pi,3} \geq \max\left(\left\lceil \frac{n_{L,3}-1}{2} \right\rceil, 1\right) = 1$, then as it is required by Lemma 5.3, $n_{\rho,3} \geq 1$. Thus, $n_{L,3} \leq 1 + n_{\pi,3} + n_{\rho,3}$ holds and, consequently, $3^{n_{L,3}} \mid \left(3 \cdot 3^{n_{\pi,3}} \cdot 3^{n_{\rho,3}}\right)$.

(3) $n_{L,3} \geq 4$

In this case, $\left\lceil \frac{n_{L,3} - 1}{2} \right\rceil > 1$.

Considering Lemma 5.3, if $n_{L,3} > n_{\pi,3} \geq \left\lceil \frac{n_{L,3} - 1}{2} \right\rceil$, then $n_{\rho,3} = n_{\pi,3}$. Consequently, if $n_{L,3}$ is even, $\left\lceil \frac{n_{L,3} - 1}{2} \right\rceil = \frac{n_{L,3}}{2}$, and then $1 + n_{\pi,3} + n_{\rho,3} = 1 + 2 \cdot n_{\pi,3} \geq 1 + n_{L,3} > n_{L,3}$. If $n_{L,3}$ is odd, $\left\lceil \frac{n_{L,3} - 1}{2} \right\rceil = \frac{n_{L,3} - 1}{2}$, and then $1 + n_{\pi,3} + n_{\rho,3} = 1 + 2 \cdot n_{\pi,3} \geq 1 + n_{L,3} - 1 = n_{L,3}$. Thus $3^{n_{L,3}} \mid \left(3 \cdot 3^{n_{\pi,3}} \cdot 3^{n_{\rho,3}}\right)$.

For $p \mid L$ and $p \neq 2, 3$, we have to show that $p^{n_{L,p}} \mid \left(p^{n_{\pi,p}} \cdot p^{n_{\rho,p}}\right)$.

We consider two cases.

(1) $n_{L,p} = 1, 2$

If $n_{\pi,p} \geq \left\lceil \frac{n_{L,p}}{2} \right\rceil = 1$, then as it is required by Lemma 5.3, $n_{\rho,p} \geq 1$. Thus, $n_{L,p} \leq n_{\pi,p} + n_{\rho,p}$ holds and, consequently, $p^{n_{L,p}} \mid \left(p^{n_{\pi,p}} \cdot p^{n_{\rho,p}}\right)$.

(2) $n_{L,p} \geq 3$

In this case, $\left\lceil \frac{n_{L,p}}{2} \right\rceil > 1$.

Considering Lemma 5.3, if $n_{L,p} > n_{\pi,p} \geq \left\lceil \frac{n_{L,p}}{2} \right\rceil$, then $n_{\rho,p} = n_{\pi,p}$. Consequently, if $n_{L,p}$ is even, $\left\lceil \frac{n_{L,p}}{2} \right\rceil = \frac{n_{L,p}}{2}$, and then $n_{\pi,p} + n_{\rho,p} = 2 \cdot n_{\pi,p} \geq n_{L,p}$. If $n_{L,p}$ is odd, $\left\lceil \frac{n_{L,p}}{2} \right\rceil = \frac{n_{L,p} + 1}{2}$, and then $n_{\pi,p} + n_{\rho,p} = 2 \cdot n_{\pi,p} \geq n_{L,p} + 1 > n_{L,p}$. Thus $p^{n_{L,p}} \mid \left(p^{n_{\pi,p}} \cdot p^{n_{\rho,p}}\right)$.

If $n_{\pi,p} \geq n_{L,p}$, considering Lemma 5.3, $n_{\rho,p} \geq n_{L,p}$. Thus $n_{\pi,p} + n_{\rho,p} \geq 2 \cdot n_{L,p} > n_{L,p}$, and, consequently, $p^{n_{L,p}} \mid \left(p^{n_{\pi,p}} \cdot p^{n_{\rho,p}}\right)$. ■

Algorithm 1: An algorithm for finding the inverse QPP(s) for a QPP $\pi(x) = q_1 \cdot x + q_2 \cdot x^2 \pmod{L}$

---

1 Factor L and q_2 as products of prime powers and find the respective exponents of each prime factor (i.e., find $n_{L,p}$'s and $n_{\pi,p}$'s for $L = \prod_{p \in \mathcal{P}} p^{n_{L,p}}$ and $q_2 = \prod_{p \in \mathcal{P}} p^{n_{\pi,p}}$);

2 Using the $n_{L,p}$'s and $n_{\pi,p}$'s obtained in step 1, determine if they satisfy the inequalities in Theorem 5.6;

3 **if** *yes* **then**

4   Check if L is an odd or an even number;

5   **if** *L is an odd number,* **then**

6   There is exactly one inverse QPP for $\pi(x)$;

7   Let the inverse QPP be $\rho(x) = r_1 \cdot x + r_2 \cdot x^2 \pmod{L}$;

8         $r_2 \equiv \{(-q_2) \cdot [(q_1 + q_2)(q_1 + 2q_2)(q_1 + 3q_2)]^{-1}\} \pmod{L}$, where $(\cdot)^{-1} \pmod{L}$ is given in Algorithm 2;

9         $r_1 \equiv \{[1 - r_2(q_1 + q_2)^2] \cdot (q_1 + q_2)^{-1}\} \pmod{L}$;

10   **return** $\rho(x)$ and the algorithm ends;

11   **else**

12   **if** *L is an even number,* **then**

13   There are exactly two inverse QPPs for $\pi(x)$;

14   Let the inverse QPPs be $\rho_1(x) = r_{1,1} \cdot x + r_{1,2} \cdot x^2 \pmod{L}$ and $\rho_2(x) = r_{2,1} \cdot x + r_{2,2} \cdot x^2 \pmod{L}$, respectively;

15             $r_{1,2} \equiv \{(-q_2) \cdot [(q_1 + q_2)(q_1 + 2q_2)(q_1 + 3q_2)]^{-1}\}\left(\bmod \frac{L}{2}\right)$;

16             $r_{1,1} \equiv \{[1 - r_{1,2} \cdot (q_1 + q_2)^2] \cdot (q_1 + q_2)^{-1}\} \pmod{L}$;

17             $(r_{2,1}, r_{2,2})$ is obtained by $r_{2,1} \equiv r_{1,1} + \frac{L}{2} \pmod{L}$, $r_{2,2} \equiv r_{1,2} + \frac{L}{2} \pmod{L}$;

18   **return** $\rho_1(x)$ and $\rho_2(x)$ and the algorithm ends;

19   **end**

20   **end**

21 **else**

22   There exists no inverse QPP for $\pi(x)$;

23   **return** no inverse QPP and the algorithm ends;

24 **end**

An algorithm for finding the inverse QPP(s) for a QPP $\pi(x) = q_1 \cdot x + q_2 \cdot x^2 \pmod{L}$ is Algorithm 1.

Now we give, without proof, a theorem from Ryu (2007), Ryu and Takeshita (2011) which allows us to determine all inverse permutation polynomials of a QPP, with the least degree d_{2-inv}. We consider a QPP described by Eq. (3.37) and

$$\rho(x) = \pi^{-1}(x) = r_1 \cdot x + r_2 \cdot x^2 + \cdots + r_{d_{2-inv}} \cdot x^{d_{2-inv}} \pmod{L} \tag{5.52}$$

its inverse polynomial of degree d_{2-inv}.

Algorithm 2: An algorithm for finding the arithmetic inverse s^{-1} for $s \pmod{L}$

$s^{-1} = 1;$
$r = 0;$
while $L \neq 0$ **do**
  $c = s \pmod{L};$
  $q = \left\lfloor \frac{s}{L} \right\rfloor;$
  $s = L;$
  $L = c;$
  $r' = s^{-1} - q \cdot r;$
  $s^{-1} = r;$
  $r = r';$
end
return s^{-1}.

Theorem 5.7 (Ryu 2007) *Let the interleaver length be* $L = \prod_{p \in \mathcal{P}} p^{n_{L,p}} \leq 2^{50}$ *and let it be* $\phi(k) = \prod_{l=k}^{2k-2} l$. *Let* $\pi(x) = q_1 x + q_2 x^2 \pmod{L}$ *be a QPP.*

Decompose $\phi(k)$ *into prime factors and denote each exponent of prime factors as* $n_{\phi(k),p}$. *Then* $\pi(x)$ *has* $\prod_{k=1}^{d_{2-inv}} \gcd(k!, L)$ *inverse polynomials with the least degree* d_{2-inv} *if and only if there is a smallest integer* d_{2-inv} *so that*

$$n_{\pi,2} \geq \begin{cases} \max\left(\left\lceil \frac{n_{L,2} - n_{\phi(d_{2-inv}+1),2}}{d_{2-inv}} \right\rceil, 1\right), & \text{if } n_{L,2} > 1 \\ 0, & \text{if } n_{L,2} = 0, 1 \end{cases} \tag{5.53}$$

$$n_{\pi,p} \geq \begin{cases} \max\left(\left\lceil \frac{n_{L,p} - n_{\phi(d_{2-inv}+1),p}}{d_{2-inv}} \right\rceil, 1\right), & \text{if } n_{L,p} > 0 \\ 0, & \text{if } n_{L,p} = 0 \end{cases} \tag{5.54}$$

We note that, if $d_{2-inv} = 2$, Theorem 5.7 is reduced to Theorem 5.6.

5.2 Upper Bounds on Minimum Distance of Turbo Codes with QPP-Based Interleavers

In 2006, in Rosnes and Takeshita (1992), Eirik Rosnes and Oscar Y. Takeshita tabulated optimum QPPs (in terms of the induced minimum distance and its corresponding multiplicity) for a large number of short interleaver lengths between 32 and 512 and for conventional nominal rate-1/3 binary turbo codes with 8-state constituent encoders (the turbo code from LTE standard 3GPP 2008) and 16-state constituent encoders (the turbo code from CCSDS standard 2002).

In 2012 Eirik Rosnes continued the study reported in Rosnes and Takeshita (1992) and gave some partial upper bounds on the minimum distance of turbo codes with QPP-based interleavers dependent on the interleaver length and an general upper bound independent on the interleaver length when QPP interleavers admit an inverse QPP interleaver.

These upper bounds can be used in the selection of lengths of good QPP-based interleavers and also to efficiently reject bad QPP candidates in a computer search. For all partial upper bounds on minimum distance, we consider that the generator matrix of the component convolutional codes of the turbo code is $G = [1, 15/13]$ (in octal form) and the turbo code is unpunctured (i.e. the turbo code used in the LTE standard 3GPP 2008).

These upper bounds are given in Theorems 5.8–5.13 taken from Rosnes (2012), but with more detailed proofs. They were obtained by identifying certain critical *interleaver patterns*. An interleaver pattern $I(\Gamma)$ is a set $\{(i, \pi(i) : i \in \Gamma)\}$, where $\pi(\cdot)$ indicates the interleaver and Γ is a subset of $\{0, 1, \ldots, L-1\}$. The interleaver pattern *size* is the same as that of Γ. There are *critical* interleaver patterns that generate small weight codewords for certain generator matrices of component codes of the turbo code. As a result, the turbo codeword has ones in the bit positions in Γ and zeros elsewhere. For this reason, the bit positions in Γ are called systematic 1-positions in the upper constituent codeword, while the interleaved bit positions in Γ are called systematic 1-positions in the lower constituent codeword. The term *systematic* refers to the information sequence part of the codeword. Figure 5.1 shows an interleaver pattern of size six. In this figure, the points $x_1, x_1 + a, x_2, x_2 + a, x_3, x_3 + a$ in the upper layer represent the positions in Γ and the points $\pi(x_1), \pi(x_1 + a), \pi(x_2), \pi(x_2 + a), \pi(x_3), \pi(x_3 + a)$ in the lower layer represent the positions in $\pi(\Gamma)$. The arrows represent the actual interleaver operation and the arcs represent a grouping of bit positions that generate an *error event* or a *fundamental path* in a constituent trellis for certain generator matrices of component codes of the turbo code. The index differences of bit positions from Γ or $\pi(\Gamma)$, are indicated in figures by horizontal lines. In Fig. 5.1, the horizontal line above the first arc (or fundamental path) in the upper constituent codeword indicates that the difference between the bit positions x_1 and $x_1 + a$ is a. In the same way, the first horizontal line below the first arc in the lower constituent codeword indicates that the difference between the bit positions $\pi(x_1)$ and $\pi(x_2)$ is b and the second horizontal line below the first arc in the lower constituent codeword indicates that the difference between the bit positions $\pi(x_1)$ and $\pi(x_3)$ is c.

All the critical interleaver patterns used in the derivation of the upper bounds in Rosnes (2012) were identified by using a tailored version of the triple impulse method (Crozier et al. 2004). This modified version gives the information words leading to low weight codewords. Consequently, knowing the information words, we know the positions of bits 1.

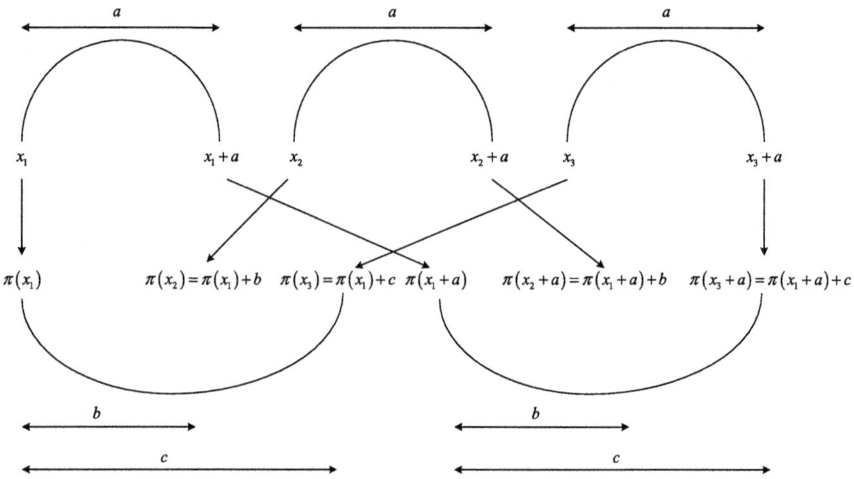

Fig. 5.1 Critical interleaver pattern of size six for QPP-based interleavers

The prime decomposition of the interleaver length is of the form $L = \prod_{i=1}^{n_L} p_{L,i}^{n_{L,p_i}}$, where $n_L \in \mathbb{N}^*$, $p_{L,i}$, with $i = \overline{1, n_L}$, are different prime numbers and $n_{L,p_i} \in \mathbb{N}^*$, $\forall i = \overline{1, n_L}$, as in Sect. 3.6.

Theorem 5.8 *The minimum distance of a conventional binary turbo code (of nominal rate 1/3) with generator matrix* $G = [1, 15/13]$ *is upper bounded by*

$$d_{min} \leq 38 + 12 \cdot l \tag{5.55}$$

for all nonnegative integers l and for all interleaver lengths L that satisfy

$$n_{L,p} \leq \begin{cases} l+4, & \text{if } p = 2 \\ 2, & \text{if } p = 7 \\ 1, & \text{otherwise} \end{cases} \tag{5.56}$$

when using QPP interleavers.

Proof Figure 5.1 displays an interleaver pattern of size six. The interleaver pattern generates an upper constituent codeword containing 3 input-weight 2 fundamental paths, and a lower constituent codeword containing 2 input-weight 3 fundamental paths. As it was explained previously, the interleaving of the systematic 1-positions of the upper constituent codeword is indicated by arrows in the figure. For the input-weight 2 fundamental paths in the upper constituent codeword, the index difference between the second and first systematic 1-positions is denoted by a. For the input-weight 3 fundamental paths in the lower constituent codeword, the index differences between the second and first systematic 1-positions and the third and first systematic 1-positions are denoted by b and c, respectively. For convenience, these index differences are also indicated in Fig. 5.1 by horizontal lines.

The six elements of permutation $\pi(\cdot)$ indicated in Fig. 5.1 are written in detail below

$$\begin{cases} x_1 \to \pi(x_1) \\ x_1 + a \to \pi(x_1 + a) \\ x_2 \to \pi(x_2) = \pi(x_1) + b \\ x_2 + a \to \pi(x_2 + a) = \pi(x_1 + a) + b \\ x_3 \to \pi(x_3) = \pi(x_1) + c \\ x_3 + a \to \pi(x_3 + a) = \pi(x_1 + a) + c \end{cases} \tag{5.57}$$

Writing $x_1 = \rho(\pi(x_1))$, $x_2 = \rho(\pi(x_2))$ and $x_3 = \rho(\pi(x_3))$, the equations corresponding to points $x_2 + a$ and $x_3 + a$ in (5.57) can be written

$$\begin{cases} \pi(\rho(\pi(x_2)) + a) = \pi(\rho(\pi(x_1)) + a) + b \pmod{L} \\ \pi(\rho(\pi(x_3)) + a) = \pi(\rho(\pi(x_1)) + a) + c \pmod{L} \end{cases} \tag{5.58}$$

Considering the equations corresponding to points x_2 and x_3 from (5.57), (5.58) becomes:

$$\begin{cases} \pi(\rho(\pi(x_1) + b) + a) = \pi(\rho(\pi(x_1)) + a) + b \pmod{L} \\ \pi(\rho(\pi(x_1) + c) + a) = \pi(\rho(\pi(x_1)) + a) + c \pmod{L} \end{cases} \tag{5.59}$$

With $\pi(x_1) = x$ in (5.59), we have

$$\begin{cases} \pi(\rho(x + b) + a) = \pi(\rho(x) + a) + b \pmod{L} \\ \pi(\rho(x + c) + a) = \pi(\rho(x) + a) + c \pmod{L} \end{cases} \tag{5.60}$$

where $x \in \mathbb{Z}_L$ denotes the first systematic 1-position (the point x_1 in Fig. 5.1). When $x = 0$, the two congruences in (5.60) are equivalent to

$$\begin{cases} \pi(\rho(b) + a) = \pi(\rho(0) + a) + b = \pi(a) + b \pmod{L} \\ \pi(\rho(c) + a) = \pi(\rho(0) + a) + c = \pi(a) + c \pmod{L} \end{cases} \Leftrightarrow \tag{5.61}$$

$$\begin{cases} q_1 \cdot \rho(b) + q_1 \cdot a + q_2 \cdot (\rho(b))^2 + 2 \cdot a \cdot q_2 \cdot \rho(b) + q_2 \cdot a^2 = \\ = \pi(a) + b \pmod{L} \\ q_1 \cdot \rho(c) + q_1 \cdot a + q_2 \cdot (\rho(c))^2 + 2 \cdot a \cdot q_2 \cdot \rho(c) + q_2 \cdot a^2 = \\ = \pi(a) + c \pmod{L} \end{cases} \Leftrightarrow \tag{5.62}$$

$$\begin{cases} \underbrace{\pi(\rho(b))}_{=b} + \pi(a) + 2 \cdot a \cdot q_2 \cdot \rho(b) = \pi(a) + b \pmod{L} \\ \underbrace{\pi(\rho(c))}_{=c} + \pi(a) + 2 \cdot a \cdot q_2 \cdot \rho(c) = \pi(a) + c \pmod{L} \end{cases} \Leftrightarrow \tag{5.63}$$

$$\begin{cases} 2 \cdot a \cdot q_2 \cdot \rho(b) = 0 \pmod{L} \\ 2 \cdot a \cdot q_2 \cdot \rho(c) = 0 \pmod{L} \end{cases} \tag{5.64}$$

With (5.52), (5.64) is written as

$$\begin{cases} 2 \cdot a \cdot b \cdot q_2 \cdot (r_1 + r_2 \cdot b + \cdots + r_{d_{2\text{-}inv}} \cdot b^{d_{2\text{-}inv}-1}) = 0 \pmod{L} \\ 2 \cdot a \cdot c \cdot q_2 \cdot (r_1 + r_2 \cdot c + \cdots + r_{d_{2\text{-}inv}} \cdot c^{d_{2\text{-}inv}-1}) = 0 \pmod{L} \end{cases} \tag{5.65}$$

The two congruences from (5.65) are satisfied if

$$\begin{cases} 2 \cdot a \cdot b \cdot q_2 = 0 \pmod{L} \\ 2 \cdot a \cdot c \cdot q_2 = 0 \pmod{L} \end{cases} \tag{5.66}$$

For $a = 2^l \cdot 7$, with $l \in \mathbb{N}$, and $b = 8$ and $c = 12$ in (5.66), we have

$$\begin{cases} 2^{l+4} \cdot 7 \cdot q_2 = 0 \pmod{L} \\ 2^{l+3} \cdot 3 \cdot 7 \cdot q_2 = 0 \pmod{L} \end{cases} \tag{5.67}$$

Then the two congruences from (5.67) are fulfilled only if

$$\begin{cases} n_{L,2} \leq n_{q_2,2} + l + 3 \\ n_{L,7} \leq n_{q_2,7} + 1 \\ n_{L,p} \leq n_{q_2,p}, \forall p \neq 2, 7 \end{cases} \tag{5.68}$$

For $n_{L,2} = 1$, the first inequality from (5.68) is always satisfied. From conditions 1.b) and 2) in Table 3.1, we have $n_{q_2,2} \geq 1$ when $n_{L,2} \geq 2$, and $n_{q_2,p} \geq 1$ when $n_{L,p} \geq 1$, for $p > 2$. Thus, the conditions from (5.68) are equivalent to

$$\begin{cases} n_{L,2} \leq l + 4 \\ n_{L,7} \leq 2 \\ n_{L,p} \leq 1, \forall p \neq 2, 7 \end{cases} \tag{5.69}$$

The conditions from (5.69) are the same as those from (5.56).

The Hamming weight of the turbo codeword generated by the interleaver pattern in Fig. 5.1 can be computed as the sum of the parity weight of the upper constituent codeword, the parity weight of the lower constituent codeword, and the input weight.

The parity weight for an input-weight 2 sequence, as those from fundamental paths of the upper constituent codeword, with $a = 2^l \cdot 7$, $l \in \mathbb{N}$, results from Table 5.1, where a constituent RSC code of LTE turbo code was considered. In Table 5.1 k is the instant time, u_k is the input bit to the RSC encoder at time k, c_k is the parity output bit at time k, and $B_1^{(k)}$, $B_2^{(k)}$, and $B_3^{(k)}$ are the contents of the three memory cells of the RSC encoder at time k. From Table 5.1 we remark that, if $l = 0$, then the parity weight (i.e. the weight of sequence c_k) is equal to 6 = 1+4+1, and if $l = 1$, the parity weight is equal to 10 = 1+4+4+1. In general, for $l \in \mathbb{N}$, the parity weight is equal to $1 + (l + 1) \cdot 4 + 1 = (l + 1) \cdot 4 + 2 = 4 \cdot l + 6$. Similarly, for an RSC encoder

Table 5.1 Determining the parity weight for an input-weight 2 sequence with the two bits 1 separated by $a = 2^l \cdot 7$ positions

k	u_k	$B_1^{(k)}$	$B_2^{(k)}$	$B_3^{(k)}$	c_k
1	1	0	0	0	1
2	0	1	0	0	1
3	0	0	1	0	1
4	0	1	0	1	1
5	0	1	1	0	0
6	0	1	1	1	0
7	0	0	1	1	1
8	0	0	0	1	0
9	0	1	0	0	1
10	0	0	1	0	1
11	0	1	0	1	1
12	0	1	1	0	0
13	0	1	1	1	0
14	0	0	1	1	1
15	1/0	0	0	1	1/0
16	0	0/1	0	0	0/1

with a primitive feedback polynomial of arbitrary degree ν and a monic feedforward polynomial, and for $a = 2^l \cdot (2^\nu - 1)$, the parity weight is equal to $(l + 1) \cdot 2^{\nu-1} + 2$.

The parity weight for an input-weight 3 sequence, as those from fundamental paths of the lower constituent codeword, with $b = 8$ and $c = 12$, results from Table 5.2, where a constituent RSC code of LTE turbo code was considered. From Table 5.2 we note that the parity weight is equal to 7.

Finally, the input weight is six, since the size of the interleaver pattern is six. Thus, the total Hamming weight of the turbo codeword generated by the interleaver pattern in Fig. 5.1 is $6 + 3 \cdot (4 \cdot l + 6) + 2 \cdot 7 = 12 \cdot l + 38$. Note that the aforementioned argument for the calculation of the Hamming weight of the turbo codeword generated by the interleaver pattern assumes that the fundamental paths of the constituent codewords do not overlap. Otherwise, the calculated weight is an upper bound on the Hamming weight of the generated codeword. ■

Theorem 5.9 *The minimum distance of a conventional binary turbo code (of nominal rate 1/3) with generator matrix $G = [1, 15/13]$ is upper bounded by*

$$d_{min} \leq 51 \tag{5.70}$$

Table 5.2 Determining the parity weight for an input-weight 3 sequence with the first and the second bits 1 separated by $b = 8$ positions, and the first and the third bits 1 separated by $c = 12$ positions

k	u_k	$B_1^{(k)}$	$B_2^{(k)}$	$B_3^{(k)}$	c_k
1	1	0	0	0	1
2	0	1	0	0	1
3	0	0	1	0	1
4	0	1	0	1	1
5	0	1	1	0	0
6	0	1	1	1	0
7	0	0	1	1	1
8	0	0	0	1	0
9	1	1	0	0	0
10	0	1	1	0	0
11	0	1	1	1	0
12	0	0	1	1	1
13	1	0	0	1	1
14	0	0	0	0	0

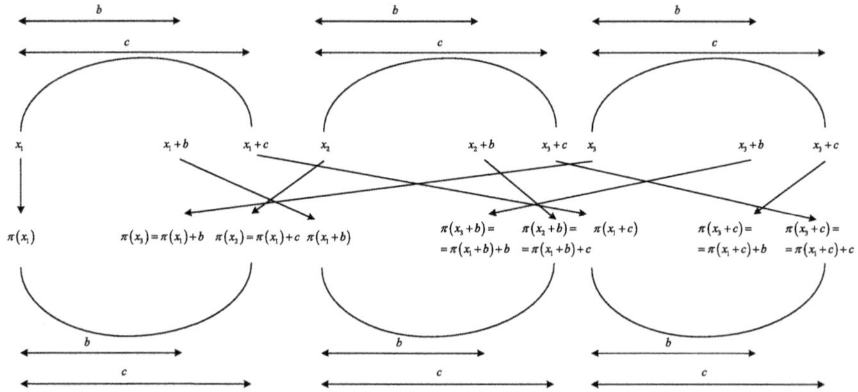

Fig. 5.2 Critical interleaver pattern of size nine for QPP-based interleavers

for all interleaver lengths L that satisfy

$$n_{L,p} \leq \begin{cases} 6, & \text{if } p = 2 \\ 1, & \text{otherwise} \end{cases} \tag{5.71}$$

when using QPP interleavers.

Proof An interleaver pattern of size nine is shown in Fig. 5.2. The interleaver pattern generates an upper and a lower constituent codeword containing 3 input-weight

3 fundamental paths. The interleaving of the systematic 1-positions (of the upper constituent codeword) is indicated by arrows. Now, let the index differences between the second and the first and the third and the first systematic 1-positions of all the fundamental paths in both the upper and lower constituent codewords be denoted by b and c, respectively. For convenience, these index differences are also indicated in Fig. 5.2 by horizontal lines.

The nine elements of permutation $\pi(\cdot)$ indicated in Fig. 5.2 are written in detail below

$$\begin{cases} x_1 \rightarrow \pi(x_1) \\ x_1 + b \rightarrow \pi(x_1 + b) \\ x_1 + c \rightarrow \pi(x_1 + c) \\ x_2 \rightarrow \pi(x_2) = \pi(x_1) + c \\ x_2 + b \rightarrow \pi(x_2 + b) = \pi(x_1 + b) + c \\ x_2 + c \rightarrow \pi(x_2 + c) = \pi(x_1 + c) + c \\ x_3 \rightarrow \pi(x_3) = \pi(x_1) + b \\ x_3 + b \rightarrow \pi(x_3 + b) = \pi(x_1 + b) + b \\ x_3 + c \rightarrow \pi(x_3 + c) = \pi(x_1 + c) + b \end{cases} \tag{5.72}$$

Writing $x = \rho(\pi(x))$, for $x = x_1$, $x = x_2$, and $x = x_3$, the equations corresponding to points $x_2 + b$, $x_2 + c$, $x_3 + b$, and $x_3 + c$ from (5.72) are written as

$$\begin{cases} \pi(\rho(\pi(x_2)) + b) = \pi(\rho(\pi(x_1)) + b) + c \pmod L \\ \pi(\rho(\pi(x_2)) + c) = \pi(\rho(\pi(x_1)) + c) + c \pmod L \\ \pi(\rho(\pi(x_3)) + b) = \pi(\rho(\pi(x_1)) + b) + b \pmod L \\ \pi(\rho(\pi(x_3)) + c) = \pi(\rho(\pi(x_1)) + c) + b \pmod L \end{cases} \tag{5.73}$$

Using the equations corresponding to points x_2 and x_3 from (5.72) in (5.73), and then replacing $\pi(x_1)$ by x, we have

$$\begin{cases} \pi(\rho(x + c) + b) = \pi(\rho(x) + b) + c \pmod L \\ \pi(\rho(x + c) + c) = \pi(\rho(x) + c) + c \pmod L \\ \pi(\rho(x + b) + b) = \pi(\rho(x) + b) + b \pmod L \\ \pi(\rho(x + b) + c) = \pi(\rho(x) + c) + b \pmod L \end{cases} \tag{5.74}$$

Thus, the interleaver pattern in Fig. 5.2 generates a turbo codeword if the four congruences from (5.74) are satisfied, where $x \in \mathbb{Z}_L$ denotes the first systematic 1-position (the point x_1 in Fig. 5.2) of the first fundamental path in the lower constituent codeword.

For $x = 0$ in (5.74), we have

$$\begin{cases} \pi(\rho(c) + b) = \pi(b) + c \pmod L \\ \pi(\rho(c) + c) = \pi(c) + c \pmod L \\ \pi(\rho(b) + b) = \pi(b) + b \pmod L \\ \pi(\rho(b) + c) = \pi(c) + b \pmod L \end{cases} \tag{5.75}$$

Similarly as in (5.61)–(5.64), we have

$$\begin{cases} \pi(b) + \underbrace{\pi(\rho(c))}_{=c} + 2 \cdot b \cdot q_2 \cdot \rho(c) = \pi(b) + c \pmod{L} \\ \pi(c) + \underbrace{\pi(\rho(c))}_{=c} + 2 \cdot c \cdot q_2 \cdot \rho(c) = \pi(c) + c \pmod{L} \\ \pi(b) + \underbrace{\pi(\rho(b))}_{=b} + 2 \cdot b \cdot q_2 \cdot \rho(b) = \pi(b) + b \pmod{L} \\ \pi(c) + \underbrace{\pi(\rho(b))}_{=b} + 2 \cdot c \cdot q_2 \cdot \rho(b) = \pi(c) + b \pmod{L} \end{cases} \Leftrightarrow \tag{5.76}$$

$$\begin{cases} 2 \cdot b \cdot q_2 \cdot \rho(c) = 0 \pmod{L} \\ 2 \cdot c \cdot q_2 \cdot \rho(c) = 0 \pmod{L} \\ 2 \cdot b \cdot q_2 \cdot \rho(b) = 0 \pmod{L} \\ 2 \cdot c \cdot q_2 \cdot \rho(b) = 0 \pmod{L} \end{cases} \Leftrightarrow \tag{5.77}$$

$$\begin{cases} 2 \cdot b \cdot c \cdot q_2 \cdot (r_1 + r_2 \cdot c + \cdots + r_{d_{2-inv}} \cdot c^{d_{2-inv}-1}) = 0 \pmod{L} \\ 2 \cdot c^2 \cdot q_2 \cdot (r_1 + r_2 \cdot c + \cdots + r_{d_{2-inv}} \cdot c^{d_{2-inv}-1}) = 0 \pmod{L} \\ 2 \cdot b^2 \cdot q_2 \cdot (r_1 + r_2 \cdot b + \cdots + r_{d_{2-inv}} \cdot b^{d_{2-inv}-1}) = 0 \pmod{L} \\ 2 \cdot b \cdot c \cdot q_2 \cdot (r_1 + r_2 \cdot b + \cdots + r_{d_{2-inv}} \cdot b^{d_{2-inv}-1}) = 0 \pmod{L} \end{cases} \tag{5.78}$$

The four congruences in (5.78) are satisfied if

$$\begin{cases} 2 \cdot b \cdot c \cdot q_2 = 0 \pmod{L} \\ 2 \cdot b^2 \cdot q_2 = 0 \pmod{L} \\ 2 \cdot c^2 \cdot q_2 = 0 \pmod{L} \end{cases} \tag{5.79}$$

For $b = 8$ and $c = 12$ in (5.79), we have

$$\begin{cases} 2^6 \cdot 3 \cdot q_2 = 0 \pmod{L} \\ 2^7 \cdot q_2 = 0 \pmod{L} \\ 2^5 \cdot 3^2 \cdot q_2 = 0 \pmod{L} \end{cases} \tag{5.80}$$

The three congruences in (5.80) are satisfied if

$$2^5 \cdot q_2 = 0 \pmod{L} \tag{5.81}$$

Equation (5.81) is equivalent to

$$\begin{cases} n_{L,2} \leq n_{q_2,2} + 5 \\ n_{L,p} \leq n_{q_2,p}, \forall p \neq 2 \end{cases} \tag{5.82}$$

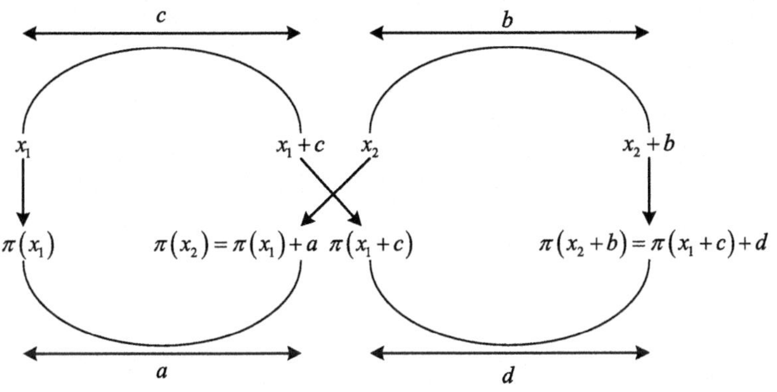

Fig. 5.3 Critical interleaver pattern of size four for QPP-based interleavers

Taking into account the comments after Eqs. (5.68), (5.82) becomes

$$\begin{cases} n_{L,2} \le 6 \\ n_{L,p} \le 1, \forall p \ne 2 \end{cases} \tag{5.83}$$

Conditions (5.83) are the same as those from (5.71).

From Table 5.2 it follows that the parity weight for the input-weight 3 error pattern, as well as those from the fundamental paths of the upper or lower constituent codewords in Fig. 5.2 for $b = 8$ and $c = 12$ is equal to 7. Then, because the input weight is nine, the Hamming weight of the codeword generated by the interleaver pattern in Fig. 5.2 is equal to $9 + 6 \cdot 7 = 51$, i.e. the upper bound given in (5.70). ∎

Figure 5.3 shows an interleaver pattern of size four. Both the upper and the lower constituent codewords generated by the interleaver pattern contain 2 input-weight 2 fundamental paths. The interleaving of the systematic 1-positions (of the upper constituent codeword) is indicated by arrows. Furthermore, the index differences between the second and the first systematic 1-positions of the first and the second fundamental paths in the lower constituent codeword are denoted by a and d, respectively. The corresponding index differences for the upper constituent codeword are denoted by c and b, respectively. These index differences are also indicated in Fig. 5.3 by horizontal lines.

The four elements of permutation $\pi(\cdot)$ indicated in Fig. 5.3 are written in detail below

$$\begin{cases} x_1 \to \pi(x_1) \\ x_1 + c \to \pi(x_1 + c) \\ x_2 \to \pi(x_2) = \pi(x_1) + a \\ x_2 + b \to \pi(x_2 + b) = \pi(x_1 + c) + d \end{cases} \tag{5.84}$$

Writing $x_2 = \rho(\pi(x_2))$, the equation corresponding to point $x_2 + b$ in (5.84) can be written as

$$\pi(\rho(\pi(x_2)) + b) = \pi(x_1 + c) + d \pmod{L} \tag{5.85}$$

Considering the equation corresponding to point x_2 from (5.84), (5.85) becomes:

$$\pi(\rho(\pi(x_1) + a) + b) = \pi(x_1 + c) + d \pmod{L} \tag{5.86}$$

With $x_1 = \rho(\pi(x_1))$ and then $\pi(x_1) = x$ in (5.86), we have

$$\pi(\rho(x + a) + b) = \pi(\rho(x) + c) + d \pmod{L} \tag{5.87}$$

where $x \in \mathbb{Z}_L$ denotes the first systematic 1-position (the point x_1 in Fig. 5.3) of the first fundamental path in the lower constituent codeword. With a primitive feedback polynomial of degree ν, all the numbers a, b, c, and d are multiples of $2^\nu - 1$. From the fact that the feedback polynomial of RSC encoder is primitive, of degree ν, and the feedforward polynomial of RSC encoder is monic, it follows that for an input-weight 2 sequence, with the index difference between the second and the first 1-positions equal to a, a multiple of $2^\nu - 1$, the parity weight is $1 + \frac{a \cdot 2^{\nu-1}}{2^\nu - 1} + 1 = 2 + \frac{a \cdot 2^{\nu-1}}{2^\nu - 1}$. An example in this sense is shown in Table 5.1 for $\nu = 3$ and $a = 14 = 2 \cdot (2^3 - 1)$. Then the Hamming weight of the codeword generated by the interleaver pattern in Fig. 5.3 is at most $4 + 4 \cdot 2 + \frac{(a+b+c+d) \cdot 2^{\nu-1}}{2^\nu - 1} = 12 + \frac{(a+b+c+d) \cdot 2^{\nu-1}}{2^\nu - 1}$.

For $x = 0$, the congruence (5.87) is equivalent to

$$\begin{gathered}
\pi(\rho(a) + b) = \pi(\rho(0) + c) + d = \pi(c) + d \pmod{L} \Leftrightarrow \\
q_1 \cdot (\rho(a) + b) + q_2 \cdot (\rho(a) + b)^2 = q_1 \cdot c + q_2 \cdot c^2 + d \pmod{L} \Leftrightarrow \\
\Leftrightarrow q_1 \cdot (\rho(a) + b) + q_2 \cdot \left((\rho(a))^2 + b^2 + 2 \cdot \rho(a) \cdot b\right) - \\
-q_1 \cdot c - q_2 \cdot c^2 - d \equiv 0 \pmod{L} \Leftrightarrow \\
\Leftrightarrow q_1 \cdot \rho(a) + q_2 \cdot (\rho(a))^2 + q_1 \cdot b + q_2 \cdot b^2 + 2 \cdot q_2 \cdot \rho(a) \cdot b - \\
-q_1 \cdot c - q_2 \cdot c^2 - d \equiv 0 \pmod{L} \Leftrightarrow \\
\Leftrightarrow \underbrace{\pi(\rho(a))}_{=a} + q_2 \cdot (b^2 - c^2) + q_1 \cdot (b - c) + \\
+2 \cdot q_2 \cdot \rho(a) \cdot b - d \equiv 0 \pmod{L} \Leftrightarrow \\
\Leftrightarrow q_2 \cdot (b^2 - c^2) + q_1 \cdot (b - c) + a - d + \\
+2 \cdot q_2 \cdot a \cdot b \cdot (r_1 + r_2 \cdot a + \cdots + r_{d_{2-inv}} \cdot a^{d_{2-inv}-1}) \equiv 0 \pmod{L}
\end{gathered} \tag{5.88}$$

If we choose $c = b$ and $d = a$, (5.88) becomes

$$2 \cdot q_2 \cdot a \cdot b \cdot (r_1 + r_2 \cdot a + \cdots + r_{d_{2-inv}} \cdot a^{d_{2-inv}-1}) \equiv 0 \pmod{L} \tag{5.89}$$

Table 5.3 Summary of conditions that make the interleaver pattern in Fig. 5.3 to be critical and the upper bounds on the weight of the corresponding codewords generated by this interleaver pattern

$(\|a'\|, \|b'\|)$	Simplified congruence (5.91)	Upper bound on the weight
(1, 1)	$2 \cdot q_2 \cdot (2^\nu - 1)^2 \equiv 0 \pmod{L}$	$12 + 2^{\nu+1}$
(1, 2)	$2^2 \cdot q_2 \cdot (2^\nu - 1)^2 \equiv 0 \pmod{L}$	$12 + 3 \cdot 2^\nu$
(2, 1)	$2^2 \cdot q_2 \cdot (2^\nu - 1)^2 \equiv 0 \pmod{L}$	$12 + 3 \cdot 2^\nu$
(2, 2)	$2^3 \cdot q_2 \cdot (2^\nu - 1)^2 \equiv 0 \pmod{L}$	$12 + 2^{\nu+2}$
(1, 3)	$2 \cdot 3 \cdot q_2 \cdot (2^\nu - 1)^2 \equiv 0 \pmod{L}$	$12 + 2^{\nu+2}$
(3, 1)	$2 \cdot 3 \cdot q_2 \cdot (2^\nu - 1)^2 \equiv 0 \pmod{L}$	$12 + 2^{\nu+2}$

In the following, let there be $a = (2^\nu - 1) \cdot a'$ and $b = (2^\nu - 1) \cdot b'$, with $a', b' \in \mathbb{Z}^*$. Then, the Hamming weight of the codeword generated by the interleaver pattern in Fig. 5.3 is (at most) $12 + 2 \cdot (|a'| + |b'|) \cdot 2^{\nu-1} = 12 + (|a'| + |b'|) \cdot 2^\nu$ and the congruence (5.89) is written as

$$2 \cdot q_2 \cdot |a'| \cdot |b'| \cdot (2^\nu - 1)^2 \cdot \big(r_1 \pm r_2 \cdot (2^\nu - 1) \cdot |a'| + \cdots + \\ + r_{d_{2-inv}} \cdot (2^\nu - 1)^{d_{2-inv}-1} \cdot (\pm|a'|)^{d_{2-inv}-1}\big) \equiv 0 \pmod{L} \tag{5.90}$$

The congruence (5.90) is true if

$$2 \cdot q_2 \cdot |a'| \cdot |b'| \cdot (2^\nu - 1)^2 \equiv 0 \pmod{L} \tag{5.91}$$

In Table 5.3, the simplified congruence in (5.91) with different values of the pair $(|a'|, |b'|)$ is listed in the second column. The last column contains an upper bound on the weight of the corresponding turbo codeword generated by the interleaver pattern in Fig. 5.3.

For $\nu = 3$ the simplified congruence (5.91) becomes

$$2 \cdot 7^2 \cdot q_2 \cdot |a'| \cdot |b'| \equiv 0 \pmod{L} \tag{5.92}$$

For different values of the pair $(|a'|, |b'|)$, from the congruence (5.92) we can get the conditions on L for which the congruence is true and thus, for which the corresponding upper bound of the minimum distance is $12 + 8 \cdot (|a'| + |b'|)$. The conditions on L for the values of the pair $(|a'|, |b'|)$ from Table 5.3 are given in Table 5.4.

We show below how the conditions on L in Table 5.4 were obtained. If $|a'| = \prod_{p \in \mathcal{P}} p^{n_{|a'|,p}}$ and $|b'| = \prod_{p \in \mathcal{P}} p^{n_{|b'|,p}}$ are the prime factorizations of $|a'|$ and $|b'|$, respectively, then the congruence (5.92) is fulfilled if

Table 5.4 Summary of conditions on L for various upper bounds on the minimum distance when $\nu = 3$ (derived from the congruence (5.92) for different values of pair $(|a'|, |b'|)$)

$(\|a'\|, \|b'\|)$	$n_{L,2}$	$n_{L,3}$	$n_{L,7}$	$n_{L,p}$, $p \neq 2, 3, 7$	Upper bound on d_{min}
(1, 1)	≤ 2	≤ 1	≤ 3	≤ 1	28
(1, 2) or (2, 1)	≤ 3	≤ 1	≤ 3	≤ 1	36
(2, 2)	≤ 4	≤ 1	≤ 3	≤ 1	44
(1, 3) or (3, 1)	≤ 2	≤ 2	≤ 3	≤ 1	44

$$\begin{cases} n_{L,2} \leq n_{q_2,2} + n_{|a'|,2} + n_{|b'|,2} + 1 \\ n_{L,7} \leq n_{q_2,7} + n_{|a'|,7} + n_{|b'|,7} + 2 \\ n_{L,p} \leq n_{q_2,p} + n_{|a'|,p} + n_{|b'|,p}, \forall p \neq 2, 7 \end{cases} \tag{5.93}$$

Taking into account the conditions for q_2 from Table 3.1, we have that $n_{q_2,2} \geq 1$ when $n_{L,2} \geq 2$ and $n_{q_2,p} \geq 1$ when $n_{L,p} \geq 1$, $\forall p > 2$. Then the conditions from (5.93) are equivalent to

$$\begin{cases} n_{L,2} \leq n_{|a'|,2} + n_{|b'|,2} + 2 \\ n_{L,7} \leq n_{|a'|,7} + n_{|b'|,7} + 3 \\ n_{L,p} \leq n_{|a'|,p} + n_{|b'|,p} + 1, \forall p \neq 2, 7 \end{cases} \tag{5.94}$$

The conditions on $n_{L,p}$ from Table 5.4 result immediately taking into account the factorizations of the values for $|a'|$ and $|b'|$ (i.e. of 1, 2 and 3).

If the QPP has an inverse QPP, then, taking into account the conditions from Theorem 5.6, the conditions from (5.93) are equivalent to

$$\begin{cases} n_{L,2} \leq n_{|a'|,2} + n_{|b'|,2} + 1 + \max\left(\left\lceil \frac{n_{L,2}-2}{2} \right\rceil, 1\right) \\ n_{L,3} \leq n_{|a'|,3} + n_{|b'|,3} + \max\left(\left\lceil \frac{n_{L,3}-1}{2} \right\rceil, 1\right) \\ n_{L,7} \leq n_{|a'|,7} + n_{|b'|,7} + 2 + \left\lceil \frac{n_{L,7}}{2} \right\rceil \\ n_{L,p} \leq n_{|a'|,p} + n_{|b'|,p} + \left\lceil \frac{n_{L,p}}{2} \right\rceil, \forall p \neq 2, 3, 7 \end{cases} \tag{5.95}$$

For $n_{L,2} \leq 2$, the first inequality from (5.95) is always fulfilled. For $n_{L,2} \geq 3$ we have $\max\left(\left\lceil \frac{n_{L,2}-2}{2} \right\rceil, 1\right) = \left\lceil \frac{n_{L,2}-2}{2} \right\rceil$ and we consider two cases.

If $n_{L,2} = 2 \cdot k$, with $k \in \mathbb{N}$ and $k \geq 2$, the first inequality from (5.95) becomes

$$2 \cdot k \leq n_{|a'|,2} + n_{|b'|,2} + 1 + k - 1 \Leftrightarrow k \leq n_{|a'|,2} + n_{|b'|,2} \Leftrightarrow$$

$$\Leftrightarrow 2 \cdot k \leq 2 \cdot (n_{|a'|,2} + n_{|b'|,2}) \tag{5.96}$$

If $n_{L,2} = 2 \cdot k + 1$, with $k \in \mathbb{N}^*$, the first inequality from (5.95) becomes

$$2 \cdot k + 1 \leq n_{|a'|,2} + n_{|b'|,2} + 1 + k \Leftrightarrow k \leq n_{|a'|,2} + n_{|b'|,2} \Leftrightarrow$$

$$\Leftrightarrow 2 \cdot k + 1 \leq 2 \cdot (n_{|a'|,2} + n_{|b'|,2}) + 1 \tag{5.97}$$

The last inequalities from (5.96) and (5.97) can be written as:

$$n_{L,2} \leq 2 \cdot (n_{|a'|,2} + n_{|b'|,2}) + 1, \text{ for } n_{L,2} \geq 3 \tag{5.98}$$

Because for $n_{L,2} \leq 2$ the first inequality from (5.95) is always fulfilled, then the first inequality from (5.95) is equivalent to

$$n_{L,2} \leq \max\left(2 \cdot (n_{|a'|,2} + n_{|b'|,2}) + 1, 2\right) \tag{5.99}$$

For $n_{L,3} \leq 1$, the second inequality from (5.95) is always fulfilled. For $n_{L,3} \geq 2$ we have $\max\left(\left\lceil \frac{n_{L,3} - 1}{2} \right\rceil, 1\right) = \left\lceil \frac{n_{L,3} - 1}{2} \right\rceil$ and we consider two cases.

If $n_{L,3} = 2 \cdot k$, with $k \in \mathbb{N}^*$, the second inequality from (5.95) becomes

$$2 \cdot k \leq n_{|a'|,3} + n_{|b'|,3} + k \Leftrightarrow k \leq n_{|a'|,3} + n_{|b'|,3} \Leftrightarrow$$

$$\Leftrightarrow 2 \cdot k \leq 2 \cdot (n_{|a'|,3} + n_{|b'|,3}) \tag{5.100}$$

If $n_{L,3} = 2 \cdot k + 1$, with $k \in \mathbb{N}^*$, the second inequality from (5.95) becomes

$$2 \cdot k + 1 \leq n_{|a'|,3} + n_{|b'|,3} + k \Leftrightarrow k \leq n_{|a'|,3} + n_{|b'|,3} - 1 \Leftrightarrow$$

$$\Leftrightarrow 2 \cdot k + 1 \leq 2 \cdot (n_{|a'|,3} + n_{|b'|,3}) - 1 \tag{5.101}$$

The last inequalities from (5.100) and (5.101) can be written as:

$$n_{L,3} \leq 2 \cdot (n_{|a'|,3} + n_{|b'|,3}), \text{ for } n_{L,3} \geq 2 \tag{5.102}$$

Because for $n_{L,3} \leq 1$ the second inequality from (5.95) is always fulfilled, then the second inequality from (5.95) is equivalent to

$$n_{L,3} \leq \max\left(2 \cdot (n_{|a'|,3} + n_{|b'|,3}), 1\right) \tag{5.103}$$

For $n_{L,7} \leq 5$, the third inequality from (5.95) is always fulfilled. For $n_{L,7} \geq 6$ we consider two cases.

If $n_{L,7} = 2 \cdot k$, with $k \in \mathbb{N}$ and $k \geq 3$, the third inequality from (5.95) becomes

$$2 \cdot k \leq n_{|a'|,7} + n_{|b'|,7} + 2 + k \Leftrightarrow k \leq n_{|a'|,7} + n_{|b'|,7} + 2 \Leftrightarrow$$

$$\Leftrightarrow 2 \cdot k \leq 2 \cdot (n_{|a'|,7} + n_{|b'|,7}) + 4 \tag{5.104}$$

If $n_{L,7} = 2 \cdot k + 1$, with $k \in \mathbb{N}$ and $k \geq 3$, the third inequality from (5.95) becomes

$$2 \cdot k + 1 \leq n_{|a'|,7} + n_{|b'|,7} + 2 + k + 1 \Leftrightarrow k \leq n_{|a'|,7} + n_{|b'|,7} + 2 \Leftrightarrow$$

$$\Leftrightarrow 2 \cdot k + 1 \leq 2 \cdot (n_{|a'|,7} + n_{|b'|,7}) + 5 \tag{5.105}$$

The last inequalities from (5.104) and (5.105) can be written as:

$$n_{L,7} \leq 2 \cdot (n_{|a'|,7} + n_{|b'|,7}) + 5, \text{ for } n_{L,7} \geq 6 \tag{5.106}$$

Because for $n_{L,7} \leq 5$ the inequality from (5.106) is always fulfilled, then the third inequality from (5.95) is equivalent to

$$n_{L,7} \leq 2 \cdot (n_{|a'|,7} + n_{|b'|,7}) + 5 \tag{5.107}$$

For $n_{L,p} \leq 1$, the fourth inequality from (5.95) is always fulfilled, $\forall p \neq 2, 3, 7$. For $n_{L,p} \geq 2$ we consider two cases.

If $n_{L,p} = 2 \cdot k$, with $k \in \mathbb{N}^*$, the fourth inequality from (5.95) becomes

$$2 \cdot k \leq n_{|a'|,p} + n_{|b'|,p} + k \Leftrightarrow k \leq n_{|a'|,p} + n_{|b'|,p} \Leftrightarrow$$

$$\Leftrightarrow 2 \cdot k \leq 2 \cdot (n_{|a'|,p} + n_{|b'|,p}) \tag{5.108}$$

If $n_{L,p} = 2 \cdot k + 1$, with $k \in \mathbb{N}^*$, the fourth inequality from (5.95) becomes

$$2 \cdot k + 1 \leq n_{|a'|,p} + n_{|b'|,p} + k + 1 \Leftrightarrow k \leq n_{|a'|,p} + n_{|b'|,p} \Leftrightarrow$$

$$\Leftrightarrow 2 \cdot k + 1 \leq 2 \cdot (n_{|a'|,p} + n_{|b'|,p}) + 1 \tag{5.109}$$

The last inequalities from (5.108) and (5.109) can be written as:

$$n_{L,p} \leq 2 \cdot (n_{|a'|,p} + n_{|b'|,p}) + 1, \text{ for } n_{L,p} \geq 2 \text{ and } p \neq 2, 3, 7 \tag{5.110}$$

Because for $n_{L,p} \leq 1$ the inequality from (5.110) is always fulfilled, then the fourth inequality from (5.95) is equivalent to

$$n_{L,p} \leq 2 \cdot (n_{|a'|,p} + n_{|b'|,p}) + 1, \forall p \neq 2, 3, 7 \tag{5.111}$$

Taking into account the factorizations of the values for $|a'|$ and $|b'|$ and the inequalities (5.99), (5.103), (5.107), and (5.111), we can obtain the conditions on L for different values of the pair $(|a'|, |b'|)$, when the QPP has an inverse QPP. For the

Table 5.5 Summary of conditions on L for various upper bounds on the minimum distance when $\nu = 3$ and when the QPP has an inverse QPP (derived from the inequalities (5.99), (5.103), (5.107), and (5.111), for different values of the pair $(|a'|, |b'|)$)

$(\lvert a'\rvert, \lvert b'\rvert)$	$n_{L,2}$	$n_{L,3}$	$n_{L,7}$	$n_{L,p}$, $p \neq 2, 3, 7$	Upper bound on d_{min}
(1, 1)	≤ 2	≤ 1	≤ 5	≤ 1	28
(1, 2) or (2, 1)	≤ 3	≤ 1	≤ 5	≤ 1	36
(2, 2)	≤ 5	≤ 1	≤ 5	≤ 1	44
(1, 3) or (3, 1)	≤ 2	≤ 2	≤ 5	≤ 1	44

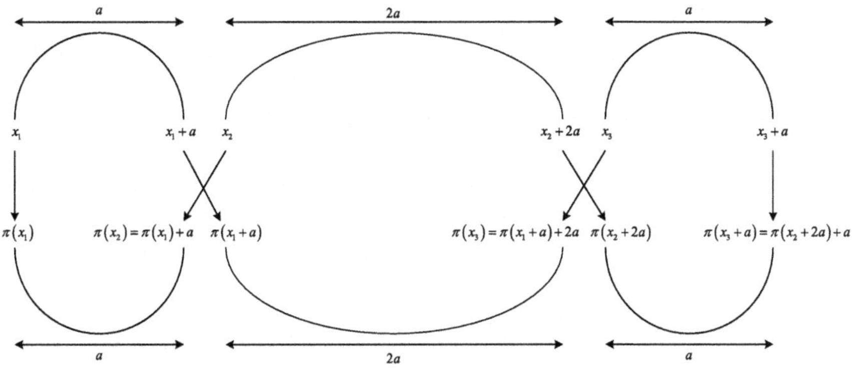

Fig. 5.4 Critical interleaver pattern of size six for QPP-based interleavers

values of the pair $(|a'|, |b'|)$ from Tables 5.3 and 5.4, the conditions on L are given in Table 5.5.

Theorem 5.10 *The minimum distance of a conventional binary turbo code (of nominal rate 1/3) using primitive feedback and monic feedforward polynomials of degree ν and a QPP with a quadratic inverse is upper bounded by*

$$d_{min} \leq 2 \cdot (2^{\nu+1} + 9) \tag{5.112}$$

Proof In Fig. 5.4, an interleaver pattern of size six is shown. Each of the upper and the lower constituent codewords generated by the interleaver pattern contain 3 input-weight 2 fundamental paths. The interleaving of the systematic 1-positions (of the upper constituent codeword) is indicated by arrows. The index difference between the second and the first systematic 1-positions of the first and the third fundamental paths in both the upper and lower constituent codewords is denoted by a. The corresponding index difference for the second, i.e. the middle fundamental path in both constituent codewords is $2a$. For convenience, these index differences are also indicated in Fig. 5.4 by horizontal lines.

The six elements of permutation $\pi(\cdot)$ indicated in Fig. 5.4 are written in detail below

$$\begin{cases} x_1 \rightarrow \pi(x_1) \\ x_1 + a \rightarrow \pi(x_1 + a) \\ x_2 \rightarrow \pi(x_2) = \pi(x_1) + a \\ x_2 + 2a \rightarrow \pi(x_2 + 2a) \\ x_3 \rightarrow \pi(x_3) = \pi(x_1 + a) + 2a \\ x_3 + a \rightarrow \pi(x_3 + a) = \pi(x_2 + 2a) + a \end{cases} \tag{5.113}$$

Writing $x_2 = \rho(\pi(x_2))$ and $x_3 = \rho(\pi(x_3))$, the equation corresponding to point $x_3 + a$ in (5.113) can be written as

$$\pi(\rho(\pi(x_3)) + a) = \pi(\rho(\pi(x_2)) + 2a) + a \pmod{L} \tag{5.114}$$

Considering the equations corresponding to points x_2 and x_3 from (5.113), (5.114) becomes:

$$\pi(\rho(\pi(x_1 + a) + 2a) + a) = \pi(\rho(\pi(x_1) + a) + 2a) + a \pmod{L} \tag{5.115}$$

With $x_1 = x$ in (5.115), we have

$$\pi(\rho(\pi(x + a) + 2a) + a) = \pi(\rho(\pi(x) + a) + 2a) + a \pmod{L} \tag{5.116}$$

where $x \in \mathbb{Z}_L$ denotes the first systematic 1-position (the point x_1 in Fig. 5.4). Now, the congruence in (5.116) is equivalent to

$$4 \cdot a^3 \cdot q_2 \cdot r_2 \cdot (1 + 2q_1 + 2aq_2 + 4xq_2) \equiv 0 \pmod{L} \tag{5.117}$$

which again is equivalent to

$$4 \cdot a^3 \cdot q_2 \cdot r_2 \cdot (1 + 2q_1 + 2aq_2) \equiv 0 \pmod{L} \tag{5.118}$$

since $4 \cdot q_2^2 \cdot r_2 \equiv 0 \pmod{L}$. This follows from Theorem 5.5, which states that $12 \cdot q_2 \cdot r_2 \equiv 0 \pmod{L}$. Indeed, if $L = 2^{n_{L,2}} \cdot \prod_{i=2}^{n_L} p_i^{n_{L,p_i}}$, with $n_{L,2} \geq 0$ and $n_{L,p_i} \geq 1, \forall i = \overline{2, n_L}$, then, because $\pi(x)$ and $\rho(x)$ are QPPs, we have $q_2 = 2^{n_{\pi,2}} \cdot \prod_{i=2}^{n_L} p_i^{n_{\pi,p_i}} \cdot L_{q_2}$, $r_2 = 2^{n_{\rho,2}} \cdot \prod_{i=2}^{n_L} p_i^{n_{\rho,p_i}} \cdot L_{r_2}$, where $n_{\pi,2}, n_{\rho,2} \geq 0$ if $n_{L,2} \in \{0, 1\}$ and $n_{\pi,2}, n_{\rho,2} \geq 1$ if $n_{L,2} \geq 2$, $n_{\pi,p_i}, n_{\rho,p_i} \geq 1, \forall i = \overline{2, n_L}$, and $L_{q_2}, L_{r_2} \in \mathbb{N}^*$ are positive integers corresponding to other prime numbers than p_i, with $i = \overline{2, n_L}$. Then, from $12 \cdot q_2 \cdot r_2 \equiv 0 \pmod{L}$, it results that $n_{\pi,2} + n_{\rho,2} + 2 \geq n_{L,2}$, $n_{\pi,p_i} + n_{\rho,p_i} \geq n_{L,p_i}$, $\forall i = \overline{2, n_L}$ with $p_i > 3$, and, if $3 \mid L$, $n_{\pi,3} + n_{\rho,3} + 1 \geq n_{L,3}$. Therefore, if $3 \nmid L$, the congruence

$4 \cdot q_2^2 \cdot r_2 \equiv 0 \pmod{L}$ results immediately because $2 \cdot n_{\pi,2} + n_{\rho,2} + 2 \geq n_{L,2}$ and $2 \cdot n_{\pi,p_i} + n_{\rho,p_i} \geq n_{L,p_i}$, $\forall i = \overline{2, n_L}$. If $3 \mid L$ we have $n_{\pi,3}, n_{\rho,3} \geq 1$ because $\pi(x)$ and $\rho(x)$ are QPPs. Then, because $n_{\pi,3} \geq 1$ and $n_{\pi,3} + n_{\rho,3} \geq n_{L,3} - 1$, we have $n_{\pi,3} + n_{\pi,3} + n_{\rho,3} \geq 1 + n_{L,3} - 1 = n_{L,3}$ and, thus, the congruence $4 \cdot q_2^2 \cdot r_2 \equiv 0 \pmod{L}$ is also true.

From Theorem 5.5, we know that $12 \cdot q_2 \cdot r_2 \equiv 0 \pmod{L}$, and it follows that $4 \cdot q_2 \cdot r_2 \equiv 0 \pmod{L}$ if 27 is not a divisor of L. Indeed, if $3 \nmid L$, the congruence $4 \cdot q_2 \cdot r_2 \equiv 0 \pmod{L}$ results immediately as we have explained above for the congruence $4 \cdot q_2^2 \cdot r_2 \equiv 0 \pmod{L}$. If $3 \mid L$, for the congruence $4 \cdot q_2 \cdot r_2 \equiv 0 \pmod{L}$ to be true, it is required that $n_{\pi,3} + n_{\rho,3} \geq n_{L,3}$. But from $12 \cdot q_2 \cdot r_2 \equiv 0 \pmod{L}$ we have $n_{\pi,3} + n_{\rho,3} \geq n_{L,3} - 1$, and because $\pi(x)$ and $\rho(x)$ are QPPs we have $n_{\pi,3}, n_{\rho,3} \geq 1$. Therefore, the inequality $n_{\pi,3} + n_{\rho,3} \geq n_{L,3}$ is always held if $n_{L,3} \leq 2$, i.e. if $27 \nmid L$. Thus, the congruence in (5.118) holds for *all* QPPs with a quadratic inverse, for a given value of L if 27 is *not* a divisor of L.

Since the feedback polynomial is primitive, we can choose $a = 2^{\nu} - 1$, and with a monic feedforward polynomial of degree ν, the parity weight for the first and the third fundamental paths in both constituent codewords in Fig. 5.4 is $2 + 2^{\nu-1}$ (see the proof of Theorem 5.8). The corresponding parity weight for the second, i.e. the middle fundamental path in both constituent codewords is $2 + 2^{\nu}$. Then the Hamming weight of the codeword generated by the interleaver pattern in Fig. 5.4 is at most $4 \cdot (2 + 2^{\nu-1}) + 2 \cdot (2 + 2^{\nu}) + 6 = 2 \cdot (2^{\nu+1} + 9)$. The equality holds if the fundamental paths do not interfere with each other as in Fig. 5.4.

Figure 5.5 presents another interleaver pattern of size six. Each of the upper and the lower constituent codewords generated by the interleaver pattern contain 3 input-weight 2 fundamental paths. The interleaving of the systematic 1-positions of the upper constituent codeword is indicated by arrows. Furthermore, the index difference between the second and the first systematic 1-positions of the first and the third fundamental paths in the upper constituent codeword, and of the second and the third fundamental paths in the lower constituent codeword is denoted by a. The corre-

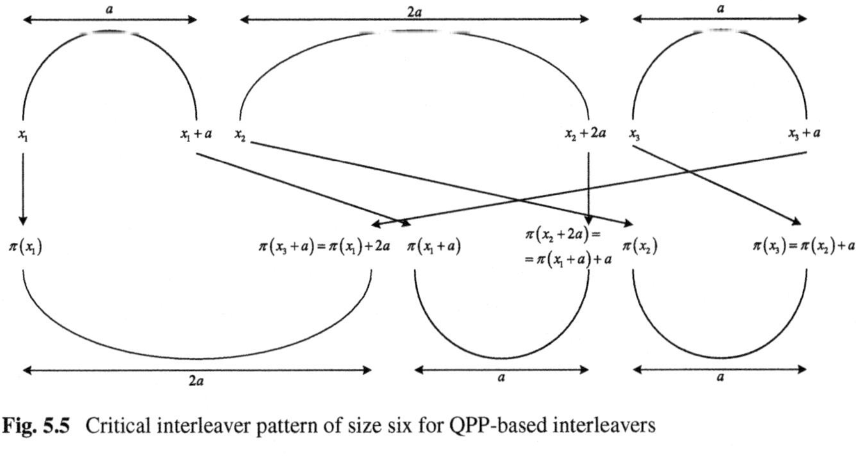

Fig. 5.5 Critical interleaver pattern of size six for QPP-based interleavers

sponding index difference for the second fundamental path in the upper constituent codeword, and for the first fundamental path in the lower constituent codeword is $2a$. These index differences are also indicated in Fig. 5.5 by horizontal lines.

The six elements of permutation $\pi(\cdot)$ indicated in Fig. 5.5 are written in detail below

$$\begin{cases} x_1 \to \pi(x_1) \\ x_1 + a \to \pi(x_1 + a) \\ x_2 \to \pi(x_2) \\ x_2 + 2a \to \pi(x_2 + 2a) = \pi(x_1 + a) + a \\ x_3 \to \pi(x_3) = \pi(x_2) + a \\ x_3 + a \to \pi(x_3 + a) = \pi(x_1) + 2a \end{cases} \tag{5.119}$$

Writing $x_2 = \rho(\pi(x_2))$ the equation corresponding to point $x_2 + 2a$ in (5.119) can be written as

$$\pi(\rho(\pi(x_2)) + 2a) = \pi(x_1 + a) + a \ (\text{mod } L) \tag{5.120}$$

Considering the equation corresponding to point x_3 from (5.119), (5.120) becomes:

$$\pi(\rho(\pi(x_3) - a) + 2a) = \pi(x_1 + a) + a \ (\text{mod } L) \tag{5.121}$$

Writing $x_3 + a = \rho(\pi(x_3 + a))$, the equation corresponding to point $x_3 + a$ from (5.119) is equivalently written as

$$x_3 + a = \rho(\pi(x_1) + 2a) \Leftrightarrow x_3 = \rho(\pi(x_1) + 2a) - a \Leftrightarrow$$

$$\Leftrightarrow \pi(x_3) = \pi(\rho(\pi(x_1) + 2a) - a) \tag{5.122}$$

With (5.122), (5.121) becomes:

$$\begin{aligned} &\pi(\rho(\pi(\rho(\pi(x_1) + 2a) - a) - a) + 2a) = \pi(x_1 + a) + a \ (\text{mod } L) \Leftrightarrow \\ &\rho(\pi(\rho(\pi(x_1) + 2a) - a) - a) + 2a = \rho(\pi(x_1 + a) + a) \ (\text{mod } L) \Leftrightarrow \\ &\rho(\pi(\rho(\pi(x_1) + 2a) - a) - a) = \rho(\pi(x_1 + a) + a) - 2a \ (\text{mod } L) \Leftrightarrow \end{aligned}$$

$$\pi(\rho(\pi(x_1) + 2a) - a) - a = \pi(\rho(\pi(x_1 + a) + a) - 2a) \ (\text{mod } L) \tag{5.123}$$

With $x_1 = x$ in (5.123), we have

$$\pi(\rho(\pi(x) + 2a) - a) = \pi(\rho(\pi(x + a) + a) - 2a) + a \ (\text{mod } L), \tag{5.124}$$

where $x \in \mathbb{Z}_L$ is the leftmost systematic 1-position (the point x_1 in Fig. 5.5) in the upper constituent codeword. Now, the congruence (5.124) is equivalent to

$$4 \cdot a^3 \cdot q_2 \cdot r_2 \cdot (1 - 2q_1 - 2aq_2 - 4xq_2) \equiv 0 \ (\text{mod } L) \tag{5.125}$$

which again is equivalent to

$$4 \cdot a^3 \cdot q_2 \cdot r_2 \cdot (1 - 2q_1 - 2aq_2) \equiv 0 \text{ (mod } L) \tag{5.126}$$

since $4 \cdot q_2^2 \cdot r_2 \equiv 0$ (mod L). This follows from Theorem 5.5, which states that $12 \cdot q_2 \cdot r_2 \equiv 0$ (mod L), as we have explained after Eq. (5.118). As for the codeword generated by the interleaver pattern in Fig. 5.4, the Hamming weight of the codeword generated by the interleaver pattern in Fig. 5.5 is at most $2 \cdot (2^{\nu+1} + 9)$.

Assume $27 \mid L$. Then, $q_2 = 3 \cdot c$, for some integer c, since, from Table 3.1, $3 \mid q_2$. Furthermore, $q_1 = 1 + 3 \cdot k$ or $2 + 3 \cdot k$, for some integer k, since, from Table 3.1, $3 \nmid q_1$. If $q_1 = 1 + 3 \cdot k$, then the congruence in (5.118) reduces to $4 \cdot a^3 \cdot q_2 \cdot r_2 \cdot (1 + 2 \cdot (1 + 3 \cdot k) + 2a \cdot 3 \cdot c) = 12 \cdot a^3 \cdot q_2 \cdot r_2 \cdot (1 + 2 \cdot k + 2a \cdot c) \equiv 0 \pmod L$, which is always true, since $12 \cdot q_2 \cdot r_2 \equiv 0$ (mod L) from Theorem 5.5. Furthermore, if $q_1 = 2 + 3 \cdot k$, then the congruence in (5.126) reduces to $4 \cdot a^3 \cdot q_2 \cdot r_2 \cdot (1 - 2 \cdot (2 + 3 \cdot k) - 2a \cdot 3 \cdot c) = 12 \cdot a^3 \cdot q_2 \cdot r_2 \cdot (-1 - 2 \cdot k - 2a \cdot c) \equiv 0 \pmod L$, which is always true, since $12 \cdot q_2 \cdot r_2 \equiv 0$ (mod L) from Theorem 5.5. Thus, there is an upper bound of $2 \cdot (2^{\nu+1} + 9)$ on the d_{min} for QPPs with a quadratic inverse for all values of L. ■

We note that Theorem 5.10 applies for all interleaver lengths L. It is achievable for a range of values L, at least for $\nu = 3$. For $\nu = 3$, the degree of all inverse permutations should be at least three, to achieve a d_{min} strictly larger than $2 \cdot (2^{\nu+1} + 9) = 50$.

Theorem 5.11 *The minimum distance of a conventional binary turbo code (of nominal rate 1/3) with generator matrix* $G = [1, 15/13]$ *is upper bounded by*

$$d_{min} \leq 38 + 12 \cdot l \tag{5.127}$$

for all nonnegative integers l *and for all interleaver lengths* L *that satisfy*

$$n_{L,p} \leq \begin{cases} 2 \cdot l + 5, & \text{if } p = 2 \\ 3, & \text{if } p = 7 \\ 1, & \text{otherwise} \end{cases} \tag{5.128}$$

when using QPP interleavers with a quadratic inverse.

Proof In this case the proof is based on interleaver pattern in Fig. 5.1. When there exists an inverse polynomial of degree two, the congruences in (5.65) reduce to

$$\begin{cases} 2 \cdot a \cdot b \cdot q_2 \cdot (r_1 + r_2 \cdot b) = 0 \text{ (mod } L) \\ 2 \cdot a \cdot c \cdot q_2 \cdot (r_1 + r_2 \cdot c) = 0 \text{ (mod } L) \end{cases} \tag{5.129}$$

For $a = 2^l \cdot 7$, $b = 8$, and $c = 12$ in (5.129), we have

$$\begin{cases} 2^l \cdot 2^4 \cdot 7 \cdot q_2 \cdot (r_1 + 2^3 \cdot r_2) = 0 \text{ (mod } L) \\ 2^l \cdot 2^3 \cdot 3 \cdot 7 \cdot q_2 \cdot (r_1 + 2^2 \cdot 3 \cdot r_2) = 0 \text{ (mod } L) \end{cases} \Leftrightarrow \tag{5.130}$$

$$\begin{cases} 2^l \cdot 112 \cdot q_2 \cdot (r_1 + 8 \cdot r_2) = 0 \pmod{L} \\ 2^l \cdot 168 \cdot q_2 \cdot (r_1 + 12 \cdot r_2) = 0 \pmod{L} \end{cases} \tag{5.131}$$

From Theorem 5.5 we have that $12 \cdot q_2 \cdot r_2 = 0 \pmod{L}$ when a QPP admits an inverse QPP. Using this congruence, the second equation from (5.130) or (5.131) is equivalent to $2^l \cdot 2^3 \cdot 3 \cdot 7 \cdot q_2 \cdot r_1 = 0 \pmod{L}$ or $2^l \cdot 168 \cdot q_2 \cdot r_1 = 0 \pmod{L}$. If $2^l \cdot 2^3 \cdot 3 \cdot 7 \cdot q_2 \cdot r_1 = 0 \pmod{L}$ and $2^2 \cdot 3 \cdot q_2 \cdot r_2 = 0 \pmod{L}$, then also $2^l \cdot 2^4 \cdot 3 \cdot 7 \cdot q_2 \cdot r_1 + 2^l \cdot 2^7 \cdot 3 \cdot 7 \cdot q_2 \cdot r_2 = 0 \pmod{L}$. The last congruence is equivalent to $2^l \cdot 2^4 \cdot 3 \cdot 7 \cdot q_2 \cdot (r_1 + 2^3 \cdot r_2) = 0 \pmod{L}$ or $2^l \cdot 3 \cdot 112 \cdot q_2 \cdot (r_1 + 8 \cdot r_2) = 0 \pmod{L}$, whence it follows that $2^l \cdot 112 \cdot q_2 \cdot (r_1 + 8 \cdot r_2) = 0 \pmod{L}$ if 9 is not a divisor of L, i.e. if $n_{L,3} \leq 1$. The congruence $2^l \cdot 2^3 \cdot 3 \cdot 7 \cdot q_2 \cdot r_1 = 0 \pmod{L}$, for $n_{L,3} \leq 1$, is satisfied if

$$\begin{cases} n_{L,2} \leq n_{q_2,2} + l + 3 \\ n_{L,7} \leq n_{q_2,7} + 1 \\ n_{L,p} \leq n_{q_2,p}, \forall p \neq 2, 3, 7 \end{cases} \tag{5.132}$$

Using Theorem 5.6, it follows that the congruences in (5.132) are satisfied for all valid values of q_2, for a given value of L, if the following conditions are satisfied

$$n_{L,2} \leq \begin{cases} l + 3 + \max\left(\left\lceil \frac{n_{L,2}-2}{2} \right\rceil, 1\right), & \text{if } n_{L,2} > 1 \\ l + 3, & \text{if } n_{L,2} = 0, 1 \end{cases} \tag{5.133}$$

$$n_{L,7} \leq 1 + \left\lceil \frac{n_{L,7}}{2} \right\rceil \tag{5.134}$$

$$n_{L,p} \leq \left\lceil \frac{n_{L,p}}{2} \right\rceil, \forall p \neq 2, 3, 7 \tag{5.135}$$

If $n_{L,2} = 2 \cdot k$, with $k \in \mathbb{N}$, (5.133) can be written as

$$2 \cdot k \leq l + 3 + \max(k - 1, 1) \tag{5.136}$$

For $k \leq 2$, (5.136) is obviously satisfied, and for $k \geq 3$ (5.136) becomes

$$2 \cdot k \leq l + k + 2 \Leftrightarrow k \leq l + 2 \Leftrightarrow 2 \cdot k \leq 2 \cdot l + 4 \tag{5.137}$$

If $n_{L,2} = 2 \cdot k + 1$, with $k \in \mathbb{N}$, (5.133) can be written as

$$2 \cdot k + 1 \leq l + 3 + \max(\lceil k - 1/2 \rceil, 1) \tag{5.138}$$

For $k \leq 2$, (5.138) is obviously satisfied, and for $k \geq 3$ it results that $\lceil k - 1/2 \rceil = k$. Then, (5.138) becomes

$$2 \cdot k + 1 \leq l + 3 + k \Leftrightarrow k \leq l + 2 \Leftrightarrow 2 \cdot k + 1 \leq 2 \cdot l + 5 \tag{5.139}$$

From (5.137) and (5.139), the first inequality in (5.128) results.

If $n_{L,7} = 2 \cdot k$, with $k \in \mathbb{N}$, (5.134) can be written as

$$2 \cdot k \leq 1 + k \Leftrightarrow k \leq 1 \Leftrightarrow 2 \cdot k \leq 2 \tag{5.140}$$

If $n_{L,7} = 2 \cdot k + 1$, with $k \in \mathbb{N}$, (5.134) can be written as

$$\begin{gathered} 2 \cdot k + 1 \leq 1 + \lceil k + 1/2 \rceil \Leftrightarrow 2 \cdot k + 1 \leq 1 + k + 1 \Leftrightarrow k \leq 1 \Leftrightarrow \\ 2 \cdot k + 1 \leq 3 \end{gathered} \tag{5.141}$$

From (5.140) and (5.141) it follows that $n_{L,7} \leq 3$ always satisfies (5.134), which is the same as the second inequality from (5.128).

If $n_{L,p} = 2 \cdot k$, with $k \in \mathbb{N}$, (5.135) can be written as

$$2 \cdot k \leq k \Leftrightarrow k \leq 0 \Leftrightarrow k = 0 \tag{5.142}$$

If $n_{L,p} = 2 \cdot k + 1$, with $k \in \mathbb{N}$, (5.135) can be written as

$$2 \cdot k + 1 \leq \lceil k + 1/2 \rceil \Leftrightarrow 2 \cdot k + 1 \leq k + 1 \Leftrightarrow k \leq 0 \Leftrightarrow 2 \cdot k + 1 \leq 1 \tag{5.143}$$

From (5.142) and (5.143) it follows that $n_{L,p} \leq 1$ always satisfies (5.135), which fits with the third inequality in (5.128).

From the proof of Theorem 5.8, the Hamming weight of the codeword generated by the interleaver pattern in Fig. 5.1 is at most $38 + 12 \cdot l$ and, thus, the theorem is proved. ■

We remark that Theorem 5.11 is useful only when $l = 0$, since the upper bound of d_{min} is at least 50 when $l \geq 1$. This follows from Theorem 5.10 (with $\nu = 3$), which gives a general upper bound of 50.

Theorem 5.12 *The minimum distance of a conventional binary turbo code (of nominal rate 1/3) with generator matrix $G = [1, 15/13]$ is upper bounded by*

$$d_{min} \leq 38 + 12 \cdot l \tag{5.144}$$

for all nonnegative integers l and for all interleaver lengths L that satisfy

$$n_{L,p} \leq \begin{cases} \left\lfloor \frac{3l}{2} \right\rfloor + 4, & \text{if } p = 2 \\ 2, & \text{if } p = 7 \\ 1, & \text{otherwise} \end{cases} \tag{5.145}$$

when using QPP interleavers with a cubic inverse.

Proof The proof is similar to the proof of Theorem 5.11. When there exists an inverse polynomial of degree three, the congruences in (5.65) reduce to

$$\begin{cases} 2 \cdot a \cdot b \cdot q_2 \cdot (r_1 + r_2 \cdot b + r_3 \cdot b^2) = 0 \pmod{L} \\ 2 \cdot a \cdot c \cdot q_2 \cdot (r_1 + r_2 \cdot c + +r_3 \cdot c^2) = 0 \pmod{L} \end{cases} \tag{5.146}$$

For $a = 2^l \cdot 7$, $b = 8$, and $c = 12$ in (5.146), we have

$$\begin{cases} 2^l \cdot 2^4 \cdot 7 \cdot q_2 \cdot (r_1 + 2^3 \cdot r_2 + 2^6 \cdot r_3) = 0 \pmod{L} \\ 2^l \cdot 2^3 \cdot 3 \cdot 7 \cdot q_2 \cdot (r_1 + 2^2 \cdot 3 \cdot r_2 + 2^4 \cdot 3^2 \cdot r_3) = 0 \pmod{L} \end{cases} \Leftrightarrow \tag{5.147}$$

$$\begin{cases} 2^l \cdot 112 \cdot q_2 \cdot (r_1 + 8 \cdot r_2 + 64 \cdot r_3) = 0 \pmod{L} \\ 2^l \cdot 168 \cdot q_2 \cdot (r_1 + 12 \cdot r_2 + 144 \cdot r_3) = 0 \pmod{L} \end{cases} \tag{5.148}$$

The two congruences from (5.147) or (5.148) are satisfied if

$$\begin{cases} n_{L,2} \leq n_{q_2,2} + l + 3 \\ n_{L,7} \leq n_{q_2,7} + 1 \\ n_{L,p} \leq n_{q_2,p}, \forall p \neq 2, 7 \end{cases} \tag{5.149}$$

For $d_{2-inv} = 3$, $\phi(d_{2-inv} + 1)$ from Theorem 5.7 is factorized as $\phi(d_{2-inv} + 1) = \phi(4) = 4 \cdot 5 \cdot 6 = 2^3 \cdot 3 \cdot 5$ and thus $n_{\phi(4),2} = 3$, $n_{\phi(4),3} = 1$, $n_{\phi(4),5} = 1$, and $n_{\phi(4),p} = 0$, $\forall p \neq 2, 3, 5,\ p \in \mathcal{P}$. Then, when a QPP admits a CPP inverse (i.e. $d_{2-inv} = 3$), from Theorem 5.7 we have

$$n_{q_2,2} \geq \begin{cases} \max\left(\left\lceil \frac{n_{L,2}-3}{3} \right\rceil, 1\right), & \text{if } n_{L,2} > 1 \\ 0, & \text{if } n_{L,2} = 0, 1 \end{cases} \tag{5.150}$$

$$n_{q_2,3} \geq \begin{cases} \max\left(\left\lceil \frac{n_{L,3}-1}{3} \right\rceil, 1\right), & \text{if } n_{L,3} > 0 \\ 0, & \text{if } n_{L,3} = 0 \end{cases} \tag{5.151}$$

$$n_{q_2,5} \geq \begin{cases} \max\left(\left\lceil \frac{n_{L,5}-1}{3} \right\rceil, 1\right), & \text{if } n_{L,5} > 0 \\ 0, & \text{if } n_{L,5} = 0 \end{cases} \tag{5.152}$$

and for $p \neq 2, 3, 5$

$$n_{q_2,p} \geq \begin{cases} \max\left(\left\lceil \frac{n_{L,p}}{3} \right\rceil, 1\right), & \text{if } n_{L,p} > 0 \\ 0, & \text{if } n_{L,p} = 0 \end{cases} \tag{5.153}$$

With (5.150)–(5.153), the conditions from (5.149) can be written as

$$n_{L,2} \leq \begin{cases} l + 3 + \max\left(\left\lceil \frac{n_{L,2}-3}{3} \right\rceil, 1\right), & \text{if } n_{L,2} > 1 \\ l + 3, & \text{if } n_{L,2} = 0, 1 \end{cases} \quad (5.154)$$

$$n_{L,3} \leq \begin{cases} \max\left(\left\lceil \frac{n_{L,3}-1}{3} \right\rceil, 1\right), & \text{if } n_{L,3} > 0 \\ 0, & \text{if } n_{L,3} = 0 \end{cases} \quad (5.155)$$

$$n_{L,5} \leq \begin{cases} \max\left(\left\lceil \frac{n_{L,5}-1}{3} \right\rceil, 1\right), & \text{if } n_{L,5} > 0 \\ 0, & \text{if } n_{L,5} = 0 \end{cases} \quad (5.156)$$

$$n_{L,7} \leq \begin{cases} 1 + \max\left(\left\lceil \frac{n_{L,7}}{3} \right\rceil, 1\right), & \text{if } n_{L,7} > 0 \\ 1, & \text{if } n_{L,7} = 0 \end{cases} \quad (5.157)$$

and for $p \neq 2, 3, 5, 7$

$$n_{L,p} \leq \begin{cases} \max\left(\left\lceil \frac{n_{L,p}}{3} \right\rceil, 1\right), & \text{if } n_{L,p} > 0 \\ 0, & \text{if } n_{L,p} = 0 \end{cases} \quad (5.158)$$

For $n_{L,2} = 3 \cdot k$, with $k \in \mathbb{N}$, in (5.154), we have

$$3 \cdot k \leq l + 3 + \max(k - 1, 1) \quad (5.159)$$

For $k \leq 1$, (5.159) is obviously satisfied. For $k \geq 2$, (5.159) is equivalent to

$$3 \cdot k \leq l + 3 + k - 1 \Leftrightarrow 2 \cdot k \leq l + 2 \Leftrightarrow k \leq \frac{l}{2} + 1 \Leftrightarrow$$

$$3 \cdot k \leq 3 \cdot \frac{l}{2} + 3 \quad (5.160)$$

For $n_{L,2} = 3 \cdot k + 1$, with $k \in \mathbb{N}$, in (5.154), we have

$$3 \cdot k + 1 \leq l + 3 + \max(k, 1) \Leftrightarrow 3 \cdot k + 1 \leq l + 3 + k \Leftrightarrow$$

$$2 \cdot k \leq l + 2 \Leftrightarrow k \leq \frac{l}{2} + 1 \Leftrightarrow 3 \cdot k + 1 \leq 3 \cdot \frac{l}{2} + 4 \quad (5.161)$$

For $n_{L,2} = 3 \cdot k + 2$, with $k \in \mathbb{N}$, in (5.154), we have

$$3 \cdot k + 2 \leq l + 3 + \max(k, 1) \Leftrightarrow 3 \cdot k + 2 \leq l + 3 + k \Leftrightarrow$$

$$2 \cdot k \leq l + 1 \Leftrightarrow k \leq \frac{l+1}{2} \Leftrightarrow 3 \cdot k + 2 \leq 3 \cdot \frac{l}{2} + \frac{7}{2} \quad (5.162)$$

For $l = 2 \cdot s$, with $s \in \mathbb{N}$, the last inequalities in (5.160), (5.161), and (5.162) become

$$\begin{cases} 3 \cdot k & \leq 3 \cdot s + 3 \\ 3 \cdot k + 1 & \leq 3 \cdot s + 4 \\ 3 \cdot k + 2 & \leq 3 \cdot s + 3 \end{cases} \Leftrightarrow \begin{cases} 3k & \in \{3 \cdot s + 3, 3 \cdot s, \dots\} \\ 3k + 1 & \in \{3 \cdot s + 4, 3 \cdot s + 1, \dots\} \\ 3k + 2 & \in \{3 \cdot s + 2, 3 \cdot s - 1, \dots\} \end{cases} \Leftrightarrow$$

$$n_{L,2} \leq 3 \cdot s + 4 = \frac{3 \cdot l}{2} + 4 \Leftrightarrow n_{L,2} \leq \left\lfloor \frac{3 \cdot l}{2} \right\rfloor + 4 \tag{5.163}$$

For $l = 2 \cdot s + 1$, with $s \in \mathbb{N}$, the last inequalities in (5.160), (5.161), and (5.162) become

$$\begin{cases} 3 \cdot k & \leq 3 \cdot s + 3 \\ 3 \cdot k + 1 & \leq 3 \cdot s + 4 \\ 3 \cdot k + 2 & \leq 3 \cdot s + 5 \end{cases} \Leftrightarrow \begin{cases} 3k & \in \{3 \cdot s + 3, 3 \cdot s, \dots\} \\ 3k + 1 & \in \{3 \cdot s + 4, 3 \cdot s + 1, \dots\} \\ 3k + 2 & \in \{3 \cdot s + 5, 3 \cdot s + 2, \dots\} \end{cases} \Leftrightarrow$$

$$n_{L,2} \leq 3 \cdot s + 5 = \frac{3 \cdot l}{2} - \frac{1}{2} + 4 \Leftrightarrow n_{L,2} \leq \left\lfloor \frac{3 \cdot l}{2} \right\rfloor + 4 \tag{5.164}$$

Thus, (5.154) is equivalent to the first inequality in (5.145).

For $n_{L,3} = 3 \cdot k$ in (5.155) or $n_{L,5} = 3 \cdot k$ in (5.156), with $k \in \mathbb{N}$, we have

$$3 \cdot k \leq \max(k, 1) \Leftrightarrow 3 \cdot k \leq k \Leftrightarrow k = 0 \Leftrightarrow n_{L,3} = 0 \text{ or } n_{L,5} = 0 \tag{5.165}$$

For $n_{L,3} = 3 \cdot k + 1$ in (5.155) or $n_{L,5} = 3 \cdot k + 1$ in (5.156), with $k \in \mathbb{N}$, we have

$$3 \cdot k + 1 \leq \max(k, 1) \Leftrightarrow k = 0 \Leftrightarrow n_{L,3} = 1 \text{ or } n_{L,5} = 1 \tag{5.166}$$

For $n_{L,3} = 3 \cdot k + 2$ in (5.155) or $n_{L,5} = 3 \cdot k + 2$ in (5.156), with $k \in \mathbb{N}$, we have

$$3 \cdot k + 2 \leq \max(k + 1, 1) \Leftrightarrow 3 \cdot k + 2 \leq k + 1 \Leftrightarrow 2 \cdot k + 1 \leq 0 \Leftrightarrow k \in \emptyset \tag{5.167}$$

Thus, (5.155) and (5.156) are equivalent to the third inequality in (5.145), for $p = 3$ and $p = 5$, respectively.

For $n_{L,7} = 3 \cdot k$, with $k \in \mathbb{N}$, in (5.157), we have

$$3 \cdot k \leq 1 + k \Leftrightarrow 2 \cdot k \leq 1 \Leftrightarrow k = 0 \Leftrightarrow n_{L,7} = 0 \tag{5.168}$$

For $n_{L,7} = 3 \cdot k + 1$, with $k \in \mathbb{N}$, in (5.157), we have

$$3 \cdot k + 1 \leq 1 + k + 1 \Leftrightarrow 2 \cdot k \leq 1 \Leftrightarrow k = 0 \Leftrightarrow n_{L,7} = 1 \tag{5.169}$$

For $n_{L,7} = 3 \cdot k + 2$, with $k \in \mathbb{N}$, in (5.157), we have

$$3 \cdot k + 2 \leq 1 + k + 1 \Leftrightarrow 2 \cdot k \leq 0 \Leftrightarrow k = 0 \Leftrightarrow n_{L,7} = 2 \tag{5.170}$$

Thus, (5.157) is equivalent to the second inequality in (5.145).

For $n_{L,p} = 3 \cdot k$, with $k \in \mathbb{N}$, in (5.158), we have

$$3 \cdot k \leq k \Leftrightarrow 2 \cdot k \leq 0 \Leftrightarrow k = 0 \Leftrightarrow n_{L,p} = 0 \tag{5.171}$$

For $n_{L,p} = 3 \cdot k + 1$, with $k \in \mathbb{N}$, in (5.158), we have

$$3 \cdot k + 1 \leq k + 1 \Leftrightarrow 2 \cdot k \leq 0 \Leftrightarrow k = 0 \Leftrightarrow n_{L,p} = 1 \tag{5.172}$$

For $n_{L,p} = 3 \cdot k + 2$, with $k \in \mathbb{N}$, in (5.158), we have

$$3 \cdot k + 2 \leq k + 1 \Leftrightarrow 2 \cdot k + 1 \leq 0 \Leftrightarrow k \in \emptyset \tag{5.173}$$

Thus, (5.158) is equivalent to the third inequality in (5.145), for $p > 7$.

From the proof of Theorem 5.8, the Hamming weight of the codeword generated by the interleaver pattern in Fig. 5.1 is at most $38 + 12 \cdot l$ and, thus, the theorem is proved. ■

Theorem 5.13 *The minimum distance of a conventional binary turbo code (of nominal rate 1/3) using primitive feedback and monic feedforward polynomials of degree* ν *and QPPs that have a cubic inverse is upper bounded by*

$$d_{min} \leq 2 \cdot (2^{\nu+1} + 9) \tag{5.174}$$

for all nonnegative integers l *and for all interleaver lengths* L *that satisfy*

$$n_{L,p} \leq \begin{cases} 4, & \text{if } p = 2 \\ \left\lfloor \dfrac{9n_{\alpha,p}}{2} \right\rfloor + 2, & \text{if } p = 3, 5 \\ \left\lceil \dfrac{9n_{\alpha,p}}{2} \right\rceil + 2, & \text{otherwise} \end{cases} \tag{5.175}$$

where $\alpha = 2^{\nu} - 1$ *is a short-hand notation for the period of the feedback polynomial of the constituent encoders.*

Proof The proof is similar to the first part of the proof of Theorem 5.10. When there exists a cubic inverse of QPP, the congruence in (5.117) (with $x = 0$) is generalized to

$$4 \cdot a^3 \cdot q_2 \cdot \left(r_2 \cdot (1 + 2q_1 + 2aq_2) + 3 \cdot r_3 \cdot a \cdot (1 + q_1 + aq_2)^2\right) \equiv 0 \pmod{L}, \tag{5.176}$$

where $a = \alpha$. Taking into account that for $\alpha = 2^{\nu} - 1$ always $n_{\alpha,2} = 0$, the congruence (5.176) is true if

$$\begin{cases} n_{L,2} \le n_{q_2,2} + n_{r_2,2} + 2 \\ n_{L,2} \le n_{q_2,2} + n_{r_3,2} + 2 \\ n_{L,3} \le n_{q_2,3} + n_{r_2,3} + 3 \cdot n_{\alpha,3} \\ n_{L,3} \le n_{q_2,3} + n_{r_3,3} + 4 \cdot n_{\alpha,3} + 1 \\ n_{L,p} \le n_{q_2,p} + n_{r_2,p} + 3 \cdot n_{\alpha,p}, \forall p \neq 2, 3 \\ n_{L,p} \le n_{q_2,p} + n_{r_3,p} + 4 \cdot n_{\alpha,p}, \forall p \neq 2, 3 \end{cases} \tag{5.177}$$

When a QPP admits a CPP inverse, from Theorem 5.7 it results that the inequalities (5.150)–(5.153) are true.

For $n_{L,2} = 0, 1$ the first two inequalities in (5.177) are always true. If 4 is a factor in L decomposition, i.e. if $n_{L,2} \ge 2$, then 2 is a factor in both r_2 and r_3, i.e. $n_{r_2,2} \ge 1$ and $n_{r_3,2} \ge 1$ (see condition 1.b) in Table 3.2). In these conditions (i.e. for $n_{L,2} \ge 2$), taking into account the inequality (5.150), the first two inequalities in (5.177) are true if

$$n_{L,2} \le \max\left(\left\lceil \frac{n_{L,2} - 3}{3} \right\rceil, 1\right) + 3, \tag{5.178}$$

which is equivalent to $n_{L,2} \le 4$, i.e. the inequality for $p = 2$ from (5.175).

In the following we consider the case when $p = 3$ or $p = 5$.

We note that for $p = 3$, the third and the fourth inequalities in (5.177) are always true when $n_{L,3} = 1$ and for $p = 5$, the last two inequalities in (5.177) are always true when $n_{L,5} = 1$. This follows from the fact that when 3 or 5 is a factor of L, then it is also a factor of q_2, i.e. $n_{q_2,3} \ge 1$ when $n_{L,3} = 1$ and $n_{q_2,5} \ge 1$ when $n_{L,5} = 1$, because $\pi(x)$ is a QPP (see condition 2) from Table 3.1).

If 3 is a factor in L, i.e. if $n_{L,3} \ge 1$, then 3 is also a factor of r_2, i.e. $n_{r_2,3} \ge 1$ (see condition 2.a) or 2.b) Table 3.2). If 5 is a double factor of L, i.e. if $n_{L,5} \ge 2$, then 5 is a factor of both r_2 and r_3, i.e. $n_{r_2,5} \ge 1$ and $n_{r_3,5} \ge 1$ (see condition 4.b) Table 3.2).

In the following, by $p_{3,5}$ we understand only the prime 3 or only the prime 5. For $p_{3,5} = 3$ and $n_{L,3} \ge 2$, we take into account the inequality (5.151) and we will write the condition for which the third and the fourth inequalities in (5.177) are true. For $p_{3,5} = 5$ and $n_{L,5} \ge 2$, we take into account the inequality (5.152) and we will write the condition for which the last two inequalities in (5.177) are true, for $p = 5$. Thus we have

$$n_{L,p_{3,5}} \le \max\left(\left\lceil \frac{n_{L,p_{3,5}} - 1}{3} \right\rceil, 1\right) + 1 + 3 \cdot n_{\alpha,p_{3,5}}, \tag{5.179}$$

If $n_{L,p_{3,5}} \ge 2$ we have that $\max\left(\left\lceil \frac{n_{L,p_{3,5}} - 1}{3} \right\rceil, 1\right) = \left\lceil \frac{n_{L,p_{3,5}} - 1}{3} \right\rceil$. We consider three cases.

If $n_{L,p_{3,5}} = 3 \cdot k$, with $k \in \mathbb{N}^*$, (5.179) becomes

$$3 \cdot k \le k + 1 + 3 \cdot n_{\alpha,p_{3,5}} \Leftrightarrow 2 \cdot k \le 3 \cdot n_{\alpha,p_{3,5}} + 1 \Leftrightarrow$$

$$\Leftrightarrow 3 \cdot k \le \left\lfloor \frac{9 \cdot n_{\alpha,p_{3,5}} + 3}{2} \right\rfloor \tag{5.180}$$

If $n_{L,p_{3,5}} = 3 \cdot k + 1$, with $k \in \mathbb{N}^*$, (5.179) becomes

$$3 \cdot k + 1 \le k + 1 + 3 \cdot n_{\alpha,p_{3,5}} \Leftrightarrow 2 \cdot k \le 3 \cdot n_{\alpha,p_{3,5}} \Leftrightarrow$$

$$\Leftrightarrow 3 \cdot k + 1 \le \left\lfloor \frac{9 \cdot n_{\alpha,p_{3,5}}}{2} \right\rfloor + 1 \tag{5.181}$$

If $n_{L,p_{3,5}} = 3 \cdot k + 2$, with $k \in \mathbb{N}^*$, (5.179) becomes

$$3 \cdot k + 2 \le k + 1 + 1 + 3 \cdot n_{\alpha,p_{3,5}} \Leftrightarrow 2 \cdot k \le 3 \cdot n_{\alpha,p_{3,5}} \Leftrightarrow$$

$$\Leftrightarrow 3 \cdot k + 2 \le \left\lfloor \frac{9 \cdot n_{\alpha,p_{3,5}}}{2} \right\rfloor + 2 \tag{5.182}$$

If $n_{\alpha,p_{3,5}} = 2 \cdot s$, with $s \in \mathbb{N}$, the last inequalities in (5.180), (5.181), and (5.182) become

$$\begin{cases} 3 \cdot k & \le 9 \cdot s + 1 \\ 3 \cdot k + 1 & \le 9 \cdot s + 1 \\ 3 \cdot k + 2 & \le 9 \cdot s + 2 \end{cases} \Leftrightarrow \begin{cases} 3 \cdot k & \in \{9 \cdot s, 9 \cdot s - 3, \ldots\} \\ 3 \cdot k + 1 & \in \{9 \cdot s + 1, 9 \cdot s - 2, \ldots\} \\ 3 \cdot k + 2 & \in \{9 \cdot s + 2, 9 \cdot s - 1, \ldots\} \end{cases} \tag{5.183}$$

Thus, when $n_{\alpha,p_{3,5}} = 2 \cdot s$, with $s \in \mathbb{N}$, (5.183) is equivalent to

$$n_{L,p_{3,5}} \le 9 \cdot s + 2 \Leftrightarrow n_{L,p_{3,5}} \le \left\lfloor \frac{9 \cdot n_{\alpha,p_{3,5}}}{2} \right\rfloor + 2, \tag{5.184}$$

i.e. the second inequality from (5.175).

If $n_{\alpha,p_{3,5}} = 2 \cdot s + 1$, with $s \in \mathbb{N}$, the last inequalities from (5.180), (5.181), and (5.182) become

$$\begin{cases} 3 \cdot k & \le 9 \cdot s + 6 \\ 3 \cdot k + 1 & \le 9 \cdot s + 5 \\ 3 \cdot k + 2 & \le 9 \cdot s + 6 \end{cases} \Leftrightarrow \begin{cases} 3 \cdot k & \in \{9 \cdot s + 6, 9 \cdot s + 3, \ldots\} \\ 3 \cdot k + 1 & \in \{9 \cdot s + 4, 9 \cdot s + 1, \ldots\} \\ 3 \cdot k + 2 & \in \{9 \cdot s + 5, 9 \cdot s + 2, \ldots\} \end{cases} \tag{5.185}$$

Thus, when $n_{\alpha,p_{3,5}} = 2 \cdot s + 1$, with $s \in \mathbb{N}$, (5.185) is equivalent to

$$n_{L,p_{3,5}} \le 9 \cdot s + 6 \Leftrightarrow n_{L,p_{3,5}} \le \frac{9 \cdot n_{\alpha,p_{3,5}}}{2} + \frac{3}{2} \Leftrightarrow$$

$$\Leftrightarrow n_{L,p_{3,5}} \le \left\lfloor \frac{9 \cdot n_{\alpha,p_{3,5}}}{2} \right\rfloor + 2, \tag{5.186}$$

i.e. the second inequality from (5.175).

We note that for $p > 5$ a factor of L, the last two inequalities from (5.177) are always true when $n_{L,p} = 1$. This follows, as for $p = 5$, from the fact that when $p > 5$

is a factor of L, then it is also a factor of q_2, i.e. $n_{q_2,p} \geq 1$, because $\pi(x)$ is a QPP (see condition 2) from Table 3.1).

If $p > 5$ is at least a double factor in L, i.e. $n_{L,p} \geq 2$, then p is a factor of both r_2 and r_3, i.e. $n_{r_2,p} \geq 1$ and $n_{r_3,p} \geq 1$ (see conditions 3) or 4.b) in Table 3.2). In these conditions, for $n_{L,p} \geq 2$, taking into account the inequality (5.153), the last two inequalities in (5.177), for $p > 5$, are true if

$$n_{L,p} \leq \max\left(\left\lceil \frac{n_{L,p}}{3} \right\rceil, 1\right) + 1 + 3 \cdot n_{\alpha,p}. \tag{5.187}$$

If $n_{L,p} \geq 2$, we have $\max\left(\left\lceil \frac{n_{L,p}}{3} \right\rceil, 1\right) = \left\lceil \frac{n_{L,p}}{3} \right\rceil$. We consider three cases.
If $n_{L,p} = 3 \cdot k$, with $k \in \mathbb{N}^*$, (5.187) becomes

$$3 \cdot k \leq k + 1 + 3 \cdot n_{\alpha,p} \Leftrightarrow 2 \cdot k \leq 3 \cdot n_{\alpha,p} + 1 \Leftrightarrow 3 \cdot k \leq \left\lfloor \frac{9 \cdot n_{\alpha,p} + 3}{2} \right\rfloor \tag{5.188}$$

If $n_{L,p} = 3 \cdot k + 1$, with $k \in \mathbb{N}^*$, (5.187) becomes

$$3 \cdot k + 1 \leq k + 1 + 1 + 3 \cdot n_{\alpha,p} \Leftrightarrow 2 \cdot k \leq 3 \cdot n_{\alpha,p} + 1 \Leftrightarrow$$

$$\Leftrightarrow 3 \cdot k + 1 \leq \left\lfloor \frac{9 \cdot n_{\alpha,p} + 5}{2} \right\rfloor \tag{5.189}$$

If $n_{L,p} = 3 \cdot k + 2$, with $k \in \mathbb{N}$, (5.187) becomes

$$3 \cdot k + 2 \leq k + 1 + 1 + 3 \cdot n_{\alpha,p} \Leftrightarrow 2 \cdot k \leq 3 \cdot n_{\alpha,p} \Leftrightarrow$$

$$\Leftrightarrow 3 \cdot k + 2 \leq \left\lfloor \frac{9 \cdot n_{\alpha,p} + 6}{2} \right\rfloor \tag{5.190}$$

If $n_{\alpha,p} = 2 \cdot s$, with $s \in \mathbb{N}$, the last inequalities in (5.188), (5.189), and (5.190) become

$$\begin{cases} 3 \cdot k & \leq 9 \cdot s + 1 \\ 3 \cdot k + 1 & \leq 9 \cdot s + 2 \\ 3 \cdot k + 2 & \leq 9 \cdot s + 3 \end{cases} \Leftrightarrow \begin{cases} 3 \cdot k & \in \{9 \cdot s, 9 \cdot s - 3, \ldots\} \\ 3 \cdot k + 1 & \in \{9 \cdot s + 1, 9 \cdot s - 2, \ldots\} \\ 3 \cdot k + 2 & \in \{9 \cdot s + 2, 9 \cdot s - 1, \ldots\} \end{cases} \tag{5.191}$$

Thus, when $n_{\alpha,p} = 2 \cdot s$, with $s \in \mathbb{N}$, (5.191) is equivalent to

$$n_{L,p} \leq 9 \cdot s + 2 \Leftrightarrow n_{L,p} \leq \left\lceil \frac{9 \cdot n_{\alpha,p}}{2} \right\rceil + 2, \tag{5.192}$$

i.e. the third inequality from (5.175).

If $n_{\alpha,p} = 2 \cdot s + 1$, with $s \in \mathbb{N}$, the last inequalities from (5.188), (5.189), and (5.190) become

$$\begin{cases} 3 \cdot k & \leq 9 \cdot s + 6 \\ 3 \cdot k + 1 & \leq 9 \cdot s + 7 \\ 3 \cdot k + 2 & \leq 9 \cdot s + 7 \end{cases} \Leftrightarrow \begin{cases} 3 \cdot k & \in \{9 \cdot s + 6, 9 \cdot s + 3, \ldots\} \\ 3 \cdot k + 1 & \in \{9 \cdot s + 7, 9 \cdot s + 4, \ldots\} \\ 3 \cdot k + 2 & \in \{9 \cdot s + 5, 9 \cdot s + 2, \ldots\} \end{cases} \tag{5.193}$$

Thus, when $n_{\alpha,p} = 2 \cdot s + 1$, with $s \in \mathbb{N}$, (5.193) is equivalent to

$$n_{L,p} \leq 9 \cdot s + 7 \Leftrightarrow n_{L,p} \leq \frac{9 \cdot n_{\alpha,p}}{2} + \frac{5}{2} \Leftrightarrow n_{L,p} \leq \left\lceil \frac{9 \cdot n_{\alpha,p}}{2} \right\rceil + 2, \tag{5.194}$$

i.e. the third inequality from (5.175).

Thus the theorem is proven. ■

We note that the upper bounds from Theorems 5.8–5.13 will appear generally with high multiplicity, i.e. there will be many codewords of weight equal to the upper bound when it is reached. This means that in many cases, when the d_{min} is slightly below the bound, with low multiplicity, the turbo code error floor performance is likely to be dominated by the bound instead of the d_{min}. This is considered in finding of good QPP interleavers by the method described in Sect. 7.5 (Trifina and Tarniceriu 2014).

In Rosnes (2012) is it remarked that the upper bounds on the d_{min} in Theorems 5.8–5.13 can be shown to hold when the dual termination (Guinand and Lodge 1994) (see Sect. 2.3) is used and the length of interleaver, L, is sufficiently large.

In Rosnes (2012) the exact minimum distances for the turbo code with all lengths from the LTE standard (3GPP 2008) and the corresponding QPP interleavers are given when using dual trellis termination (Guinand and Lodge 1994).

Some d_{min}-optimal LTE QPPs (3GPP 2008), reported in Table IV from Rosnes (2012), are given in Tables 5.6, 5.7, 5.8 and 5.9. For each QPP, the exact multiplicity $N_{d_{min}}$, i.e. the number of minimum-weight codewords, is also enlisted.

LTE QPPs from Table 5.6 reach the upper bound of d_{min} equal to 36 (from the third row in Table 5.5), because for the lengths $312 = 2^3 \cdot 3 \cdot 13, 344 = 2^3 \cdot 43, 440 = 2^3 \cdot 5 \cdot 11$, and $488 = 2^3 \cdot 61$, we have $n_{L,2} = 3$, $n_{L,3} \leq 1$, and $n_{L,p} \leq 1$, for $p \neq 2, 3$.

LTE QPPs from Table 5.7 reach the upper bound of d_{min} equal to $38 = 38 + 12 \cdot l$, for $l = 0$, from Theorem 5.8, because for the lengths $496 = 2^4 \cdot 31$, $624 = 2^4 \cdot 3 \cdot 13, 656 = 2^4 \cdot 41, 688 = 2^4 \cdot 43, 752 = 2^4 \cdot 47, 816 = 2^4 \cdot 3 \cdot 17, 848 = 2^4 \cdot 53$, $880 = 2^4 \cdot 5 \cdot 11$, $912 = 2^4 \cdot 3 \cdot 19$, $944 = 2^4 \cdot 59$, and $976 = 2^4 \cdot 61$, we have $n_{L,2} = 4 \leq l + 4$, for $l = 0$, and $n_{L,p} \leq 1$, for $p \neq 2, 7$.

LTE QPPs from Table 5.8 reach the upper bound of d_{min} equal to $50 = 38 + 12 \cdot l$, for $l = 1$, from Theorem 5.8, because for the lengths $1696 = 2^5 \cdot 53$, $1760 = 2^5 \cdot 5 \cdot 11$, and $1952 = 2^5 \cdot 61$, we have $n_{L,2} = 5 \leq l + 4$, for $l = 1$, and $n_{L,p} \leq 1$, for $p \neq 2, 7$. We note that LTE QPPs from Table 5.8 do not reach the upper bound of d_{min} equal to 50 from Theorem 5.9 because they have no QPP inverses, but CPP inverses (given in the third column in Table 5.8).

Table 5.6 Some d_{min}-optimal LTE QPPs that reach the upper bound of d_{min} equal to 36 (from the third row in Table 5.5)

L	$\pi(x)$	d_{min}	$N_{d_{min}}$
312	$19x + 78x^2$	36	1654
344	$193x + 86x^2$	36	1846
440	$91x + 110x^2$	36	2422
488	$91x + 122x^2$	36	2710

Table 5.7 Some d_{min}-optimal LTE QPPs that reach the upper bound of d_{min} equal to 38 from Theorem 5.8, for $l = 0$

L	$\pi(x)$	d_{min}	$N_{d_{min}}$
496	$157x + 62x^2$	38	906
624	$41x + 234x^2$	38	1162
656	$185x + 82x^2$	38	1226
688	$21x + 86x^2$	38	1290
752	$23x + 94x^2$	38	1418
816	$127x + 102x^2$	38	1546
848	$239x + 106x^2$	38	1610
880	$137x + 110x^2$	38	1674
912	$29x + 114x^2$	38	1738
944	$21x + 86x^2$	38	1290
976	$59x + 122x^2$	38	1866

Table 5.8 Some d_{min}-optimal LTE QPPs that reach the upper bound of d_{min} equal to 50 from Theorem 5.8, for $l = 1$

L	$\pi(x)$	$\pi^{-1}(x)$	d_{min}	$N_{d_{min}}$
1696	$55x + 954x^2$	$663x + 530x^2 + 424x^3$	50	3264
1760	$27x + 110x^2$	$163x + 990x^2 + 1320x^3$	50	3392
1952	$59x + 610x^2$	$579x + 1586x^2 + 488x^3$	50	3776

Table 5.9 Some d_{min}-optimal LTE QPPs that reach the upper bound of d_{min} equal to 51 from Theorem 5.9

L	$\pi(x)$	d_{min}	$N_{d_{min}}$
4288	$33x + 134x^2$	51	4484
4544	$357x + 142x^2$	51	4476
5056	$39x + 158x^2$	51	5300
5312	$41x + 166x^2$	51	5244
5568	$43x + 174x^2$	51	5500
6080	$47x + 190x^2$	51	6012

Table 5.10 Some d_{min}-optimal LTE QPPs that reach the upper bound of d_{min} equal to 50, in the class of QPPs with inverse QPPs, from Theorem 5.10

L	$\pi(x)$	$\pi^{-1}(x)$	d_{min}	$N_{d_{min}}$
2496	$181x + 468x^2$	$1117x + 1404x^2$	50	9748
2624	$27x + 164x^2$	$1555x + 2132x^2$	50	10259
2752	$143x + 172x^2$	$943x + 1548x^2$	50	10771
3264	$443x + 204x^2$	$2483x + 2652x^2$	50	12819
3392	$51x + 212x^2$	$3259x + 1060x^2$	50	13331
3456	$451x + 192x^2$	$2539x + 3264x^2$	50	6802
3520	$257x + 220x^2$	$2753x + 1540x^2$	50	13843
3648	$313x + 228x^2$	$1993x + 3420x^2$	50	14355
3712	$271x + 232x^2$	$2671x + 232x^2$	50	14611
3776	$179x + 236x^2$	$443x + 1180x^2$	50	14867
3904	$363x + 244x^2$	$1667x + 3172x^2$	50	15379
3968	$375x + 248x^2$	$455x + 2232x^2$	50	15635
6144	$263x + 480x^2$	$5303x + 5856x^2$	50	12179

Table 5.11 d_{min} - optimal QPPs with improved error performance for two particular interleaver lengths

L	$\pi(x)$	$\pi^{-1}(x)$	d_{min}	$N_{d_{min}}$
1504	$49x + 658x^2$	$353x + 470x^2 + 1128x^3$	50	3241
2048	$21x + 128x^2$	$1853x + 1408x^2$	50	9198

LTE QPPs from Table 5.9 reach the upper bound of d_{min} equal to 51 from Theorem 5.9, because for the lengths $4288 = 2^6 \cdot 67$, $4544 = 2^6 \cdot 71$, $5056 = 2^6 \cdot 79$, $5312 = 2^6 \cdot 83$, $5568 = 2^6 \cdot 3 \cdot 29$, and $6080 = 2^6 \cdot 5 \cdot 19$, we have $n_{L,2} = 6$ and $n_{L,p} \leq 1$, for $p \neq 2$.

LTE QPPs from Table 5.10 reach the upper bound of d_{min} equal to 50, in the class of QPPs with inverse QPPs, from Theorem 5.10. The inverse QPPs for the reported QPPs are given in the third column in Table 5.10.

In Rosnes (2012), for some lengths, there are also tabulated d_{min}-optimal or improved QPPs found by computer search. To quickly reject bad QPPs during the search, the special version of the triple impulse method (Crozier et al. 2004), mentioned at the beginning of this section, was used. Furthermore, the size of the search space can be reduced by using Theorems 5.8–5.13. For instance, Theorems 5.8 and 5.9 can be used to terminate the search when a QPP has been found that led to a minimum distance equal to one of the upper bounds provided in the theorems. Furthermore, Theorems 5.10 and 5.11 can be used to terminate the search within the class of QPPs with a QPP inverse when a QPP achieving one of the upper bounds provided by the theorems has been found. Similarly, Theorems 5.12 and 5.13 can be used to terminate the search within the class of QPPs with a CPP inverse. In the final stage, the best QPP candidate can be checked using the exhaustive algorithm for

Table 5.12 Some d_{min}-optimal QPPs with improved error rate performance which do not have an inverse QPP

L	$\pi(x)$	$\pi^{-1}(x)$	d_{min}	$N_{d_{min}}$
2496	$119x + 702x^2$	$215x + 390x^2 + 218x^3$	51	3034
2624	$125x + 1066x^2$	$677x + 246x^2 + 2296x^3$	51	2556
2752	$21x + 430x^2$	$557x + 1634x^2 + 344x^3$	51	2853
3008	$143x + 94x^2$	$1951x + 470x^2 + 1128x^3$	51	2940
3264	$55x + 102x^2$	$2359x + 2958x^2 + 2856x^3$	51	3396
3392	$81x + 106x^2$	$513x + 2862x^2 + 2968x^3$	51	3324
3520	$27x + 110x^2$	$1923x + 2750x^2 + 3080x^3$	51	3452
3648	$43x + 114x^2$	$403x + 1026x^2 + 2280x^3$	51	3580
3776	$1359x + 826x^2$	$1151x + 354x^2 + 1416x^3$	51	3708
3904	$1283x + 854x^2$	$2715x + 2806x^2 + 2440x^3$	51	3836

computing the distance spectrum of a turbo code from Rosnes and Ytrehus (2005) or Garello et al. (2001).

In Table 5.11, which is a part of Table V from Rosnes (2012), two d_{min}-optimal QPPs are given for two medium interleaver lengths. For each QPP, the exact multiplicity $N_{d_{min}}$ is also listed. These polynomials were selected based on error-rate performance through simulations. The inverse polynomial, denoted by $\pi^{-1}(x)$, is tabulated in the third column of the table. We can verify that the minimum distance for the QPP of length 2048 from Table 5.11 reaches the general upper bound given by Theorem 5.10 above because this QPP has an inverse QPP and for $\nu = 3$, $d_{min} \le 2 \cdot (2^{3+1} + 9) = 50$ results. The minimum distance for the QPP of length 1504 from Table 5.11 reaches the partial upper bound given in Theorem 5.12 above because this QPP has an inverse CPP, $1504 = 2^5 \cdot 47$ and for $l = 1$, it follows that $\left\lfloor \frac{3 \cdot l}{2} \right\rfloor + 4 = 5$, and $d_{min} \le 38 + 12 \cdot l = 50$.

Table 5.12, which is a part of Table VI from Rosnes (2012), gives some d_{min}-optimal improved QPPs for medium-to-long interleaver lengths. For each QPP, the corresponding exact multiplicity $N_{d_{min}}$ is also listed. In the third column of the table, an inverse polynomial, denoted by $\pi^{-1}(x)$, is tabulated. Because for the lengths $2496 = 2^6 \cdot 3 \cdot 13$, $2624 = 2^6 \cdot 41$, $2752 = 2^6 \cdot 43$, $3008 = 2^6 \cdot 47$, $3264 = 2^6 \cdot 3 \cdot 17$, $3392 = 2^6 \cdot 53$, $3520 = 2^6 \cdot 5 \cdot 11$, $3648 = 2^6 \cdot 3 \cdot 19$, $3776 = 2^6 \cdot 59$ and $3904 = 2^6 \cdot 61$, the exponents of the factors from their prime decomposition meet the equality $n_{L,p} = \begin{cases} 6, \text{ if } p = 2 \\ 1, \text{ otherwise.} \end{cases}$, the minimum distances for QPPs lengths from Table 5.12 reach the partial upper bound given in Theorem 5.9, namely 51. From Table 5.12 we note that QPPs of the same lengths as those from Table 5.10 have significantly smaller multiplicities $N_{d_{min}}$. This is due to the fact that the QPPs from Table 5.12 have inverse CPPs, while QPPs from Table 5.10 have inverse QPPs.

Deriving a general upper bound on the d_{min} (if it exists) that holds for any interleaver length without any constraints on the minimum degree of the inverse polynomials is still an open problem. Even finding a general upper bound when the minimum degree of the inverse polynomials is three is also an open problem. At the end of the paper (Rosnes 2012) the author has remarked that the QPP $39x + 760x^2 \pmod{9728}$ (which has a CPP inverse) gives an estimated d_{min} of 56, which indicates that 51 is probably *not* a universal bound on the d_{min} when the minimum degree of the inverse polynomials is three. This fact is also indicated by the QPP $59x + 1680x^2 \pmod{6144}$, which has a CPP inverse, a minimum distance of 51 and a very low multiplicity of 94.

5.3 Upper Bound on Minimum Distance of Turbo Codes with Interleavers Based on PPs of Any Degree

In this section we give an upper bound on the minimum distance of an interleaver based on a PP of any degree, for lengths multiples of 8, that is of the form $L = 2^3 \cdot M$, with M a positive integer (Ryu et al. 2015).

Firstly, we define the notion of PLPP.

Definition 5.14 A PLPP interleaver is an interleaver $\pi(x)$ of length L so that

$$\pi(x) = \begin{cases} P_{1,0} \cdot x + P_{0,0}, & \text{for } x \pmod{R} = 0 \\ P_{1,1} \cdot x + P_{0,1}, & \text{for } x \pmod{R} = 1 \\ \cdots\cdots\cdots\cdots\cdots & \\ P_{1,R-1} \cdot x + P_{0,R-1}, & \text{for } x \pmod{R} = R-1, \end{cases} \tag{5.195}$$

which can be also represented in the following form

$$\pi(x) = \begin{cases} P_{1,0} \cdot Ry + P'_{0,0}, & \text{for } x = Ry \\ P_{1,1} \cdot Ry + P'_{0,1}, & \text{for } x = Ry + 1 \\ \cdots\cdots\cdots\cdots\cdots\cdots\cdots\cdots\cdots\cdots\cdots & \\ P_{1,R-1} \cdot Ry + P'_{0,R-1}, & \text{for } x = Ry + R - 1, \end{cases} \tag{5.196}$$

with $1 \le R < L$, $R \mid L$ and $P'_{0,l} = P_{0,l} + P_{1,l} \cdot l$, $\forall l = \overline{0, R-1}$.

The efficient address generation method in Ryu (2012a) can be used for implementing a PLPP.

The following lemma from Ryu (2012a) is useful to show that all PPs are equivalent to PLPPs when the interleaver lengths are of the form $L = 2^3 \cdot M$, with M a positive integer.

Lemma 5.15 (Ryu 2012a) *Let* $\pi(x) = \sum_{i=1}^{d} q_i x^i \pmod{L}$ *be a PP, where* $\gcd(q_1, L) = 1$ *and all factors of L divide* q_i, $\forall i = \overline{2, d}$. *Then,* $\pi(x)$ *is a PP. Let*

$\pi'(x) = \sum_{i=1}^{d} i q_i x^{i-1}$ *be the formal derivative of* $\pi(x)$ *and let also* R *be an integer so that* $L \mid (q_i R^2), \forall i = \overline{2, d}$. *Then* $\pi(x)$ *is decomposed into a PLPP so that* $P_{1,l} = \pi'(l)$ *and* $P'_{0,l} = \pi(l)$, *with* $l = 0, 1, \cdots, R-1$.

Proof For $x = Ry + l$, the ith degree term of $\pi(x)$ can be written as

$$q_i x^i = q_i (Ry + l)^i = q_i \cdot \sum_{m=0}^{i} C_i^m \cdot (Ry)^m \cdot l^{i-m} =$$

$$= q_i \cdot l^i + q_i \cdot i \cdot (Ry) \cdot l^{i-1} + q_i \cdot \sum_{m=2}^{i} C_i^m \cdot (Ry)^m \cdot l^{i-m} \tag{5.197}$$

Since $L \mid (q_i R^2), \forall i = \overline{2, d}$, it follows that $L \mid (q_i R^n)$, where $2 \le i \le d$ and $n \ge 3$. Thus $q_i x^i \pmod L = i q_i l^{i-1} \cdot Ry + q_i \cdot l^i \pmod L$. Then, $\pi(Ry + l) = (Ry) \cdot \sum_{i=1}^{d} (i q_i l^{i-1}) + \sum_{i=1}^{d} (q_i l^i) = \pi'(l) \cdot Ry + \pi(l)$. Consequently $P_{1,l} = \sum_{i=1}^{d} i q_i l^{i-1} = \pi'(l)$ and $P'_{0,l} = \sum_{i=1}^{d} q_i l^i = \pi(l), l = 0, 1, \cdots, R-1$. ■

Lemma 5.16 (Ryu 2012b) *Let* $\pi(x) = \sum_{i=1}^{d} q_i x^i \pmod L$ *be a PP. Assume that* $L = 2^3 \cdot M$, *with* M *a positive integer. Then,* $\pi(x)$ *is equivalent to a PLPP and* $R \le L/2$.

Proof We choose a positive integer R so that $L \mid R^2$ and $R \mid L$. Assume that the prime decomposition of L is $L = 2^{n_{L,2}} \cdot \prod_{i=2}^{n_L} p_{L,i}^{n_{L,i}}$, with $n_{L,2} \ge 3$, $n_L \in \mathbb{N}^*$ and $n_{L,i} \ge 1$, $\forall i = \overline{2, n_L}$. Then, a valid choice for R is the value $2^{n_{R,2}} \cdot \prod_{i=2}^{n_L} p_i^{n_{R,i}}$, where $n_{R,2} = \lceil n_{L,2}/2 \rceil$ and $n_{R,i} = \begin{cases} n_{L,i}, & \text{if } n_{L,i} = 1, \\ \lceil n_{L,i}/2 \rceil, & \text{if } n_{L,i} \ge 2 \end{cases}, \forall i = \overline{2, n_L}$.

For this choice of R, as $2 \cdot \lceil n_{L,i}/2 \rceil \ge n_{L,i}$, for $n_{L,i} \ge 2$, we have $L \mid R^2$, and as $\lceil n_{L,i}/2 \rceil < n_{L,i}$, for $n_{L,i} \ge 2$, we have $R \mid L$.

Because $L \mid R^2$, we have $L \mid (q_i R^2)$, for $2 \le i \le d$, and according to Lemma 5.15 it follows that $\pi(x)$ can be decomposed into a PLPP so that $P_{1,l} = \pi'(l)$ and $P'_{0,l} = \pi(l)$, with $l = 0, 1, \cdots, R-1$.

The fact that $R \le L/2$ follows from $n_{R,2} \le n_{L,2} - 1$ and $n_{R,i} \le n_{L,i}, \forall i = \overline{2, n_L}$. ■

Lemma 5.17 gives the upper bound of minimum distance for an interleaver based on PP of any degree, for lengths multiples of 8.

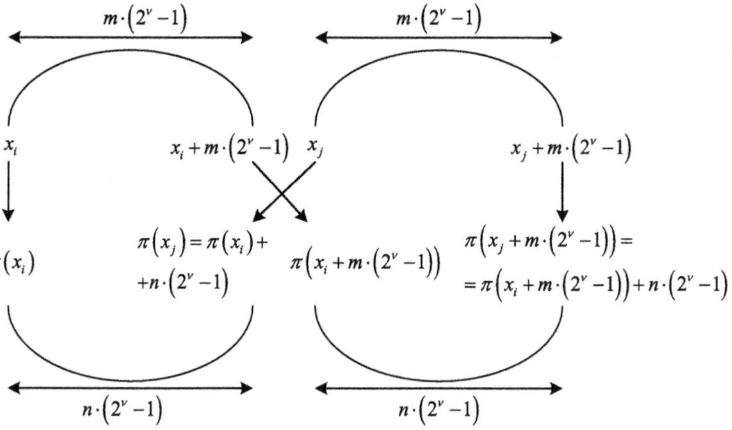

Fig. 5.6 Critical interleaver pattern of size 4

Lemma 5.17 (Ryu et al. 2015) *Let there be a PP of arbitrary degree modulo a length multiple of 8 and let the first coefficients of the equivalent PLPP be equal for all* l, *i.e.,* $P_{1,0} = P_{1,1} = \ldots = P_{1,R-1} = P$. *Let* m *and* n *be positive integers and* $R|(m \cdot (2^\nu - 1))$, *where* ν *is the degree of the primitive feedback and monic feedforward polynomials of RSC codes, which are component codes of a conventional turbo code. Under these conditions, there exists a critical interleaver pattern of size* 4 *as shown in Fig. 5.6 and the minimum distance of the turbo code with this PP (or PLPP) interleaver is upper bounded by* $(m + n) \cdot 2^\nu + 12$.

Proof Consider the codeword generated by the interleaver pattern shown in Fig. 5.6. For both constituent codes the interleaver pattern contains two fundamental paths with input sequences of weight 2. The difference between 1-positions in the two input sequences of weight 2 is $m \cdot (2^\nu - 1)$ for the upper constituent code, and $n \cdot (2^\nu - 1)$ for the lower constituent code, respectively.

The four elements of permutation $\pi(\cdot)$, indicated in Fig. 5.6, are written in detail below

$$\begin{cases} x_i \to \pi(x_i) \\ x_i + m \cdot (2^\nu - 1) \to \pi(x_i + m \cdot (2^\nu - 1)) \\ x_j \to \pi(x_j) = \pi(x_i) + n \cdot (2^\nu - 1) \\ x_j + m \cdot (2^\nu - 1) \to \pi(x_j + m \cdot (2^\nu - 1)) = \\ = \pi(x_i + m \cdot (2^\nu - 1)) + n \cdot (2^\nu - 1) \end{cases} \tag{5.198}$$

Since the distance between the first two points for the upper constituent code is $m \cdot (2^\nu - 1)$ and $R|(m \cdot (2^\nu - 1))$, the two points are in the same ith component LPP of the PLPP. Thus, the two points are mapped into $\pi(x_i) = Px_i + R_i$ and $\pi(x_i + m \cdot (2^\nu - 1)) = P(x_i + m \cdot (2^\nu - 1)) + R_i$, respectively.

Since the input sequences for the upper and lower constituent codes are mapped by an interleaver, there is a point in the input for the upper constituent code that is mapped into the point $Px_i + R_i + n \cdot (2^\nu - 1)$ in the input for the lower constituent code. Let us call it x_j, where j is the number of the corresponding component LPP of the PLPP. Then, $\pi(x_j) = Px_j + R_j = Px_i + R_i + n \cdot (2^\nu - 1)$. Since the distance between the last two points for the upper constituent code is $m \cdot (2^\nu - 1)$ and $R|(m \cdot (2^\nu - 1))$, the two points are in the same jth component LPP of the PLPP.

From those above, we have

$$\begin{aligned} P(x_j + m \cdot (2^\nu - 1)) + R_j = Px_j + R_j + Pm \cdot (2^\nu - 1) = \\ = Px_i + R_i + n \cdot (2^\nu - 1) + Pm \cdot (2^\nu - 1) = \\ = P(x_i + m \cdot (2^\nu - 1)) + R_i + n \cdot (2^\nu - 1) \end{aligned} \tag{5.199}$$

But (5.199) is the equation corresponding to the fourth point from (5.198). Thus, an input sequence of weight 4, shown by the interleaver pattern from Fig. 5.6, exists for PLPP with R LPPs, when $R|(m \cdot (2^\nu - 1))$.

It is easy to check that the parity weight of codeword generated by the upper constituent code is $2 \cdot (m \cdot 2^{\nu-1} + 2)$ (see the proof of Theorem 5.8). Then, the weight of the corresponding turbo codeword is at most $2 \cdot (m \cdot 2^{\nu-1} + 2) + 2 \cdot (n \cdot 2^{\nu-1} + 2) + 4 = (m + n) \cdot 2^\nu + 12$. ■

In Table 5.13 upper bounds on the minimum distance for turbo codes with PPs when $\nu = 3$ are shown. The result in Lemma 5.17 is similar to Tables 5.4 and 5.5 (Tables II and III in Rosnes 2012). However, Lemma 5.17 can also be applied to higher order PPs.

Reference Trifina et al. (2017) gives up to five degree PPs of short LTE lengths (from 40 to 512) with optimum minimum distances, when using LTE turbo codes with dual trellis termination. Four CPPs and two 4-PPs obtained in Trifina et al. (2017) reaching the upper bound of minimum distance equal to 36 (their PLPP representations have $R = 2$) are given in Table 5.14. Many PPs of degree greater than two better than QPPs in terms of TUB(FER) at high SNR for AWGN channel (see Eq. (2.22)), considering only the first term in the distances spectra, were found in Trifina et al. (2017). For the PPs found in Trifina et al. (2017), the authors have computed the number of component LPPs from their PLPP representation under the constraint that the coefficients of linear terms of the LPPs are equal to each other (R). From the results given in Tables I and II from Trifina et al. (2017) it can be observed that if some PPs of a certain length have minimum distances close to each other, then, for those with greater values of R for their PLPP representation, the multiplicities are smaller. Thus, an important conclusion resulting from Trifina et al. (2017) is that the value of R highlights a tradeoff between the error rate performance and implementation complexity. For very good error rate performance the values of R, and thus the complexity, should increase, and for lower implementation complexity the values of R should decrease, but in this case the error rate performance is compromised.

Table 5.13 Upper Bounds (UBs) for the minimum distance of turbo codes using PP based interleavers. Generator matrix of recursive systematic convolutional codes is as in the LTE turbo code (3GPP 2008)

$R \in$	m	n	UB (d_{min})
$R_1 = \{1, 7\}$	1	1	28
		2	36
		3	44
		4	52
		5	60
		6	68
		7	76
		8	84
$R_2 = R_1 \cup \{2, 14\}$	2	1	36
		2	44
		3	52
		4	60
		5	68
		6	76
		7	84
$R_3 = R_1 \cup$ $\cup \{3, 21\}$	3	1	44
		2	52
		3	60
		4	68
		5	76
		6	84
$R_4 = R_2 \cup$ $\cup \{4, 28\}$	4	1	52
		2	60
		3	68
		4	76
		5	84
$R_5 = R_1 \cup$ $\cup \{5, 35\}$	5	1	60
		2	68
		3	76
		4	84
$R_6 = R_1 \cup R_2 \cup$ $\cup R_3 \cup \{6, 42\}$	6	1	68
		2	76
		3	84
$R_7 = R_1 \cup$ $\cup \{49\}$	7	1	76
		2	84
$R_8 = R_4 \cup$ $\cup \{8, 56\}$	8	1	84

The upper bound in Lemma 5.17 shows some interesting points, specifically, for $R = 4$. In Rosnes (2012) (see Theorems 5.9 and 5.10 in this chapter), it was shown that the upper bound on the minimum distance of turbo codes with QPP based interleavers, when QPPs have a quadratic inverse, is 50 and for some QPPs with non-QPP inverse, 51. In it is shown that for every PP, not just for QPPs, the upper bound on the minimum distance is 52 when $R = 4$. Although the PPs that achieve the upper bound 52 have not been identified Lemma 5.17 shows that the upper bound

Table 5.14 Four d_{min} - optimal CPPs and two d_{min} - optimal 4-PPs for some short LTE lengths, which reach the upper bound of minimum distance equal to 36 for $R = 2$ (from Table 5.13)

L	$\pi(x)$	d_{min}	$N_{d_{min}}$
288	$211x + 36x^2 + 24x^3$	36	1511
304	$67x + 38x^2 + 76x^3$	36	1607
320	$47x + 40x^2 + 80x^3$	36	1702
368	$57x + 46x^2 + 92x^3$	36	1990
256	$183x + 96x^2 + 64x^3 + 16x^4$	36	1422
288	$79x + 120x^2 + 0x^3 + 6x^4$	36	1511

will increase only with 1 or 2 for any degree of polynomials when $R = 4$, i.e., no large difference in terms of minimum distance.

Finally, we mention that, to avoid the limitation of PPs previously shown, dithered LPP interleavers are proposed in Ryu et al. (2015). These interleavers are, actually, PLPPs as in (5.195), but the coefficients of the component LPPs are relaxed ("dithered") to allow improving the performance of a specific turbo code. In Ryu et al. (2015) it is shown that for some short lengths, good dithered LPP interleavers were found, which outperform good QPPs from 3GPP (2008), Rosnes and Takeshita (1992) or Trifina and Tarniceriu (2014) or CPPs from Trifina and Tarniceriu (2013).

References

3GPP TS 36.212 V8.3.0, 3rd Generation partnership project, multiplexing and channel coding (Release 8) (2008), http://www.etsi.org/

S. Crozier, P. Guinand, A. Hunt, Computing the minimum distance of turbo-codes using iterative decoding techniques, in *22th Biennial Symposium on Communication*, Kingston, Ontario, Canada, 31 May–3 June 2004, pp. 306–308

R. Garello, P. Pierleoni, S. Benedetto, Computing the free distance of turbo codes and serially concatenated codes with interleavers: algorithms and applications. IEEE J. Sel. Areas Commun. **19**(5), 800–812 (2001)

P. Guinand, J. Lodge, Trellis termination for turbo encoders, in *17th Biennial Symposium Communication*, Queen's University, Kingston, Canada, 30 May–1 June 1994, pp. 389–392

G.H. Hardy, E.M. Wright, *An Introduction to the Theory of Numbers*, 4th edn. (Oxford University Press, Clarendon, 1975)

E. Rosnes, On the minimum distance of turbo codes with quadratic permutation polynomial interleavers. IEEE Trans. Inf. Theory **58**(7), 4781–4795 (2012)

E. Rosnes, O.Y. Takeshita, Optimum distance quadratic permutation polynomial-based interleavers for turbo codes, in *IEEE International Symposium Information Theory (ISIT)*, Seattle, USA, 9–14 July 2006, pp. 1988–1992

E. Rosnes, Ø. Ytrehus, Improved algorithms for the determination of turbo-code weight distributions. IEEE Trans. Commun. **53**(1), 20–26 (2005)

J. Ryu, Permutation polynomial based interleavers for turbo codes over integer rings. Ph.D. thesis. Ohio State University (2007), https://etd.ohiolink.edu/rws_etd/document/get/osu1181139404/inline

J. Ryu, Efficient address generation for permutation polynomial based interleavers over integer rings. IEICE Trans. Fundam. **E95–A**(1), 421–424 (2012a)

J. Ryu, Permutation polynomials of higher degrees for turbo code interleavers. IEICE Trans. Commun. **E95–B**(12), 3760–3762 (2012b)

J. Ryu, O.Y. Takeshita, On quadratic inverses for quadratic permutation polynomials over integers rings. IEEE Trans. Inf. Theory **52**(3), 1254–1260 (2006)

J. Ryu, O.Y. Takeshita, On inverses for quadratic permutation polynomials over integers rings, 10 February 2011, http://arxiv.org/abs/1102.2223

J. Ryu, L. Trifina, H. Balta, The limitation of permutation polynomial interleavers for turbo codes and a scheme for dithering permutation polynomials. AEÜ Int. J. Electron. Commun. **69**(10), 1550–1556 (2015)

Telemetry channel coding, consultative committee for space data systems (CCSDS), CCSDS 101.0-B-6, Blue Book (2002)

L. Trifina, J. Ryu, D. Tarniceriu, Up to five degree permutation polynomial interleavers for short length LTE turbo codes with optimum minimum distance, in *IEEE International Symposium on Signals Circuits System (ISSCS)*, Iasi, Romania, 13–14 July 2017

L. Trifina, D. Tarniceriu, Analysis of cubic permutation polynomials for turbo codes. Wirel. Pers. Commun. **69**(1), 1–22 (2013)

L. Trifina, D. Tarniceriu, Improved method for searching interleavers from a certain set using Garello's method with applications for the LTE standard. Ann. Telecommun. **69**(5–6), 251–272 (2014)

Chapter 6
Parallel Turbo Decoding for Permutation Polynomial Interleavers

6.1 Preliminaries

The very good error correction properties of turbo codes make them an attractive choice for all types of communication systems. However, in applications requiring high data rate, there is a problem related to the overall large delay of the iterative decoder. In fact, each frame must be processed through several iterations before estimating the corresponding data. Consequently, the implementation of the decoder should be optimized so that the delay should be kept as low as possible.

A direct way to increase the throughput of an iterative decoder consists in using many parallel iterative decoders. The use of M iterative decoders leads to an increase of M times of the throughput, but the complexity also increases M times. Particularly, the memory used, that becomes a significant portion of the overall complexity for the medium-to-long interleaver lengths, will increase linearly with M.

A more efficient solution to increase the throughput of an iterative decoder is a parallelized structure where the operations for decoding a turbo-codeword are performed at the same time by different processors.

The turbo decoder iterates for a certain number of iterations between two decoders with a soft-input soft-output (SISO) decoding algorithm (Trifina and Munteanu 2008). At a given iteration, decoder 1 accepts at its input what decoder 2 supplies at its output in the previous iteration (except for the first iteration, when there is no input from decoder 2). Actually, decoder 1 sends its output to decoder 2, which accepts it as input for the current iteration. Between the two decoders, an interleaver and a de-interleaver rearrange the information corresponding to the permutation π and to the inverse permutation π^{-1}, respectively. In parallel concatenation the information exchanged (called extrinsic information) belongs to input bits of both encoders.

When convolutional codes are constituent codes of the turbo encoder, the algorithm carried in each SISO decoder is usually the Bahl–Cocke–Jelinek–Raviv (BCJR) (Bahl et al. 1974). Ideally it requires two recursions on the code trellis, a forward one, from the beginning to the end of the block, and a backward one, from the end to

L. Trifina and D. Tarniceriu, *Permutation Polynomial Interleavers for Turbo Codes*, Signals and Communication Technology,
https://doi.org/10.1007/978-981-13-2625-7_6

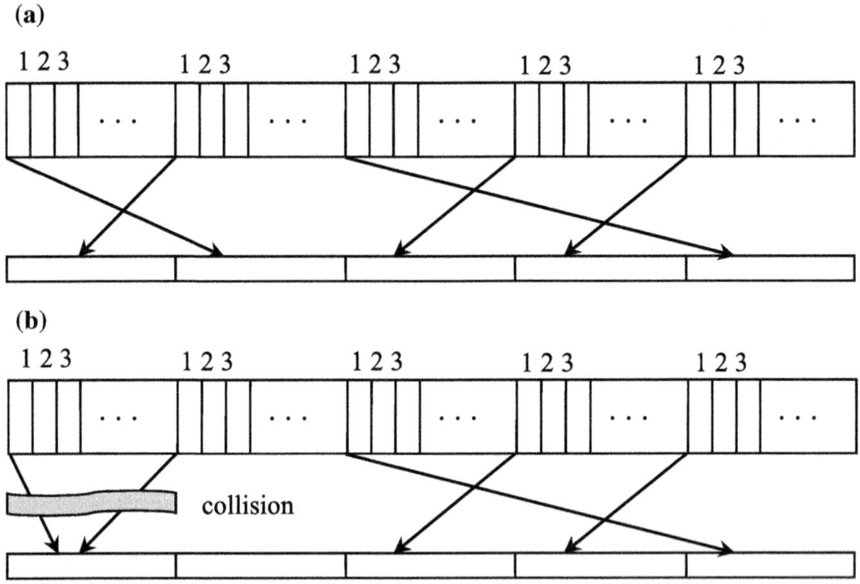

Fig. 6.1 **a** Avoiding collisions; **b** A collision

the beginning of the block. Practically, in low delay implementations the algorithm is "windowed" by dividing the entire block into sub-blocks and making some shorter recursions within sub-blocks, all of them associated to distinct processors working in parallel.

Since every processor performs the same algorithm, all of them access the memory at the same time. Then, collisions can occur, weakening the effectiveness of implementation. It is assumed that the memory is divided into a number of banks. A collision occurs when two (or more) processors access the same memory bank at the same time. There are no collisions when processors access different memory banks at the same time. If there are P processors and each of them accesses the memory once at a time, then there will be at most P simultaneous accesses to memory. To ensure an implementation without collisions, there must be at least P memory banks. A schematic view of collisions is shown in Fig. 6.1.

The memory access is performed when the SISO decoder reads input variables and when it writes output variables. Actually, the output variables are an update of input variables and it is assumed that the update of a given variable is written on the same memory element where it has been stored. It is also assumed that the cost for reading and writing operations of corresponding updates coincide, thus the reading and writing operations are considered equivalent in terms of memory access.

The problem of parallelism consists in that there are two schemes of reading/writing operations, one associated with decoder 1, which reads and writes variables in natural order, the other one associated with decoder 2, which reads and writes in an interleaved (permuted) order. In other words, the two semi-iterations of turbo decoding algorithm impose different constraints on processing variables in the memory banks. This is the main problem of parallelism in turbo decoders.

Thus, the interleaver has the main role in this scenario. If the interleaver is defined by the identity permutation, i.e., $\pi(i) = i, \forall i \in \mathbb{Z}_L$, the problem of parallelism becomes considerably easier, since both decoder 1 and decoder 2 work in the same way, at least in what concerns the memory access scheme. Both decoders write/read in a natural order. But in this case the turbo code performance is weaker compared to a well designed interleaver.

Many researchers have tried to find good interleavers that allow a parallel implementation of the turbo decoder without collisions. Out of these we mention: block or multi-stage interleavers for parallel decoding (Giulietti et al. 2002), divisible interleavers for parallel decoding (Kwak and Lee 2002), inter-window randomization (IWR) interleavers for parallel decoding (Nimbalker et al. 2003). For more references concerning collision-free interleaver design we refer the reader to Benedetto et al. (2006).

When a specific permutation must be implemented in a parallel decoder, as it is the case for standards as LTE (3GPP 2008) or Digital Video Broadcasting (DVB) (DVB-RCS 2003), or if a decoder must work with multiple interleavers and/or with different types of parallel processing, then particular collision-free interleavers as those mentioned above are not viable. However, the solution of collision-free interleavers has been successfully applied to LTE standard (3GPP 2008) since QPP interleavers are contention-free for any number of processors divisible by interleaver length (Takeshita 2006). We present the proof for this property of any degree PP interleavers in Sect. 6.2. Almost regular permutation (ARP) interleavers (Berrou et al. 2004) can be designed so that they have high degree parallel processing and good error correction performance.

Another method to solve the collision problem is to use a scheme which maps the inputs to be read or written in the memory, in a desired order so that collisions are avoided.

In Tarable and Benedetto (2004, 2005) and Tarable et al. (2004), the authors proposed an algorithm based on the permutation decomposition which allows to implement a parallel decoder based on an arbitrary interleaver with any parallelism degree.

A method with the same approach to avoid collisions was proposed in Nieminen (2014), but it uses a butterfly network of 2-by-2 crossbar switches for the mapping scheme between processors and memories. The use of a butterfly network of 2-by-2 crossbar switches involves that the number of processors used in parallel iterative decoding is a power of 2, but in this case the complexity is smaller than that for the solution previously mentioned. Actually, the solution is optimal in terms of address information. In Nieminen (2014) the control bits needed for routing the inputs in a butterfly network are obtained in a quite complicated way. In Nieminen (2017) it is shown that the control bits are obtained in an easier way for QPP interleavers. The same easy way to obtain the control bits is proved to be valid for any degree PP interleavers and ARP interleavers in Trifina and Tarniceriu (2017). This solution is presented in Sect. 6.3 of this chapter.

6.2 Maximum Contention-Free Property of Any Degree PP Interleavers

In this section we show that PP interleavers are without collision for any number of processors divisible by interleaver length (i.e. they are "MCF"). The condition for collision-free interleavers for window size W and permutation $\pi(\cdot)$ describing the interleaver is Nimbalker et al. (2004, 2008)

$$\lfloor \pi(j + tW)/W \rfloor \neq \lfloor \pi(j + vW)/W \rfloor \text{ and} \tag{6.1}$$

$$\lfloor \pi^{-1}(j + tW)/W \rfloor \neq \lfloor \pi^{-1}(j + vW)/W \rfloor, \tag{6.2}$$

where $0 \leq j < W$, $0 \leq t < v < L/W$, and $\pi^{-1}(\cdot)$ is the inverse permutation. In Nimbalker et al. (2004, 2008), it is shown that the number of collision-free interleavers $A(W, M)$ is lower and upper bounded by

$$W! \cdot (M!)^W + M! \cdot (W!)^M - M! \cdot W! < A(W, M) < (M!)^W \cdot (W!)^M \tag{6.3}$$

The fact that the upper bound is smaller than $(MW)!$ (the total number of interleavers of length MW) shows that the fraction of collision-free interleavers is small. This upper bound is represented in Fig. 6.2 on a logarithmic scale. It is divided by to the total number of interleavers for the first 17 lengths from LTE standard (3GPP 2008) when the number of processors for parallel decoding is $M = 2$, $M = 4$ or $M = 8$.

The fact that all PP interleavers are MCF is given by the following theorem (Takeshita 2006).

Theorem 6.1 (MCF Property of PP interleavers) *Let* $\pi(x) = q_0 + q_1 \cdot x + q_2 \cdot x^2 + \cdots + q_d \cdot x^d \pmod L$, $0 \leq x \leq L - 1$, *be a permutation polynomial interleaver of degree d. Then,* $\pi(x)$ *generates an MCF interleaver.*

Proof Firstly, we check condition (6.1) for this interleaver.

Let there be

$$Q_t = \left\lfloor \frac{\pi(j + tW)}{W} \right\rfloor \text{ and } Q_v = \left\lfloor \frac{\pi(j + vW)}{W} \right\rfloor \tag{6.4}$$

Then

$$\begin{aligned} \pi(j + tW) &= Q_t \cdot W + \left(\pi(j + tW) \pmod W\right) \text{ and} \\ \pi(j + vW) &= Q_v \cdot W + \left(\pi(j + vW) \pmod W\right) \end{aligned} \tag{6.5}$$

We have to show that $Q_t \neq Q_v$ for $t - v \neq 0 \pmod M$ and any $0 \leq j < W$.

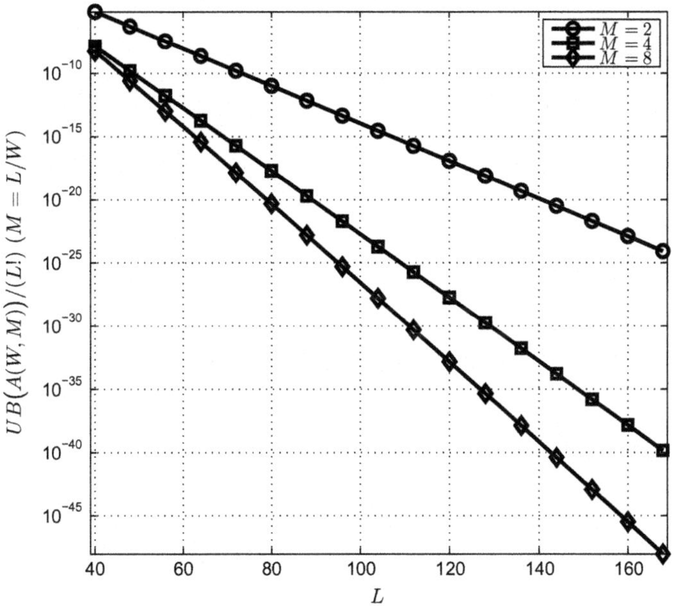

Fig. 6.2 The upper bound of the number of collision-free interleavers from (6.3) ($UB(A(W, M))$) divided by to the total number of interleavers of length $L = MW$, for the first 17 lengths of LTE standard, when the number of processors is $M = 2$, $M = 4$ or $M = 8$

Assume that $Q_t = Q_v$. Then

$$Q_t - Q_v = \frac{\pi(j + tW) - \big(\pi(j + tW) \pmod W\big)}{W} - \\ - \frac{\pi(j + vW) - \big(\pi(j + vW) \pmod W\big)}{W} = 0 \quad (6.6)$$

But

$$\pi(j + tW) = q_0 + q_1 \cdot j + q_2 \cdot j^2 + \cdots + q_d \cdot j^d \pmod W \text{ and} \\ \pi(j + vW) = q_0 + q_1 \cdot j + q_2 \cdot j^2 + \cdots + q_d \cdot j^d \pmod W, \quad (6.7)$$

since $W = L/M$ is integer and, in this case, for x and y integers, from $x \equiv y \pmod L$ it follows that $x \equiv y \pmod W$. Thus

$$\pi(j + tW) \pmod W = \pi(j + vW) \pmod W \quad (6.8)$$

and the absolute value of (6.6) can be simplified as

$$|Q_t - Q_v| = \frac{|\pi(j + tW) - \pi(j + vW)|}{W} = 0 \tag{6.9}$$

As $0 \leq j < W$ and $0 \leq tW < vW < L$, it follows that $0 \leq j + tW < j + vW < L + W$. Obviously, we have $(j + tW) \pmod L \neq (j + vW) \pmod L$. Therefore, $\pi(j + tW) \neq \pi(j + vW)$, because $\pi(x)$ generates a permutation polynomial and in this way a contradiction exists in (6.9).

To verify condition (6.2), we note that the permutation polynomials form a finite group $\mathcal{G}$ under the operation of function composing, i.e. $\pi(\pi(x))$ is a permutation polynomial and the inverse function $\pi^{-1}(x)$ can be found by a sufficient number of function composition of $\pi(x)$ to itself. Then, it is sufficient to show that each element from $\mathcal{G}$ which includes the inverse function $\pi^{-1}(x)$ satisfies (6.1). This implies the same steps as in the first part of the theorem proof, replacing $\pi(j + tW)$ by $\pi(\pi(j + tW))$ and $\pi(j + vW)$ by $\pi(\pi(j + tW))$. We have:

$$\begin{aligned}\pi\big(\pi(j + tW)\big) = q_0 + q_1 \cdot \pi(j + tW) + q_2 \cdot \big(\pi(j + tW)\big)^2 + \cdots + \\ +q_d \cdot \big(\pi(j + tW)\big)^d \pmod W\end{aligned} \tag{6.10}$$

or, using (6.7),

$$\begin{aligned}\pi\big(\pi(j + tW)\big) = q_0 + q_1 \cdot \pi(j) + q_2 \cdot \big(\pi(j)\big)^2 + \cdots + \\ +q_d \cdot \big(\pi(j)\big)^d \pmod W\end{aligned} \tag{6.11}$$

Similarly

$$\begin{aligned}\pi\big(\pi(j + vW)\big) = q_0 + q_1 \cdot \pi(j) + q_2 \cdot \big(\pi(j)\big)^2 + \cdots + \\ +q_d \cdot \big(\pi(j)\big)^d \pmod W = \pi\big(\pi(j + tW)\big) \pmod W\end{aligned} \tag{6.12}$$

Therefore, (6.9) becomes

$$|Q_t - Q_v| = \frac{\big|\pi\big(\pi(j + tW)\big) - \pi\big(\pi(j + vW)\big)\big|}{W} = 0 \tag{6.13}$$

As $(j + tW) \pmod L \neq (j + vW) \pmod L$, and $\pi(x)$ generates a permutation polynomial, it follows that $\pi(j + tW) \neq \pi(j + vW)$ and thus $\pi\big(\pi(j + tW)\big) \neq \pi\big(\pi(j + vW)\big)$. So, the assumption that $Q_t = Q_v$ is not true.

Finally, we use the mathematical induction method to show that every function obtained by successively composing $\pi(x)$ (which after a number of steps generate inverse function $\pi^{-1}(x)$) generates a MCF interleaver. ■

6.3 Parallel Access by Butterfly Networks for Any Degree Permutation Polynomials

Let the interleaver length be factorized as $L = M \cdot W$ for positive integers M and W. In this section we denote by I_L, $L \in \mathbb{N}^*$, the set $\{0, 1, \ldots, L-1\}$. Let functions $a_j(k) : I_M \times I_W \to I_L$ be defined so that $a_j(k) \neq a_i(k)$, $\forall k \in I_W$, $\forall i, j \in I_M$ with $j \neq i$, and $\forall x \in I_L$, there is a unique function a_{j_x}, so that $a_{j_x}(k_x) = x$. For a parallel turbo decoding implementation, the linear and interleaved accesses to the memory banks at time $k \in I_W$ are defined by the address vectors $\big(a_0(k), a_1(k), \ldots, a_{M-1}(k)\big)$ and $\big(\pi(a_0(k)), \pi(a_1(k)), \ldots, \pi(a_{M-1}(k))\big)$, respectively. To avoid the collisions at a certain moment in the same memory bank, there must exist a function $F_M : I_L \to I_W$, so that

(1) $F_M\big(a_i(k)\big) \neq F_M\big(a_j(k)\big)$ (linear parallel access)

(2) $F_M\big(\pi(a_i(k))\big) \neq F_M\big(\pi(a_j(k))\big)$ (interleaved parallel access)

$\forall i, j \in I_M$, $i \neq j$, and $k \in I_W$. If there exists a function F_M for an interleaver over I_L, then we say that the interleaver is *contention-free* for parameters M and W. We note that this definition is different from that given by Eqs. (6.1) and (6.2). In this case, a parallel turbo decoding implementation is possible with M processors and M memory banks, each of them with W memory cells.

The linear parallel access for any degree PP interleaver with functions $a_j(k) = j \cdot W + k$ was proved in Theorem 6.1 (Takeshita 2006).

In Theorem 6.3 it is proved that any PP interleaver of length $L = 2^n \cdot W$, with n and W positive integers, is contention-free by the function

$$F_{2^n}(x) = x \pmod{2^n}, \tag{6.14}$$

and the function (6.14) provides exactly the same mapping of addresses as a $2^n \times 2^n$ butterfly network does.

In Fig. 6.3 a 8×8-butterfly network and eight memories are shown. The 8×8-butterfly network consists of twelve 2-by-2 crossbar switches (i.e. $\log_2(8) = 3$ columns multiplied by $8/2 = 4$ switches per column). Each 2-by-2 crossbar switch is controlled by one bit. If the bit is zero, a direct connection is performed, and if the bit is one, a cross connection is performed. In Fig. 6.3 the twelve control bits $x_0, x_1, \ldots, x_{11}$ are set to zero. Hence all 2-by-2 crossbar switches perform a direct connection. To get from an input bit i to an output bit j, three control bits are required (Lawrie 1975). If the binary representations of decimal numbers i and j are $(i_2 i_1 i_0)_2$ and $(j_2 j_1 j_0)_2$, respectively, then the control bits cb_k, with $k = 0, 1, 2$, are computed by $cb_k = i_k \oplus j_k$, where $\oplus$ is the modulo 2 operator. The bit with index 0 in a binary representation is the least significant bit. The first control bit used is cb_0, then cb_1, and then cb_2, i.e. the three control bits are applied from the left to the right side of the butterfly network.

Before proving the main theorem in this section we give the definition of a uniformly p^n-dyadic vector, where p is a prime number.

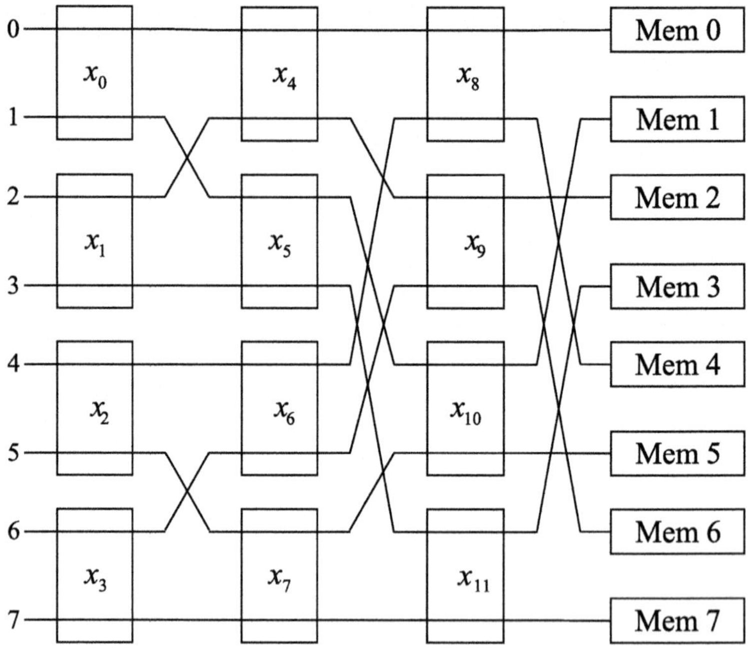

Fig. 6.3 8 × 8-butterfly network with eight memories

Definition 6.2 Let p be a prime number. A vector $\big(a_0(k), a_1(k), \ldots, a_{p^n-1}(k)\big)$ of integer components $a_l \geq 0, l \in I_{p^n}$, is uniformly p^n-dyadic if

$$a_{p^q k+i} \neq a_{p^q k+j} \pmod{p^q}, \tag{6.15}$$

$\forall i, j \in I_{p^q}$ with $i \neq j, \forall k \in I_{p^{n-q}}$, and $\forall q = 1, 2, \ldots, n$.

For example, the vector $\big(a_0(k), a_1(k), \ldots, a_{3^2-1}(k)\big) = (0, 1, 2, 3, 4, 5, 6, 7, 8)$ is uniformly 9-dyadic because $a_0(k) \neq a_1(k) \neq a_2(k) \pmod 3$ (i.e. $0 \neq 1 \neq 2 \pmod 3$), $a_3(k) \neq a_4(k) \neq a_5(k) \pmod 3$ (i.e. $3 \neq 4 \neq 5 \pmod 3$), $a_6(k) \neq a_7(k) \neq a_8(k) \pmod 3$ (i.e. $6 \neq 7 \neq 8 \pmod 3$), and $a_0(k) \neq a_1(k) \neq \cdots \neq a_8(k) \pmod{3^2}$ (i.e. $0 \neq 1 \neq \cdots \neq 8 \pmod 9$). Vector $\big(a_0(k), a_1(k), \ldots, a_{3^2-1}(k)\big) = (3, 7, 11, 6, 4, 5, 9, 10, 2)$ fulfills the conditions $a_{3k+i} \neq a_{3k+j} \pmod 3$, $\forall i, j \in I_3$ with $i \neq j, \forall k \in I_3$, but it is not uniformly 9-dyadic because $a_2(k) = a_8(k) \pmod{3^2}$ (i.e. $11 = 2 \pmod 9$). Vector $\big(a_0(k), a_1(k), \ldots, a_{3^2-1}(k)\big) = (17, 7, 11, 6, 4, 5, 9, 10, 3)$ also fulfills the conditions $a_{3\cdot 0+i} \neq a_{3\cdot 0+j} \pmod{3^2}$, $\forall i, j \in I_9$ with $i \neq j$, but it is not uniformly 9-dyadic because $a_0(k) = a_2(k) \pmod 3$ (i.e. $17 = 11 = 2 \pmod 3$).

The classes on I_L modulo p^n are denoted by

$$U_i(p) = \{x \in I_L | x = i \pmod{p^n}\} \tag{6.16}$$

for all $i \in I_{p^n}$. The Cartesian product of the sets $U_i(p)$ is denoted by U^{p^n}, i.e., $U^{p^n} = U_0(p) \times U_1(p) \times U_2(p) \times \cdots \times U_{p^n-1}(p)$. Obviously, from (6.16) it results that the sets $U_i(p)$ are disjoint.

For the interleaver length $L = 2^n \cdot W$ vectors $A_a(k)$ and $A_\pi(k)$ are written as

$$A_a(k) = \big(a_0(k), a_1(k), \ldots, a_{2^n-1}(k)\big) \tag{6.17}$$

and

$$A_\pi(k) = \big(\pi(a_0(k)), \pi(a_1(k)), \ldots, \pi(a_{2^n-1}(k))\big) \tag{6.18}$$

for all $k \in I_W$ and components $a_l(k) \in I_L$.

The main theorem is given below.

Theorem 6.3 (Contention-free (CF) property of PP interleavers by the function (6.14) Nieminen 2017) *Let n and W be positive integers and $L = 2^n \cdot W$. Let $\pi(\cdot)$ be a PP interleaver on I_L of arbitrary degree, as in (3.1). Then it holds that*

(1) $A_a(k)$ is uniformly 2^n-dyadic
if and only if
(2) $A_\pi(k)$ is uniformly 2^n-dyadic
if and only if
(3) there exists the transition matrix B_k of a $2^n \times 2^n$ butterfly network so that $B_k A_a(k) \in U^{2^n}$
if and only if
(4) there exists the transition matrix C_k of a $2^n \times 2^n$ butterfly network so that $C_k A_\pi(k) \in U^{2^n}$

for al $k \in I_W$.

From Theorem 6.3 it follows that any PP interleaver on I_L, with $L = 2^n \cdot W$, is contention-free by the function F_{2^n} given in (6.14). On the base of this theorem we may construct many kinds of vectors of component functions a_j to access the extrinsic memories without collisions for PP interleavers on $I_{2^n M}$. We have only to check that the designed vectors of component functions the a_j are uniformly 2^n-dyadic according to (6.15) for $p = 2$.

To prove the theorem, three lemmas are firstly stated and proved.

Lemma 6.4 *Let n and W be positive integers and $L = 2^n W$. Assume that $A = (a_0, a_1, \ldots, a_{2^n-1})$ is a vector whose components $a_l \in I_L$, $\forall l \in I_{2^n}$. Then A is uniformly 2^n-dyadic if and only if there exists the transition matrix B of a $2^n \times 2^n$ butterfly network so that*

$$BA \text{ belongs to } U^{2^n}. \tag{6.19}$$

The assumption that $a_l \in I_L$, $\forall l \in I_{2^n}$, can be replaced by the requirement that a_l are non-negative integers, i.e. $a_l \geq 0$, $\forall l \in I_{2^n}$.

Proof "⇒" For the direct proof, we assume that vector A is uniformly 2^n-dyadic. We show that, if a_l is applied to the input pin l of the $2^n \times 2^n$ butterfly network, then it is mapped to the output $U_i(2)$ with $i = a_l \pmod{2^n}$. Let $l_0, l_1, \ldots, l_{n-1}$ and $a_{l;0}, a_{l;1}, \ldots, a_{l;n-1}$, be the least n significant bits of l and a_l, respectively. From Lawrie (1975) the n control bits cb_j, with $j \in I_n$, which make an interconnection along the $2^n \times 2^n$ butterfly network from the input pin l to the output pin $a_l \pmod{2^n}$ are determined by the modulo 2 sums $cb_j = (l_j \oplus a_{l;j})$, with $j \in I_n$.

We begin with the least significant bit $a_{l;0}$ of the component a_l. We denote the original vector by A_0, that is, $A^0 = A = (a_0, a_1, \ldots, a_{2^n-1})$. Since $A^0 = A$ is uniformly 2^n-dyadic by assumption, from (6.15) for $p = 2$ and $q = 1$, we have that $a_{2k} \neq a_{2k+1} \pmod 2$, $\forall k \in I_{2^{n-1}}$. We note that if $l = 2k$ then $l_0 = 0$ and the first control bit is $a_{l;0}$ and if $l = 2k+1$ then $l_0 = 1$ and the first control bit is $a_{l;0} \oplus 1$. We derive the vector $A^1 = (a_0^1, a_1^1, \ldots, a_{2^n-1}^1)$ from A^0 by the following rule: assign $a_{2k}^1 = a_{2k}$ and $a_{2k+1}^1 = a_{2k+1}$ if $a_{2k} = 0 \pmod 2$ (i.e. if $cb_0 = a_{2k;0} \oplus 0 = 0 \oplus 0 = a_{2k+1;0} \oplus 1 = 1 \oplus 1 = 0$), otherwise (i.e. if $cb_0 = a_{2k;0} \oplus 0 = 1 \oplus 0 = a_{2k+1;0} \oplus 1 = 0 \oplus 1 = 1$) $a_{2k}^1 = a_{2k+1}$ and $a_{2k+1}^1 = a_{2k}$, $\forall k \in I_{2^{n-1}}$. We denote by B_0 the permutation matrix which permutes A^0 to get A^1 according to this rule, that is, $A^1 = B_0 A^0$.

In Table 6.1 the two least significant bits are shown for the components a_l^0 and a_l^1, for an arbitrary sequence of four successive components a_l, $l = 4k, 4k+1, 4k+2, 4k+3$, with $k \in I_{2^{n-2}}$. By the fact that A is uniformly 4-dyadic, from (6.15) for $p = 2$ and $q = 2$, we have that $a_{4k} \neq a_{4k+1} \neq a_{4k+2} \neq a_{4k+3} \pmod 4$. For the case shown in Table 6.1 this means that $a_{4k;1} \neq a_{4k+3;1}$ and $a_{4k+1;1} \neq a_{4k+2;1}$. From the column $a_l^1 \pmod 4$ in Table 6.1 we have that $a_{4k;0}^1 = a_{4k+2;0}^1 = 0, a_{4k+1;0}^1 = a_{4k+3;0}^1 = 1$. Thus, because $a_{4k} \neq a_{4k+1} \neq a_{4k+2} \neq a_{4k+3} \pmod 4$, it results that $a_{4k;1}^1 \neq a_{4k+2;1}^1$ and $a_{4k+1;1}^1 \neq a_{4k+3;1}^1$. Equations $a_{4k;0}^1 = a_{4k+2;0}^1 = 0, a_{4k+1;0}^1 = a_{4k+3;0}^1 = 1, a_{4k;1}^1 \neq a_{4k+2;1}^1$, and $a_{4k+1;1}^1 \neq a_{4k+3;1}^1$ can be shown to be also fulfilled for the other three cases of sequences $a_{4k;0}, a_{4k+1;0}, a_{4k+2;0}, a_{4k+3;0}$, consisting in the least significant bit of the four successive components a_l, namely 0101, 1001, and 1010. Thus we have that $a_{4k+i}^1 \neq a_{4k+2+i}^1 \pmod 4$ and $a_{4k+i}^1 = a_{4k+2+i}^1 = i \pmod 2$, $\forall\, i \in I_2$. Now we derive the vector $A^2 = (a_0^2, a_1^2, \ldots, a_{2^n-1}^2)$ from A^1 based on second least significant bits $a_{l;1}$ of the components a_l, as follows. For $i \in I_2$, we assign $a_{4k+i}^2 = a_{4k+i}^1$ and $a_{4k+2+i}^2 = a_{4k+2+i}^1$ if $a_{4k+i}^1 = i \pmod 4$ (i.e. if $cb_1 = a_{4k;1} \oplus 0 = 0 \oplus 0 = a_{4k+2;1} \oplus 0 = 0 \oplus 0 = a_{4k+1;1} \oplus 1 = 1 \oplus 1 = a_{4k+3;1} \oplus 1 = 1 \oplus 1 = 0$) and otherwise (i.e. if $cb_1 = a_{4k;1} \oplus 0 = 1 \oplus 0 = a_{4k+2;1} \oplus 0 = 1 \oplus 0 = a_{4k+1;1} \oplus 1 = 0 \oplus 1 = a_{4k+3;1} \oplus 1 = 0 \oplus 1 = 1$) $a_{4k+i}^2 = a_{4k+2+i}^1$ and $a_{4k+i}^2 = a_{4k+2+i}^1$, $\forall k \in I_{2^{n-2}}$. We note that condition $a_{4k+i}^1 = i \pmod 4$ is equivalent to $a_{4k;1} = a_{4k+1;1} = 0$ and $a_{4k+2;1} = a_{4k+3;1} = 1$, and condition $a_{4k+i}^1 \neq i \pmod 4$ is equivalent to $a_{4k;1} = a_{4k+1;1} = 1$ and $a_{4k+2;1} = a_{4k+3;1} = 0$. So it follows that $a_{4k+j}^2 = j \pmod 4$, $\forall k \in$

Table 6.1 The two least significant bits for the components a_l^0 and a_l^1, for an arbitrary sequence of four successive components a_l

l	$a_l^0 \pmod 4$	$a_l^1 \pmod 4$
$4k$	$a_{4k;1}0$	$a_{4k;1}0$
$4k+1$	$a_{4k+1;1}1$	$a_{4k+1;1}1$
$4k+2$	$a_{4k+2;1}1$	$a_{4k+3;1}0$
$4k+3$	$a_{4k+3;1}0$	$a_{4k+2;1}1$

$I_{2^{n-2}}$ and $\forall j \in I_4$. We denote by B^1 the permutation matrix which permutes A^1 to get A^2. Then we have $A^2 = B_1 A^1 = B_1 B_0 A^0$.

We continue in the same way by assigning the components of the vectors A^q, with $q = 3, 4, \ldots, n$. In general, $\forall k \in I_{2^{n-q}}$ and $\forall i \in I_{2^{q-1}}$, the components of vector A^{q-1} satisfy conditions $a^{q-1}_{2^q k+i} \neq a^{q-1}_{2^q k+2^{q-1}+i} \pmod{2^q}$, since vector A is uniformly 2^q-dyadic, and $a^{q-1}_{2^q k+i} = a^{q-1}_{2^q k+2^{q-1}+i} = i \pmod{2^{q-1}}$, by the construction of vector A^{q-1}. Therefore for $i \in I_{2^{q-1}}$, we choose the components of A^q from A^{q-1} by $a^q_{2^q k+i} = a^{q-1}_{2^q k+i}$ and $a^q_{2^q k+2^{q-1}+i} = a^{q-1}_{2^q k+2^{q-1}+i}$ if $a^q_{2^{q-1}k+i} = i \pmod{2^q}$ (i.e. if $cb_{q-1} = 0$) and otherwise (i.e. if $cb_{q-1} = 1$) $a^q_{2^q k+i} = a^{q-1}_{2^q k+2^{q-1}+i}$ and $a^q_{2^q k+2^{q-1}+i} = a^{q-1}_{2^q k+i}$, $\forall k \in I_{2^{n-q}}$. Then the components of vector A^n will satisfy conditions $a^n_i = i \pmod{2^n}$, i.e. $a^n_i \in U_i(2)$, $\forall i \in I_{2^n}$. As an example, in Table 6.2 the components a^q_l for $q = 1, 2, 3$, for the 8-dyadic vector of eight components $A = (a_0, a_1, \ldots, a_7) = (7, 4, 1, 6, 0, 3, 5, 2)$ at the input of an 8×8 butterfly network (Fig. 6.4) and the corresponding control bits at each of the three columns of switches are given. In parenthesis binary representations are shown. For each of the three columns consisting in four 2-by-2 crossbar switches the values at the input of the top switch (x_0, x_4, and x_8 in Fig. 6.4) are colored in blue, the values at the input of the second switch (x_1, x_5, and x_9 in Fig. 6.4) are coloured in red, the values at the input of the third switch (x_2, x_6, and x_{10} in Fig. 6.4) are coloured in green, and the values at the input of the fourth (bottom) switch (x_3, x_7, and x_{11} in Fig. 6.4) are coloured in magenta. For each switch, the control bits in Table 6.2 and the routing paths in Fig. 6.4 are coloured in the same way. The corresponding permutation matrices B_0, B_1, B_2, and $B = B_2 B_1 B_0$ are given below.

$$B_0 = \begin{bmatrix} 0&1&0&0&0&0&0&0 \\ 1&0&0&0&0&0&0&0 \\ 0&0&0&1&0&0&0&0 \\ 0&0&1&0&0&0&0&0 \\ 0&0&0&0&1&0&0&0 \\ 0&0&0&0&0&1&0&0 \\ 0&0&0&0&0&0&0&1 \\ 0&0&0&0&0&0&1&0 \end{bmatrix}, \quad B_1 = \begin{bmatrix} 1&0&0&0&0&0&0&0 \\ 0&0&0&1&0&0&0&0 \\ 0&0&1&0&0&0&0&0 \\ 0&1&0&0&0&0&0&0 \\ 0&0&0&0&1&0&0&0 \\ 0&0&0&0&0&0&0&1 \\ 0&0&0&0&0&0&1&0 \\ 0&0&0&0&0&1&0&0 \end{bmatrix},$$

$$B_2 = \begin{bmatrix} 0&0&0&0&1&0&0&0 \\ 0&1&0&0&0&0&0&0 \\ 0&0&0&0&0&0&1&0 \\ 0&0&0&0&0&0&0&1 \\ 1&0&0&0&0&0&0&0 \\ 0&0&0&0&0&1&0&0 \\ 0&0&1&0&0&0&0&0 \\ 0&0&0&1&0&0&0&0 \end{bmatrix}, \quad B = \begin{bmatrix} 0&0&0&0&1&0&0&0 \\ 0&0&1&0&0&0&0&0 \\ 0&0&0&0&0&0&0&1 \\ 0&0&0&0&0&1&0&0 \\ 0&1&0&0&0&0&0&0 \\ 0&0&0&0&0&0&1&0 \\ 0&0&0&1&0&0&0&0 \\ 1&0&0&0&0&0&0&0 \end{bmatrix} \tag{6.20}$$

As it was previously shown by the control bits, one choice of two new components of A^q amounts to the permutation by one 2-by-2 crossbar switch in the qth column of

Table 6.2 The components a_l^q for $q = 1, 2, 3$, for an arbitrary 8-dyadic vector of eight components $a_l, l = 0, 1, \ldots, 7$, at the input of an 8×8 butterfly network and the corresponding control bits at each of the three columns of switches. Binary representations are shown in parenthesis

l	$a_l = a_l^0$	cb_0	a_l^1	cb_1	a_l^2	cb_2	a_l^3
0 (000)	7 (111)	1	4 (100)	0	4 (100)	1	0 (000)
1 (001)	4 (100)	1	7 (111)	1	1 (001)	0	1 (001)
2 (010)	1 (001)	1	6 (110)	0	6 (110)	1	2 (010)
3 (011)	6 (110)	1	1 (001)	1	7 (111)	1	3 (011)
4 (100)	0 (000)	0	0 (000)	0	0 (000)	1	4 (100)
5 (101)	3 (011)	0	3 (011)	1	5 (101)	0	5 (101)
6 (110)	5 (101)	1	2 (010)	0	2 (010)	1	6 (110)
7 (111)	2 (010)	1	5 (101)	1	3 (011)	1	7 (111)

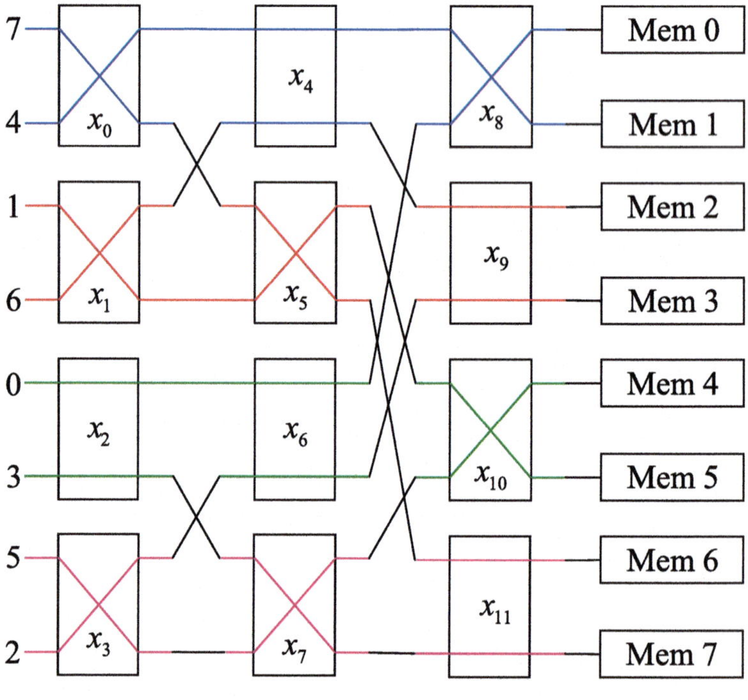

Fig. 6.4 Routing the 8-dyadic vector $A = (7, 4, 1, 6, 0, 3, 5, 2)$ by an 8×8-butterfly network with eight memories

the $2^n \times 2^n$ butterfly network. It holds that $A^q = B_{q-1}A^{q-1}$. Therefore we have that $A^n = B_{n-1}B_{n-2}\ldots B_0 A$. The matrix $B = B_{n-1}B_{n-2}\ldots B_0$ is the required transition matrix of a $2^n \times 2^n$ butterfly network and thus (6.19) holds.

"$\Leftarrow$" For the inverse proof we assume that (6.19) holds with some transition matrix of a $2^n \times 2^n$ butterfly network and we prove that vector A is uniformly 2^n-dyadic. We suppose that there exists four integers q, k, i, j, with $q \le n$, $i, j \in I_{2^q}$, $i \neq j$, and

$k \in I_{2^{n-q}}$, so that (6.15) does not hold for $p = 2$, i.e. $a_{2^q k+i} = a_{2^q k+j} \pmod{2^q}$. This means that the least significant q bits of values $a_{2^q k+i}$ and $a_{2^q k+j}$ are identical. By assumption (6.19) we have that the inputs $a_{2^q k+i}$ and $a_{2^q k+j}$ end up into two sets $U_{i_1}(2)$ and $U_{j_1}(2)$, with $i_1 \neq j_1 \pmod{2^n}$. Since $a_{2^q k+i} \in U_{i_1}(2)$ and $a_{2^q k+j} \in U_{j_1}(2)$, from (6.16) with $p = 2$, we have $a_{2^q k+i} = i_1 \pmod{2^n}$ and $a_{2^q k+j} = j_1 \pmod{2^n}$. Since the least significant q bits of values $a_{2^q k+i}$ and $a_{2^q k+j}$ are identical, we have $i_{1;0} = j_{1;0} = o_0^q, i_{1;1} = j_{1;1} = o_1^q, \ldots, i_{1;q-1} = j_{1;q-1} = o_{q-1}^q$, where we have denoted by $o_0^q, o_1^q, \ldots, o_{q-1}^q$ these identical least significant q bits.

We consider now the $(k+1)$th group, $k \in I_{2^{n-q}}$, of the 2^q inputs in the $2^n \times 2^n$ butterfly network. We note that the values $a_{2^q k+i}$ and $a_{2^q k+j}$ belong to this set of inputs. Since these values get to the outputs i_1 and j_1 of the network, from Lawrie (1975) it results that the first q control bits for input $a_{2^q k+i}$ are $cb_{i;0} = i_0 \oplus o_0^q$, $cb_{i;1} = i_1 \oplus o_1^q, \ldots, cb_{i;q-1} = i_{q-1} \oplus o_{q-1}^q$ and the first q control bits for input $a_{2^q k+j}$ are $cb_{j;0} = j_0 \oplus o_0^q$, $cb_{j;1} = j_1 \oplus o_1^q, \ldots, cb_{j;q-1} = j_{q-1} \oplus o_{q-1}^q$. In the previous equations we used the fact that the term $2^q k$ does not affect the least significant q bits of the input positions $2^q k + i$ and $2^q k + j$. Without loss of generality we may assume $i < j$. If $i = 2k_1$ and $j = 2k_1 + 1$, with $k_1 \in I_{2^{q-1}}$, then the values $a_{2^q k+i}$ and $a_{2^q k+j}$ are at the inputs of the same 2-by-2 crossbar switch in the first column of switches of the network. Moreover, we have $i_0 = 0$ and $j_0 = 1$ and the control bit of the previously mentioned switch is $cb_{i;0} = o_0^q$ for the input $a_{2^q k+i}$ and $cb_{j;0} = o_0^q \oplus 1$ for the input $a_{2^q k+j}$. If $i \in \{4k_1, 4k_1+1\}$ and $j \in \{4k_1+2, 4k_1+3\}$, where $k_1 \in I_{2^{q-2}}$, then the values $a_{2^q k+i}$ and $a_{2^q k+j}$ get at the inputs of the same 2-by-2 crossbar switch in the second column of switches of the network. Moreover, we have $i_1 = 0$ and $j_1 = 1$ and the control bit of the previously mentioned switch is $cb_{i;1} = o_1^q$ for the input $a_{2^q k+i}$ and $cb_{j;1} = o_1^q \oplus 1$ for the input $a_{2^q k+j}$. In general, if $i \in \{2^{q_1} k_1, 2^{q_1} k_1 + 1, \ldots, 2^{q_1} k_1 + 2^{q_1-1} - 1\}$ and $j \in \{2^{q_1} k_1 + 2^{q_1-1}, 2^{q_1} k_1 + 2^{q_1-1} + 1, \ldots, 2^{q_1} k_1 + 2^{q_1} - 1\}$, where $k_1 \in I_{2^{q-q_1}}$, $q_1 \in \{1, 2, \ldots, q\}$, then the values $a_{2^q k+i}$ and $a_{2^q k+j}$ get at the inputs of the same 2-by-2 crossbar switch in the q_1th column of switches of the network. Moreover, we have $i_{q_1-1} = 0$ and $j_{q_1-1} = 1$ and the control bit of the previously mentioned switch is $cb_{i;q_1-1} = o_{q_1-1}^q$ for the input $a_{2^q k+i}$ and $cb_{j;q_1-1} = o_{q_1-1}^q \oplus 1$ for the input $a_{2^q k+j}$. We see that in all previous cases we have two different control bits for the same 2-by-2 crossbar switch. But this is not possible because, by the construction of the butterfly network, each switch has only one control bit. Thus the assumption $a_{2^q k+i} = a_{2^q k+j} \pmod{2^q}$ is not valid and the vector A is uniformly 2^n-dyadic. We note that if $a_{2^q k+i} \neq a_{2^q k+j} \pmod{2^q}$, $\forall i, j \in I_{2^q}, i \neq j$, and $\forall k \in I_{2^{n-q}}$, then, according to direct proof, the routing to the outputs i_1 and j_1 of the $2^n \times 2^n$ butterfly network is possible. Thus the lemma is proved. ■

Before proceeding further, we prove a property of a PP of any degree.

Lemma 6.5 *Consider a PP of degree d as in (3.1). Then,* $\forall N \in I_L$, *so that* $N \mid L$, *and* $\forall x \in I_L$, *we have* $\pi(x) \pmod{N} = \pi\big(x \pmod{N}\big) \pmod{N}$.

Proof Let x be

$$x = x_N + k \cdot N, \tag{6.21}$$

so that $x_N \in I_N$ and $k \in \mathbb{N}$, i.e. $x_N = x \pmod{N}$.

Then we can write the ith power of x from (6.21) modulo N, with $i \in I_d$, as

$$\begin{aligned} x^i \pmod N &= (x_N + k \cdot N)^i \pmod N = \\ &= \sum_{j=0}^{i} C_i^j \cdot (x_N)^j \cdot (k \cdot N)^{i-j} \pmod N = \\ &= \sum_{j=0}^{i-1} C_i^j \cdot (x_N)^j \cdot (k \cdot N)^{i-j} \pmod N + (x_N)^i \pmod N = \\ &= (x_N)^i \pmod N \end{aligned} \tag{6.22}$$

The last equality is true since $C_i^j \in \mathbb{N}$ and $i - j > 0$, $\forall j = 0, 1, \ldots, i-1$, the indexes in the sum from the second line of (6.22). With (6.22) and (3.1) the lemma results immediately. ■

Now we restate Lemmas 3, 4, and 5 from Nieminen (2017) for a PP of any degree (Trifina and Tarniceriu 2017). In fact, in Lemmas 6.6 and 6.7 we prove more general results for any degree PP.

Lemma 6.6 *Let n and M be positive integers and $L = p^n \cdot M$, where p is any prime number. Assume that π is a PP of arbitrary degree d on I_L, as in (3.1). Let x and y be in I_M. Then*

$$x \neq y \pmod{p^n}, \tag{6.23}$$

if and only if

$$\pi(x) \neq \pi(y) \pmod{p^n}. \tag{6.24}$$

Proof Let $n_{\max}$ be the greatest positive integer so that $p^{n_{\max}} \mid L$. From Theorem 3.8 we have that π is a PP of degree d on $I_{p^{n_{\max}}}$. If $n_{\max} > 1$, from Theorem 3.7 we have that π is a PP of degree d on I_p and $\pi'(x) \neq 0 \pmod p$, $\forall x \in I_p$. With the previous considerations, from Theorem 3.7 it also results that π is a PP of degree d on I_{p^q}, $\forall q \in \mathbb{N}$ with $q \geq 2$. Taking into account Lemma 6.5 this means that Eqs. (6.23) and (6.24) imply each other, $\forall n \in \mathbb{N}^*$, so that $p^n \mid L$. ■

For $p = 2$, Lemma 6.6 gives the same result as Lemma 3 from Nieminen (2017), but for a PP of any degree.

Lemma 6.7 *Let n and M be positive integers and $L = p^n \cdot M$, where p is any prime number. Assume that π is a PP of arbitrary degree d on I_L, as in (3.1). Assume that a_i are in I_L for every i in I_{p^n}. Then $A_a = (a_0, a_1, \ldots, a_{p^n-1})$ is uniformly p^n-dyadic if and only if $A_\pi = \big(\pi(a_0), \pi(a_1), \ldots, \pi(a_{p^n-1})\big)$ is uniformly p^n-dyadic.*

Proof As Lemma 4 from Nieminen (2017), Lemma 6.7 is a direct consequence of Lemma 6.6.

If we detail Eq. (6.15) for any $q = 1, 2, \ldots, n$, we have

$$a_{pk} \neq a_{pk+1} \neq \cdots \neq a_{pk+p-1} \pmod p, \forall k \in I_{p^{n-1}}, \tag{6.25}$$

$$a_{p^2k} \neq a_{p^2k+1} \neq \cdots \neq a_{p^2k+p^2-1} \pmod{p^2}, \forall k \in I_{p^{n-2}}, \tag{6.26}$$

a.s.o.

$$a_{p^{n-1}k} \neq a_{p^{n-1}k+1} \neq \cdots \neq a_{p^{n-1}k+p^{n-1}-1} \pmod{p^{n-1}}, \forall k \in I_p, \tag{6.27}$$

$$a_0 \neq a_1 \neq \cdots \neq a_{p^n-1} \pmod{p^n}. \tag{6.28}$$

Taking into account Lemma 6.6, Eqs. (6.25)–(6.28) hold if and only if

$$\pi(a_{pk}) \neq \pi(a_{pk+1}) \neq \cdots \neq \pi(a_{pk+p-1}) \pmod{p}, \forall k \in I_{p^{n-1}}, \tag{6.29}$$

$$\pi(a_{p^2k}) \neq \pi(a_{p^2k+1}) \neq \cdots \neq \pi(a_{p^2k+p^2-1}) \pmod{p^2}, \forall k \in I_{p^{n-2}}, \tag{6.30}$$

a.s.o.

$$\pi(a_{p^{n-1}k}) \neq \pi(a_{p^{n-1}k+1}) \neq \cdots \neq \pi(a_{p^{n-1}k+p^{n-1}-1}) \pmod{p^{n-1}}, \forall k \in I_p, \tag{6.31}$$

$$\pi(a_0) \neq \pi(a_1) \neq \cdots \neq \pi(a_{p^n-1}) \pmod{p^n}. \tag{6.32}$$

But Eqs. (6.29)–(6.32) mean that vector A_π is uniformly p^n-dyadic. Thus the lemma is proved. ■

For $p = 2$, Lemma 6.7 gives the same result as Lemma 4 from Nieminen (2017), but for a PP of any degree.

Now we can prove Theorem 6.3.

Proof of Theorem 6.3. The equivalences (1) ⇔ (3) and (2) ⇔ (4) follow from Lemma 6.4. Finally, the equivalence (1) ⇔ (2) follows from Lemma 6.7. ■

The next lemma allows an easy deriving of control bits of a $2^n \times 2^n$ butterfly network for a PP of any degree.

Lemma 6.8 *Let n and M be positive integers and $L = 2^n \cdot M$. Assume that π is a PP of arbitrary degree d on I_L, as in (3.1). Then for any $x \in I_L$, we have*

$$\pi(x + k2^{n-1}) = \pi(x) + \big(k \pmod 2\big)2^{n-1} \pmod{2^n}, \forall k \in I_L. \tag{6.33}$$

Proof We have

$$\pi(x + k2^{n-1}) = \sum_{i=1}^{d} q_i \cdot (x + k2^{n-1})^i \pmod{2^n} =$$

$$= \sum_{i=1}^{d} q_i \cdot \left(\sum_{j=0}^{i} C_i^j \cdot x^j \cdot (k2^{n-1})^{i-j} \right) \pmod{2^n} =$$

$$= \sum_{i=1}^{d} q_i \cdot \left(x^i + \sum_{j=0}^{i-1} C_i^j \cdot x^j \cdot (k2^{n-1})^{i-j} \right) \pmod{2^n} =$$

$$= \pi(x) + \sum_{i=1}^{d} q_i \cdot \left(\sum_{j=0}^{i-1} C_i^j \cdot x^j \cdot (k2^{n-1})^{i-j} \right) \pmod{2^n} =$$

$$= \pi(x) + q_1 \cdot k \cdot 2^{n-1} + \sum_{i=2}^{d} q_i \cdot \left(\sum_{j=0}^{i-1} C_i^j \cdot x^j \cdot (k2^{n-1})^{i-j} \right) \pmod{2^n}. \quad (6.34)$$

If $k = 0 \pmod 2$, i.e. $k = 2 \cdot l$, with $l \in \mathbb{N}$, from (6.34) we have

$$\pi(x + k2^{n-1}) = \pi(x) \pmod{2^n}, \quad (6.35)$$

i.e. (6.33) for $k = 0 \pmod 2$.

If $k = 1 \pmod 2$, i.e. $k = 2 \cdot l + 1$, with $l \in \mathbb{N}$, from (6.34) we have

$$\pi(x + k2^{n-1}) = \pi(x) + q_1 \cdot 2^{n-1} + \sum_{i=2}^{d} q_i \cdot \left(\sum_{j=0}^{i-1} C_i^j \cdot x^j \cdot (2^{n-1})^{i-j} \right) \pmod{2^n}. \quad (6.36)$$

For $i - j \geq 2$ and $n \geq 2$ we have $(i-j)(n-1) \geq n$, and thus $(2^{n-1})^{i-j} = 0 \pmod{2^n}$. Then, for $n \geq 2$, (6.36) reduces to

$$\pi(x + k2^{n-1}) = \pi(x) + q_1 \cdot 2^{n-1} + \sum_{i=2}^{d} q_i \cdot C_i^{i-1} \cdot x^{i-1} \cdot 2^{n-1} \pmod{2^n} =$$
$$= \pi(x) + q_1 \cdot 2^{n-1} + \sum_{i=2}^{d} q_i \cdot i \cdot x^{i-1} \cdot 2^{n-1} \pmod{2^n} =$$
$$= \pi(x) + 2^{n-1} \cdot \left(q_1 + \sum_{i=2}^{d} q_i \cdot i \cdot x^{i-1} \right) \pmod{2^n}. \quad (6.37)$$

For $x = 0 \pmod 2$, (6.37) becomes

$$\pi(x + k2^{n-1}) = \pi(x) + 2^{n-1} \cdot \left(q_1 + \sum_{i=2}^{d} q_i \cdot i \cdot (2l)^{i-1} \right) \pmod{2^n} =$$
$$= \pi(x) + 2^{n-1} \cdot q_1 \pmod{2^n} = \pi(x) + 2^{n-1} \pmod{2^n}, \quad (6.38)$$

i.e. (6.33) for $k = 1 \pmod 2$. The last equality in (6.38) is true because π is PP on I_{2^n}, and thus, considering Theorem 3.6, $q_1 = 1 \pmod 2$.

For $x = 1 \pmod 2$, (6.37) becomes

$$\pi(x + k2^{n-1}) = \pi(x) + 2^{n-1} \cdot \left(q_1 + \sum_{i=2}^{d} q_i \cdot i \cdot (2l+1)^{i-1} \right) \pmod{2^n} =$$
$$= \pi(x) + 2^{n-1} \cdot \left(q_1 + \sum_{i=2}^{d} q_i \cdot i \right) \pmod{2^n} =$$
$$= \pi(x) + 2^{n-1} \cdot (q_1 + 2q_2 + 3q_3 + \cdots + dq_d) \pmod{2^n} =$$
$$= \pi(x) + 2^{n-1} \cdot (q_1 + 3q_3 + 5q_5 + 7q_7 + \cdots) +$$

$$+\underbrace{2^{n-1}\cdot\left(2q_2+4q_4+6q_6+\cdots\right)}_{=0\ (\mathrm{mod}\ 2^n)} (\mathrm{mod}\ 2^n) =$$

$$= \pi(x) + 2^{n-1}\cdot\left(q_1+q_3+q_5+q_7+\cdots\right)+$$

$$+\underbrace{2^{n-1}\cdot\left(2q_3+4q_5+6q_7+\cdots\right)}_{=0\ (\mathrm{mod}\ 2^n)} (\mathrm{mod}\ 2^n) =$$

$$= \pi(x) + 2^{n-1}\cdot \underbrace{q_1}_{=1\ (\mathrm{mod}\ 2)} + 2^{n-1}\cdot\underbrace{\left(q_3+q_5+q_7+\cdots\right)}_{=0\ (\mathrm{mod}\ 2)} (\mathrm{mod}\ 2^n) =$$

$$= \pi(x) + 2^{n-1}\ (\mathrm{mod}\ 2^n), \tag{6.39}$$

i.e. (6.33) for $k = 1\ (\mathrm{mod}\ 2)$. The last equality in (6.38) is true because π is PP on I_{2^n}, and thus $q_1 = 1\ (\mathrm{mod}\ 2)$ and $\left(q_3+q_5+q_7+\cdots\right) = 0\ (\mathrm{mod}\ 2)$ from Theorem 3.6.

Equations (6.37)–(6.39) are valid for $n \geq 2$. For $n = 1$, (6.36) becomes

$$\pi(x+k) = \pi(x) + q_1 + \textstyle\sum_{i=2}^{d} q_i \cdot \left(\sum\limits_{j=0}^{i-1} C_i^j \cdot x^j\right) (\mathrm{mod}\ 2) =$$

$$= \pi(x) + q_1 + \textstyle\sum_{i=2}^{d} q_i + \sum_{i=2}^{d} q_i \cdot \left(\sum\limits_{j=1}^{i-1} C_i^j \cdot x^j\right) (\mathrm{mod}\ 2) =$$

$$= \pi(x) + \sum_{i=1}^{d} q_i + \sum_{i=2}^{d} q_i \cdot \left(\sum_{j=1}^{i-1} C_i^j \cdot x^j\right) (\mathrm{mod}\ 2) =$$

$$= \pi(x) + 1 + \sum_{i=2}^{d} q_i \cdot \left(\sum_{j=1}^{i-1} C_i^j \cdot x^j\right) (\mathrm{mod}\ 2). \tag{6.40}$$

The last equality in (6.40) is true because π is PP on I_2, and thus $\sum\limits_{i=1}^{d} q_i = 1\ (\mathrm{mod}\ 2)$ from Lemma 3.1.

For $x = 0\ (\mathrm{mod}\ 2)$, (6.40) becomes

$$\pi(x+k) = \pi(x) + 1\ (\mathrm{mod}\ 2), \tag{6.41}$$

i.e. (6.33) for $k = 1\ (\mathrm{mod}\ 2)$ and $n = 1$.

For $x = 1\ (\mathrm{mod}\ 2)$, (6.40) becomes

$$\pi(x+k) = \pi(x) + 1 + \sum_{i=2}^{d} q_i \cdot \left(\sum_{j=1}^{i-1} C_i^j\right) (\mathrm{mod}\ 2) =$$

$$= \pi(x) + 1 + \sum_{i=2}^{d} q_i \cdot \left(\sum_{j=0}^{i} C_i^j - 1 - 1\right) (\mathrm{mod}\ 2) =$$

$$= \pi(x) + 1 + \sum_{i=2}^{d} q_i \cdot \left((1+1)^i - 2\right) (\mathrm{mod}\ 2) =$$

$$= \pi(x) + 1 + \sum_{i=2}^{d} q_i \cdot \left(2^i - 2\right) \pmod 2 = \pi(x) + 1 \pmod 2 \tag{6.42}$$

i.e. (6.33) for $k = 1 \pmod 2$ and $n = 1$. Thus the proof is completed. ■

The usefulness of Lemma 6.8 is shown in the following. Let there be $x \in I_{2^n}$. Let $x = \left(x_{n-1}x_{n-2}\ldots x_1x_0\right)_2$ and $\pi(x) \pmod{2^n} = \left(\pi_{x,n-1}\pi_{x,n-2}\ldots\pi_{x,1}\pi_{x,0}\right)_2$ be the writing in base 2 of x and $\pi(x) \pmod{2^n}$, respectively. The bit with index 0 is the least significant bit and the bit with index $n-1$ is the most significant bit. From Lawrie (1975) we know that to get from an input pin x of a $2^n \times 2^n$ butterfly network to the output pin $\pi(x) \pmod{2^n}$, the control bits are obtained by equation

$$cb_{x,j} = x_j \oplus \pi_{x,j}, \forall j = 0, 1, \ldots, n-1, \forall x \in I_{2^n}, \tag{6.43}$$

where the control bits $cb_{x,j}$ for $j = 0, 1, \ldots, n-1$ are taken from the left to the right of the butterfly network. The decimal value of the control bits $(cb_{x,n-1}cb_{x,n-2}\ldots cb_{x,1}cb_{x,0})$ for the input pin x and the output pin $\pi(x) \pmod{2^n}$ is denoted by cb_x.

From (6.33), we have

$$\begin{aligned}
cb_{x+2^{n-1}} &= \left(\pi(x+2^{n-1}) \pmod{2^n}\right)_2 \oplus \left(x + 2^{n-1} \pmod{2^n}\right)_2 = \\
&= \left((\pi(x)+2^{n-1}) \pmod{2^n}\right)_2 \oplus \left(x + 2^{n-1} \pmod{2^n}\right)_2 = \\
&= \Big((\pi_{x,n-1}\pi_{x,n-2}\cdots\pi_{x,1}\pi_{x,0})_2 + (1\underbrace{0\cdots00}_{n-1 \text{ bits } 0})_2 \pmod{2^n}\Big)\oplus \\
&\oplus \Big((x_{n-1}x_{n-2}\cdots x_1x_0)_2 + (1\underbrace{0\cdots00}_{n-1 \text{ bits } 0})_2 \pmod{2^n}\Big)_2 = \\
&= \left((\pi_{x,n-1}\oplus 1)\pi_{x,n-2}\cdots\pi_{x,1}\pi_{x,0}\right)_2 \oplus \left((x_{n-1}\oplus 1)x_{n-2}\cdots x_1x_0\right)_2 = \\
&= (\pi_{x,n-1}\pi_{x,n-2}\cdots\pi_{x,1}\pi_{x,0})_2 \oplus (x_{n-1}x_{n-2}\cdots x_1x_0)_2.
\end{aligned} \tag{6.44}$$

Using Eq. (6.44) for $n = 1, 2, \ldots$, we have

$$\begin{aligned}
cb_{0,0} = cb_{1,0} = \cdots = cb_{n-1,0} &= \pi_{0,0}, \\
cb_{0,1} = cb_{2,1} = cb_{4,1} = \cdots = cb_{2^n-2,1} &= \pi_{0,1}, \\
cb_{1,1} = cb_{3,1} = cb_{5,1} = \cdots = cb_{2^n-1,1} &= \pi_{1,1}, \\
cb_{0,2} = cb_{4,2} = cb_{8,2} = \cdots = cb_{2^n-4,2} &= \pi_{0,2}, \\
cb_{1,2} = cb_{5,2} = cb_{9,2} = \cdots = cb_{2^n-3,2} &= \pi_{1,2}, \\
cb_{2,2} = cb_{6,2} = cb_{10,2} = \cdots = cb_{2^n-2,2} &= \pi_{2,2}, \\
cb_{3,2} = cb_{7,2} = cb_{11,2} = \cdots = cb_{2^n-1,2} &= \pi_{3,2},
\end{aligned}$$

....

Table 6.3 The interleaved addresses for 5-PP interleaver $\pi(x) = 65x + 38x^2 + 16x^3 + 10x^4 + 12x^5 \pmod{112}$

k	$\pi(16k+i)$ for $i = 0, 1, \ldots, 15$															
	0	1	2	3	4	5	6	7	8	9	10	11	12	13	14	15
0	0	29	58	103	52	97	46	91	8	37	82	31	76	25	70	99
1	16	61	10	55	4	49	78	107	40	101	34	95	28	57	86	19
2	80	13	74	7	36	65	110	59	104	53	98	15	44	89	38	83
3	32	77	106	23	68	17	62	11	56	85	2	47	108	41	102	35
4	64	93	26	87	20	81	14	43	72	5	66	111	60	105	22	51
5	96	45	90	39	84	1	30	75	24	69	18	63	92	9	54	3
6	48	109	42	71	100	33	94	27	88	21	50	79	12	73	6	67

$$cb_{0,n-1} = cb_{2^{n-1},n-1} = \pi_{0,n-1}, cb_{1,n-1} = cb_{2^{n-1}+1,n-1} = \pi_{1,n-1}, \ldots$$
$$cb_{2^{n-1}-1,n-1} = cb_{2^n-1,n-1} = \pi_{2^{n-1}-1,n-1}. \tag{6.45}$$

From (6.45) it results that, of all $n \cdot 2^n$ control bits, only $2^n - 1$ values need to be stored.

In the following we exemplify the determination of control bits of a 16×16 butterfly network for a 5-PP interleaver.

Example 6.1 We consider the five degree PP (5-PP) interleaver $\pi(x) = 65x + 38x^2 + 16x^3 + 10x^4 + 12x^5 \pmod{112}$ of length $L = 112$ found by the method in Trifina and Tarniceriu (2014). This 5-PP interleaver also leads to optimum minimum distance as that given in Trifina et al. (2017), when using LTE turbo codes (3GPP 2008), but offers better performance since it is optimized by distance spectrum with nine terms not only by the first term. For maximum degree of parallelism $M = 2^4 = 16$. For address vectors $a_i(k) = 16k + i \ \forall i \in I_{16}$ and $\forall k \in I_7$, the interleaved addresses are mapped to the memories $\pi(i) = 65i + 38i^2 + 16i^3 + 10i^4 + 12i^5 \pmod{16} = i + 6i^2 + 10i^4 + 12i^5 \pmod{16} \ \forall i \in I_{16}$. The interleaved address vectors $\pi(a_i(k))$ are given in Table 6.3 and the physical interleaved address vectors are given in Table 6.4. $\pi(i)$ in Table 6.4 shows the index of the memory accessed for physical interleaved addresses. In the last row "cb" gives the control bits for the 16×16 butterfly network. We observe that these control bits follow Eq. (6.45) for $n = 4$. For example, at time $k = 4$, the value $\pi(16k + i)$ for $i = 3$, i.e. $\pi(67) = 87$, is routed from the input $i = 3$ of the 16×16-butterfly network to the output $\pi(3) \pmod{16} = 7$. The control bits for this routing path are given by the value $(4)_{10} = (0100)_2$ (the value from the last row in the column corresponding to $i = 3$ in Table 6.4), i.e. the first control bit is equal to 0, the second control bit is equal to 0, the third control bit is equal to 1, and the last control bit is equal to 0. ■

Table 6.4 The physical interleaved addresses for 5-PP interleaver $\pi(x) = 65x + 38x^2 + 16x^3 + 10x^4 + 12x^5 \pmod{112}$

k	$\lfloor \pi(16k+i)/16 \rfloor$ for $i = 0, 1, \ldots, 15$															
	0	1	2	3	4	5	6	7	8	9	10	11	12	13	14	15
0	0	1	3	6	3	6	2	5	0	2	5	1	4	1	4	6
1	1	3	0	3	0	3	4	6	2	6	2	5	1	3	5	1
2	5	0	4	0	2	4	6	3	6	3	6	0	2	5	2	5
3	2	4	6	1	4	1	3	0	3	5	0	2	6	2	6	2
4	4	5	1	5	1	5	0	2	4	0	4	6	3	6	1	3
5	6	2	5	2	5	0	1	4	1	4	1	3	5	0	3	0
6	3	6	2	4	6	2	5	1	5	1	3	4	0	4	0	4
$\pi(i) \pmod{16}$	0	13	10	7	4	1	14	11	8	5	2	15	12	9	6	3
cb	0	12	8	4	0	4	8	12	0	12	8	4	0	4	8	12

References

3GPP TS 36.212 V8.3.0, 3rd generation partnership project, Multiplexing and channel coding (Release 8) (2008), http://www.etsi.org

L.R. Bahl et al., Optimal decoding of linear codes for minimizing symbol error rate. IEEE Trans. Inf. Theory **20**(2), 284–287 (1974)

S. Benedetto et al., Design issues on the parallel implementation of versatile, high-speed iterative decoders, in *4rd International Symposium on Turbo Codes and Related Topics, Munich, Germany* 3–7 April 2006

C. Berrou et al., Designing good permutations for turbo codes: towards a single model, in *IEEE International Conference on Communications (ICC), Paris, France*, vol. 1 (2004), pp. 341–345

ETSI EN 301 790 V1.3.1, Digital video broadcasting (DVB); Interaction channel for satellite distribution systems (2003), http://www.broadcasting.ru/pdf-standard-specifications/interactivity/dvb-rcs/en301790.v1.3.1.pdf

A. Giulietti, L. Van der Perre, M. Strum, Parallel turbo coding interleavers: avoiding collisions in accesses to storage elements. Electron. Lett. **38**(5), 232–234 (2002)

J. Kwak, K. Lee, Design of dividable interleaver for parallel decoding in turbo codes. Electron. Lett. **38**(22), 1362–1364 (2002)

D.H. Lawrie, Access and alignment of data in an array processor. IEEE Trans. Comput. **24**(12), 1145–1155 (1975)

E. Nieminen, A contention-free parallel access by butterfly networks for turbo interleavers. IEEE Trans. Inf. Theory **60**(1), 237–251 (2014)

E. Nieminen, On quadratic permutation polynomials, turbo codes, and butterfly networks. IEEE Trans. Inf. Theory **63**(9), 5793–5801 (2017)

A. Nimbalker et al., Inter-window shuffle interleavers for high throughput turbo decoding, in *3rd International Symposium on Turbo Codes and Related Topics, Brest, France*, 1–5 September 2003, pp. 355–358

A. Nimbalker et al., Contention-free interleavers, in *IEEE International Symposium on Information Theory (ISIT), Chicago, Illinois, USA*, June 27–July 2 2004, p. 54

A. Nimbalker et al., Contention-free interleavers for high-throughput turbo decoding. IEEE Trans. Commun. **56**(8), 1258–1267 (2008)

O.Y. Takeshita, Maximum contention-free interleavers and permutation polynomials over integers rings. IEEE Trans. Inf. Theory **52**(3), 1249–1253 (2006)

A. Tarable, S. Benedetto, Mapping interleaving laws to parallel turbo decoder architectures. IEEE Commun. Lett. **8**(3), 162–164 (2004)

A. Tarable, S. Benedetto, Further results on mapping functions, in *IEEE ITSOC Information Theory Workshop Coding Complex (ITW), Rotorua, New Zealand*, August 28–September 1 2005, pp. 221–225

A. Tarable, S. Benedetto, G. Montorsi, Mapping interleaving laws to parallel turbo and LDPC decoder architectures. IEEE Trans. Inf. Theory **50**(9), 2002–2009 (2004)

L. Trifina, V. Munteanu, *Coduri Turbo* (Politehnium, Iasi, 2008)

L. Trifina, D. Tarniceriu, Improved method for searching interleavers from a certain set using Garello's method with applications for the LTE standard. Ann. Telecommun. **69**(5–6), 251–272 (2014)

L. Trifina, D. Tarniceriu, Parallel access by butterfly networks for any degree permutation polynomial and ARP interleavers. Submitted for possible publication (under review) (2017)

L. Trifina, J. Ryu, D. Tarniceriu, Up to five degree permutation polynomial interleavers for short length LTE turbo codes with optimum minimum distance, in *IEEE International Symposium on Signals, Circuits and Systems (ISSCS), Iasi, Romania*, 13–14 July 2017

Chapter 7
Methods to Search Permutation Polynomial Interleavers for Turbo Codes

7.1 Preliminaries

The main problem in finding an interleaver for a turbo code consists in that the number of interleavers of a certain length is very large. The number of all interleavers of length L is $L!$ and the number of all true different PP of degree d interleavers, with $d = \overline{1, 5}$, was given in Chap. 4. Therefore finding metrics or methods that reduce the searching complexity is a key issue in designing a turbo code.

In Sects. 7.2 and 7.3, we present two metrics to search generic QPP interleavers proposed by Takeshita (2007). *Generic interleavers* are those that are not tailored to a specific component convolutional code of turbo code. The most part of Sect. 7.2 is from Section II in Takeshita (2007) and Sect. 7.3 is from Section III in Takeshita (2007). In Takeshita (2007) an interleaver is represented by an *interleaver-code*, which is the geometric representation of an interleaver by pairs of coordinates $(x, \pi(x))$ forming *points* in the set $\mathbb{Z}_L^2$.

In Sects. 7.4 and 7.5 we present some methods of searching PP interleavers adapted to a specific component code of the turbo code. These methods aim to increase the minimum distance or to improve the distance spectrum for turbo codes with PP interleavers. Numerical results are given for the component code of the turbo code from the LTE standard, whose generator matrix in octal form is $G = [1, 15/13]$.

7.2 The Spread Factor of a QPP Interleaver

A well-known measure of merit in turbo coding applications is the *spread factor* (Divsalar and Pollara 1995). The spread factor of an interleaver is defined as

$$D_E = \min_{\substack{i \neq j, \\ i, j \in \mathbb{Z}_L}} \left\{ \delta(p_i, p_j) \right\}, \tag{7.1}$$

L. Trifina and D. Tarniceriu, *Permutation Polynomial Interleavers for Turbo Codes*, Signals and Communication Technology,
https://doi.org/10.1007/978-981-13-2625-7_7

where $\delta(p_i, p_j)$ is the L_1 or Manhattan metric between the points $p_i = (i, \pi(i))$ and $p_j = (j, \pi(j))$:

$$\delta(p_i, p_j) = |i - j| + |\pi(i) - \pi(j)|. \tag{7.2}$$

The spread factor was initially proposed by Dolinar and Divsalar. The definition in (7.1) was introduced in Crozier (2000) to avoid self-terminating information sequences which lead to low-weight turbo codewords. In Divsalar and Pollara (1995) the authors have proposed a construction of linear interleavers achieving spread factors D_E equal to or close to $\sqrt{2L}$. The minimum distance of turbo codes obtained only for weight-two self-terminating information sequences grows approximately as $\sqrt{2L}$ when LPP interleavers are used. Further results show that the true minimum distance grows asymptotically, at most logarithmically, along with the interleaver length for all interleavers (Breiling 2004; Perotti and Benedetto 2004).

Takeshita gave in Takeshita (2007) an easily modified definition for the spreading factor beside that from Crozier (2000), in the following way:

$$D = \min_{\substack{i \neq j, \\ i,j \in \mathbb{Z}_L}} \left\{ \delta_L(p_i, p_j) \right\}, \tag{7.3}$$

where $\delta_L(p_i, p_j)$ is the Lee metric (Lee 1958) between points $p_i = (i, \pi(i))$ and $p_j = (j, \pi(j))$:

$$\delta_L(p_i, p_j) = |i - j|_L + |\pi(i) - \pi(j)|_L \tag{7.4}$$

and

$$|i - j|_L = \min \left\{ (i - j) \pmod{L}, (j - i) \pmod{L} \right\} \tag{7.5}$$

The upper bound of D, denoted by $\mathrm{ub}_D(L)$, was proved in Boutillon and Gnaedig (2005) to be $\sqrt{2L}$. According to definitions of $\delta(p_i, p_j)$ and $\delta_L(p_i, p_j)$ in (7.2) and (7.4), respectively, we have $D \leq D_E$. An upper bound of D_E was found in Takeshita (2007). It is denoted by $\mathrm{ub}_{D_E}(L)$, and it is close to $\mathrm{ub}_D(L)$. It is given below

$$\mathrm{ub}_{D_E}(L) = \begin{cases} \dfrac{2(L-1)}{\sqrt{2L} - 1}, L = 2p^2, p = 2, 3, 4, \ldots \\ \dfrac{2(L-1)}{\sqrt{2L-1} - 1}, L = p^2 + (p-1)^2, p = 2, 3, 4, \ldots \end{cases} \tag{7.6}$$

In Fig. 7.1 the difference $\mathrm{ub}_D(L) - \mathrm{ub}_{D_E}(L)$, for the lengths in (7.6) smaller than 10,000, is plotted. It can be seen that this difference goes to 1 as L grows.

The definition of a *maximum-spread interleaver* is given below.

Definition 7.1 A maximum-spread interleaver is an interleaver which achieves a spread factor D equal to the upper bound $\sqrt{2L}$, where L is the interleaver length. In fact the maximum spread can be equal to $\lfloor \sqrt{2L} \rfloor$, because the spread factor must be an integer.

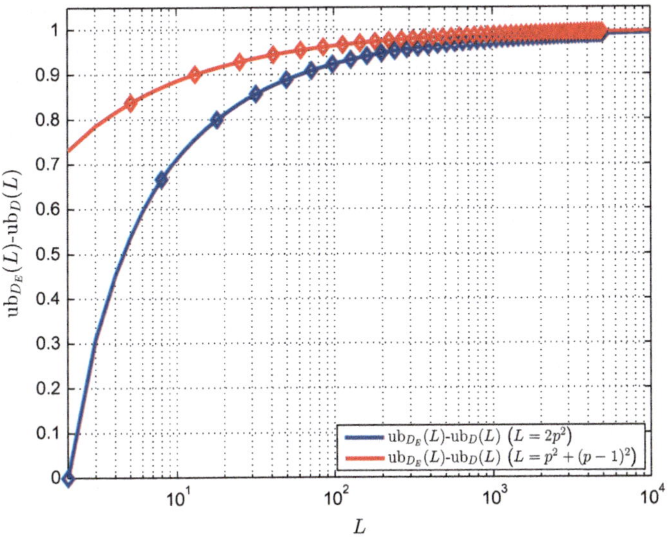

Fig. 7.1 The difference $\mathrm{ub}_D(L) - \mathrm{ub}_{D_E}(L)$

In Dolinar and Divsalar (1995) have stated that LPP interleavers either achieve or closely approximate a spread factor of $\sqrt{2L}$ for any L. When L is twice a perfect square

$$L = 2n^2, n = 1, 2, 3, \ldots \tag{7.7}$$

then all maximum-spread interleavers are of the form

$$\pi(x) = q_1 \cdot x \ (\mathrm{mod}\ L = 2n^2) \tag{7.8}$$

These maximum-spread interleavers are obtained for

$$q_1 = l \cdot \sqrt{2L} \pm 1, \tag{7.9}$$

for positive integers $l < \sqrt{L/2}$ relatively prime to $\sqrt{L/2}$. The resulting interleavers are not appropriate for turbo coding because of their high regularity. The linear interleaver asymptote in Takeshita and Costello (2000) refers to this matter. It implies the existence of low-weight codewords of input-weight four and a high multiplicity, close to L. Consequently, they proposed an algorithm for construction of a *semi-random interleaver* (Dolinar and Divsalar 1995), named *S-random interleaver* with parameter S. The algorithm for generating S-random interleavers is Algorithm 1.

The searching time for the above algorithm increases with S, and it is not guaranteed to finish successfully. However, it was observed that choosing $S < \sqrt{L/2}$ usually produces a solution in reasonable time.

Algorithm 1: Algorithm for the generation of S-random interleavers

input : The interleaver length L and the value of parameter S.
output: The set $\{\pi(0), \pi(1), \ldots, \pi(L-1)\}$, describing the S-random interleaver of length L.

```
Select randomly an arbitrary integer i from the set Z_L ;
π(0) ← i;
k ← 1;
A_k ← Z_L − {i};
while (A_k ≠ ∅) do
    repeat
        bool_valid_i_pres ← true ;
        Select randomly an arbitrary integer i_pres from the set A_k ;
        for j ← max{k − S, 0} to k − 1 do
            if (|π(k) − π(j)| ≥ S) then
                bool_valid_i_pres ← false ;
                break ;
            end
        end
    until (bool_valid_i_pres = true);
    π(k) = i_pres ;
    k ← k + 1 ;
    A_k ← A_{k−1} − {i_pres} ;
end
```

The algorithm for S-random interleaver sacrifices the spread factor D, which typically is $D = S + 1 \leq \sqrt{L/2} + 1$, i.e., smaller than about 50% of the upper bound $\text{ub}_D(L) = \lfloor \sqrt{2L} \rfloor$. Turbo codes using S-random interleavers have very good error rate performance, becoming typical benchmark interleavers. Their main drawbacks consists in expensive storage of a sequence of S integers that characterize the interleaver. Due to the construction algorithm based on a pseudorandom number generator, the compression capability of the sequence is very low. This leads to the imposibility to reproduct accurately other simulation results with S-random interleavers because, in general, only parameter S is reported in literature. However, for a given parameter S, different S-random interleavers perform similarly for error rates that are not very low, which also reflects a good minimum distance of the associated turbo codes. Thus the repeatability problem is not so critical.

In Crozier (2000) has proposed two interleaver construction algorithms for maximization of the spread factor, but avoiding or minimizing the regularity of linear interleavers. They are *high-spread* construction and the *dithered-diagonal* construction. For interleaver sizes given in (7.7) the dithered-diagonal interleavers are maximum-spread and have large spread factors for others (Crozier 2000). $n = \sqrt{L/2}$ integer parameters are needed to define dithered-diagonal interleavers. In Crozier (2000) it is shown a spectacular error performance that exceeds S-random interleavers for $L = 512$. The progress of dithered-diagonal interleaver construction consists in achieving a large spread factor D combined with sufficient irregularity leading to

Table 7.1 The three stages for generating DRP interleavers

1. The input vector v_{in} is dithered (permuted locally), using a small read dither vector r, of length R (the vector r is a permutation of indexes $0, 1, \ldots, R-1$)
2. The resulting vector v_a is permuted using a relativelly prime interleaver (or an interleaver with cyclic shift) to obtain a good spread
3. The resulting vector v_b is dithered using a small write dither vector w, of length W, to generate the output vector v_{out}

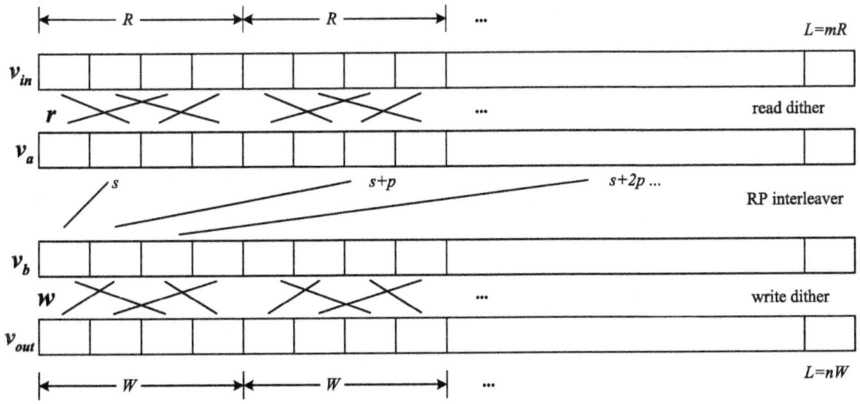

Fig. 7.2 Design approach of a DRP interleaver

very high error performance. The number of integer parameters characterizing the interleaver is much smaller compared with S-random interleavers.

In Crozier and Guinand (2001), Stewart Crozier and Paul Guinand proposed *DRP interleavers*, which are among the best known ones in terms of error rate performance. The approach to design DRP interleavers consists of three stages (Crozier and Guinand 2001) given in Table 7.1.

The DRP interleaver length L must be a multiple of both R and W.

The above approach is depicted in Fig. 7.2.

The equations which describe the DRP interleaver are:

$$v_a(i) = v_{in}\big(\pi_a(i)\big),\ v_b(i) = v_a\big(\pi_b(i)\big),\ v_{out}(i) = v_b\big(\pi_c(i)\big),$$

$$i = 0, 1, \ldots, L-1 \tag{7.10}$$

$$\pi_a(i) = R \cdot \lfloor i/R \rfloor + r(i \pmod R)), i = 0, 1, \ldots, L-1 \tag{7.11}$$

$$\pi_b(i) = (s + i \cdot p) \pmod L, i = 0, 1, \ldots, L-1 \tag{7.12}$$

$$\pi_c(i) = W \cdot \lfloor i/W \rfloor + w(i \pmod W)), i = 0, 1, \ldots, L-1 \tag{7.13}$$

$$v_{out}(i) = v_{in}\big(\pi(i)\big), i = 0, 1, \ldots, L-1 \tag{7.14}$$

$$\pi(i) = \pi_a\big(\pi_b\big(\pi_c(i)\big)\big), i = 0, 1, \ldots, L-1 \tag{7.15}$$

Thus, DRP interleavers require $R + W + 2$ integers for their specification, namely values $r(0), r(1), \ldots, r(R-1)$, $w(0), w(1), \ldots, w(W-1)$, s, and p. With the above equations, for $R = W = 4$ they require 10 integers and for $R = W = 8$ they require 18 integers for their generation.

It is shown that DRP interleavers are very efficient in terms of storage requirements. Let M be the least common multiple of R and W. It can be shown that the following equation holds for a DRP interleaver:

$$\pi\big((i + M) \text{ (mod } L)\big) = \big(\pi(i) + M \cdot p\big) \text{ (mod } L), i = 0, 1, \ldots, L-1 \tag{7.16}$$

It follows that the interleaver indexes can be recursively computed by equation:

$$\pi(i) = \big(\pi(i-1) + P(i \text{ (mod } M)\big) \text{ (mod } L), i = 1, 2, \ldots, L-1, \tag{7.17}$$

where $\pi(0)$ is arbitrary, and the M index increments in vector P are defined by (7.16) and

$$P(i \text{ (mod } M)) = \big(\pi(i) - \pi(i-1)\big) \text{ (mod } L), i = 1, 2, \ldots, M. \tag{7.18}$$

Thus, DRP interleavers require a much smaller number of integer parameters for their specification compared to previous interleavers. If $M = R = W$, DRP interleavers require $M + 1$ integers for their generation, namely $\pi(0)$, and values $P(0)$, $P(1)$, ..., $P(M-1)$. Thus, considering Eq. (7.17), for $M = R = W = 4$ they require 5 integers and for $M = R = W = 8$ they require 9 integers for their generation.

Figure 7.3 shows an exhaustive search over true different QPP interleavers for the largest achievable spread $D_{\max}(L)$ for $2 \leq L \leq 4096$. From Fig. 7.3 we see that the majority QPP interleavers, more exactly 83.2%, with the largest spreads lie between $\text{ub}_D(L) = \sqrt{2L}$ and $\sqrt{L}$ (which means about 70% of the upper bound). For $2 \leq L \leq 4096$, there are 1190 values of L for which there exist true QPPs (Number of LPPs/QPPs/CPPs 2015) (i.e. roughly 29% of lengths values). Due to several algebraic and geometric properties of QPP interleavers, explained in Sect. 7.3, the exhaustive search is efficiently completed in a very short time using Theorem 7.10.

A few of the polynomials for some interleaver lengths (smaller than or equal to 1024) are given in Table 7.2. Some of the QPPs in Table 7.2 result in very good turbo codes. Section 7.5 presents QPPs that do not simply attempt maximization of the spread factor. To compare these interleavears with those found by methods from Sect. 7.5, in Table 7.2 we also give the first term in the distance spectra of turbo codes of 1/3 nominal coding rate using these interleavers, generator matrix $G = [1, 15/13]$ and post-interleaver trellis termination. In Table 7.2 the nonlinearity degree ζ and the refined nonlinearity degree ζ' are also given for the reported QPPs. These metrics are defined in the next section.

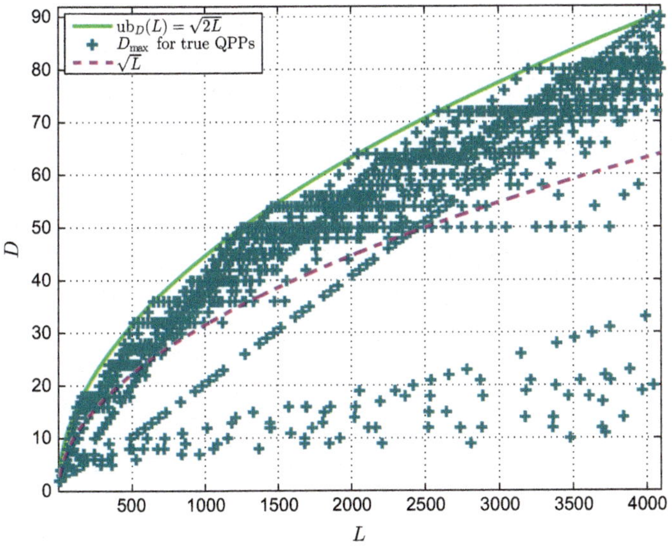

Fig. 7.3 Maximum achievable spreads $D_{\max}(L)$ with true QPP interleavers for lengths in the range $2 \leq L \leq 4096$

Table 7.2 QPP interleavers with the largest spread (LS-QPP) (Takeshita 2007) and the first term in the distance spectra of turbo codes of 1/3 nominal coding rate using these interleavers, generator matrix $G = [1, 15/13]$ and post-interleaver trellis termination

L	$\pi(x)$	D	ζ	ζ'	$d_{min}/N_{d_{min}}/w_{d_{min}}$
40	$x + 10x^2$	4	2	2	11/1/1
80	$9x + 20x^2$	10	2	2	14/1/2
128	$15x + 32x^2$	16	2	2	16/1/2
160	$19x + 40x^2$	16	2	2	19/2/4
256	$15x + 32x^2$	16	4	3	16/1/2
320	$19x + 40x^2$	20	4	3	20/3/6
400	$17x + 100x^2$	20	2	2	23/1/2
408	$25x + 102x^2$	24	2	2	25/1/1
512	$31x + 64x^2$	32	4	3	27/1/1
640	$39x + 80x^2$	32	4	3	31/4/8
752	$31x + 188x^2$	32	2	2	28/2/4
800	$17x + 80x^2$	32	5	5	31/3/9
1024	$123x + 256x^2$	34	2	2	27/1/2

From Fig. 7.3 we note that the fraction of maximum-spread QPP interleavers is very small. The range between $\mathrm{ub}_D(L) = \sqrt{2L}$ and $\sqrt{L}$ contains the most of QPP interleavers with the largest spreads. We note that the value $\sqrt{L}$ is about 70% of the upper bound and S-random interleavers typically achieve only 50% of $\mathrm{ub}_D(L)$.

The minimization of the number of low-weight codewords caused by self-terminating weight-2 input sequences or by short bursts of self-terminating input sequences is useful and it is related to the maximization of the spread factor D. To understand if we should always maximize the spread factor, the following observations are done in Takeshita (2007):

- many linear interleavers are maximum-spread, but undergo high-multiplicity low-weight codewords (Takeshita and Costello 2000).
- at least one dithered diagonal interleaver (Crozier 2000) is maximum-spread and assures a significant error performance.
- DRP interleavers which maximize error performance are typically not maximum-spread (Crozier and Guinand 2001).

Considering the second observation, from an error rate perspective the best interleavers may be those that achieve or closely approximate a maximum-spread interleaver and simultaneously have a large degree of "randomness". The same principle was used to find the Welch–Costas permutation-based interleavers from Trifina et al. (2006a, b). Interleavers with a certain degree of regularity or of low entropy, such DRP or PP interleavers, require a decrease of the spread factor to increase their entropy or "randomness".

The following theorem, given without proof, provides an infinite sequence of *maximum-spread* QPP interleavers (Takeshita 2007).

Theorem 7.2 (Theorem 2 from Takeshita 2007) *The following sequence is an infinite sequence of QPPs that generate maximum-spread interleavers:*

$$\pi(x) = \left(2^k - 1\right) \cdot x + 2^{k+1} \cdot x^2 \left(\text{mod } 2^{2k-1}\right), k = 1, 2, 3, \ldots \quad (7.19)$$

There are true QPPs in (7.19) only when $k > 3$. For $k = 1$ and $k = 2$, the second coefficient of $\pi(x)$ in (7.19) is $q_2 = 4 \ (\text{mod } 2) = 0$ and $q_2 = 8 \ (\text{mod } 8) = 0$. For $k = 3$, we have $q_2 = 16 \ (\text{mod } 32) = 32/2$ and since $\pi(x) = 16 \cdot x + 16 \cdot x^2 \ (\text{mod } 32)$ is a QNP (see Theorem 4.4), then $\pi(x)$ in (7.19) can be reduced to a LPP. For $k > 3$, $\pi(x)$ in (7.19) cannot be reduced to a LPP because it has a degree of nonlinearity $\zeta > 1$ (see Sect. 7.3). The first six terms of maximum-spread QPP sequence in Theorem 7.2, that are not reducible to first-degree polynomials, are shown in Table 7.3.

Table 7.3 QPP interleavers with maximum spread (MS-QPP) (Takeshita 2007)

k	L	$\pi(x)$	$D = \text{ub}_D(L)$	ζ	ζ'	ϵ
4	128	$15x + 32x^2$	16	2	2	64
5	512	$31x + 64x^2$	32	4	3	128
6	2048	$63x + 128x^2$	64	8	4	256
7	8192	$127x + 256x^2$	128	16	7	512
8	32768	$255x + 512x^2$	256	32	12	1024
9	131072	$511x + 1024x^2$	512	64	23	2048

7.3 Ω′ Metric for Searching QPP Inter-leavers

The following concepts, well known in the context of geometrically uniform codes (Forney 1991), are useful for finding the metric in this section.

A symmetry of a metric space $\mathcal{T} = (\mathbb{Z}_L^2, \delta_L)$ is a mapping of $\mathcal{T}$ to itself so that the distance between points is preserved. The set of symmetries which present interest are those obtained by translations of the space $\mathbb{Z}_L^2$, meaning circular "slides" in the vertical, horizontal directions, and their combinations. The symmetries obtained by translations are exactly those tied by the multiplicity of codewords in a turbo code. Other possible symmetries, but not allowed for our situation, are rotations and reflections. The algebraic equivalent of a translation $\mathcal{A}(k_0, k_1) : \mathbb{Z}_L^2 \to \mathbb{Z}_L^2$ that circularly "slides" to the right by k_0 and upwards by k_1 is given by:

$$\mathcal{A}(k_0, k_1) : (x_0, x_1) \to (x_0 + k_0, x_1 + k_1), k_0, k_1 \in \mathbb{Z}_L \tag{7.20}$$

The set of symmetry functions forms a group $\mathcal{G}$ under function composition. As translations are the only allowed symmetries, $\mathcal{G}$ is a commutative group isomorphic to $\mathcal{G}_{c,L}^2$ (the Cartesian product of two cyclic groups of order L). An isometry of an interleaver code Q is a symmetry $\mathcal{A}$ of $\mathcal{T}$ inducing $\mathcal{A} : Q \to G$ so that $Q = G$. The set of isometries of Q form a subgroup $\mathcal{H}$ of $\mathcal{G}$. Two points $p_{x_1} \in Q$ and $p_{x_2} \in Q$ are equivalent when there exists an isometry $\mathcal{A}$ of $\mathcal{F}$ so that p_{x_1} is mapped into p_{x_2}.

We give below the definitions for the *orbit of a point* from the interleaver code and those of the *nonlinearity degree* and of the *degree of shift-invariance* of an interleaver.

Definition 7.3 The orbit of a point $p_x \in Q$ is the set of points $\mathcal{O}_{p_x}$ equivalent under the action of the isometry group $\mathcal{H}$.

The next two definitions are valid because there is just one way to express Q as the disjoint union of a family of orbits and all orbits have the same size.

Definition 7.4 The number of distinct orbits represents the degree of nonlinearity ζ of an interleaver.

Definition 7.5 The degree of shift-invariance ϵ of an interleaver represents the size of the orbits.

Because the number of points from Q is equal to L, all orbits have the same size and there is just one way to express Q as a disjoint union of a family of orbits, it results that

$$\zeta = L/\epsilon \tag{7.21}$$

The next theorem gives a formula for computing the degree of nonlinearity of a QPP interleaver.

Theorem 7.6 (Theorem 3 from Takeshita 2007) *The degree of nonlinearity of a QPP interleaver given by* $\pi(x) = q_1 \cdot x + q_2 \cdot x^2 \pmod L$ *is*

$$\zeta = L/\gcd(2q_2, L) \tag{7.22}$$

Proof If $\mathcal{A}(k_0, k_1)$ is an isometry of the interlevear code Q for a QPP interleaver described by $\pi(x) = q_1 \cdot x + q_2 \cdot x^2$ then, from (7.20), we must have $\pi(x + k_0) = \pi(x) + k_1 \pmod{L}$. Developing it, we have

$$\begin{aligned} q_2 \cdot (x + k_0)^2 + q_1 \cdot (x + k_0) = q_2 \cdot x^2 + q_1 \cdot x + k_1 \pmod{L} \Leftrightarrow \\ \Leftrightarrow q_2 \cdot x^2 + (q_1 + 2 \cdot q_2 \cdot k_0) \cdot x + q_1 \cdot k_0 + q_2 \cdot k_0^2 - k_1 = \\ = q_2 \cdot x^2 + q_1 \cdot x \pmod{L} \Leftrightarrow \end{aligned}$$

$$\Leftrightarrow (2 \cdot q_2 \cdot k_0) \cdot x + q_1 \cdot k_0 + q_2 \cdot k_0^2 - k_1 = 0 \pmod{L} \tag{7.23}$$

Thus, we just need to ensure that $2 \cdot q_2 \cdot k_0 = 0 \pmod{L}$. This is a linear congruence and its solution is given by Theorem 57 from Hardy and Wright (1975)

$$k_0(i) = \frac{L \cdot i}{\gcd(2q_2, L)}, i = 0, 1, 2, \ldots, \gcd(2q_2, L) - 1 \tag{7.24}$$

With k_0 determined from (7.24), k_1 results from (7.23) as

$$k_1(i) = q_1 \cdot k_0(i) + q_2 \cdot \left(k_0(i)\right)^2 \tag{7.25}$$

There are exactly $\gcd(2q_2, L)$ distinct solutions, meaning that each point in Q belongs to an orbit of size $\gcd(2q_2, L)$, i.e., the degree of shift invariance is $\epsilon(Q) = \gcd(2q_2, L)$. Using (7.21), the degree of nonlinearity for QPPs is $\zeta(Q) = L/\gcd(2q_2, L)$. ■

Considering the proof of Theorem 7.6, and because $p_0 = (0, 0) \in Q$, under the assumption that a QPP has $q_0 = 0$, the set

$$\mathcal{O}_{p_0} = \left\{\left(k_0(i), k_1(i)\right) | i = 0, 1, \ldots, \gcd(2q_2, L) - 1\right\} \tag{7.26}$$

is exactly one of the orbits in Q.

The next theorem gives a lower bound for the Lee metric between two different points of the interleaver code.

Theorem 7.7 (Theorem 5 from Takeshita 2007) *Let there be $p_{x_1}, p_{x_2} \in \mathcal{O}_{(0,0)}$ and $p_{x_1} \neq p_{x_2}$. A lower bound on the distance $\delta_L(p_{x_1}, p_{x_2})$ is $2L/\gcd(2q_2, L)$.*

Proof From (7.24) it follows that the minimum distance in the set $\left\{k_0(i) | i = 0, 1, \ldots, \gcd(2q_2, L) - 1\right\}$ is $L/\gcd(2q_2, L)$. From the proof of Theorem 6.1 it follows that $\pi(j + tW) \neq \pi(j + vW)$, for $0 \leq j < W, 0 \leq t < v < L/W$, where $W \mid L$. Then, because $L/\gcd(2q_2, L)$ is a valid value for W, it follows that $L/\gcd(2q_2, L)$ is the minimum distance in the set $\left\{k_1(i) | i = 0, 1, \ldots, \gcd(2q_2, L) - 1\right\}$. Because the points from the same orbit are equivalent under the isometry group of translations, it results that there is a pair of values $(k_0(i), k_1(i))$, with $i \in \{0, 1, 2, \ldots, \gcd(2q_2, L) - 1\}$, so that for two points $p_{x_1}, p_{x_2} \in \mathcal{O}_{(0,0)}$, we have $x_2 = x_1 + k_0(i)$ and $\pi(x_2) = \pi(x_1) + k_1(i)$. Then

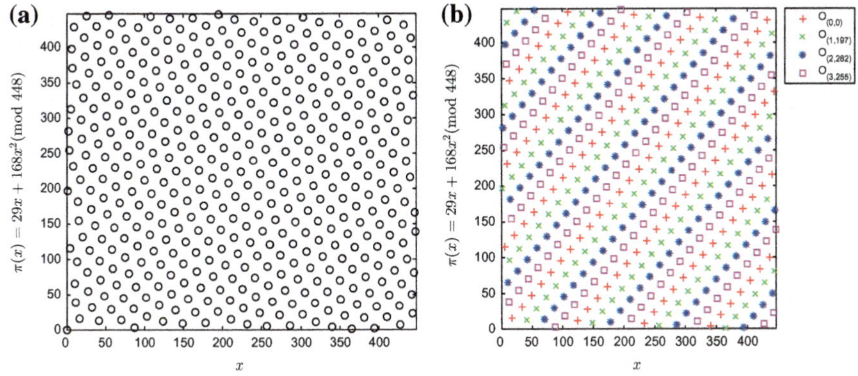

Fig. 7.4 The interleaver $\pi(x) = 29x + 168x^2 \pmod{448}$ and its four orbits

$$\delta_L(p_{x_1}, p_{x_2}) = |x_2 - x_1|_L + |\pi(x_2) - \pi(x_1)|_L = |k_1(i)|_L + |k_2(i)_L \leq \\ \leq \frac{L}{\gcd(2q_2, L)} + \frac{L}{\gcd(2q_2, L)} = \frac{2 \cdot L}{\gcd(2q_2, L)} \tag{7.27}$$

This is even the result in the theorem. ■

To find the other orbits except (7.26), we only need a representative from each one.

Theorem 7.8 (Theorem 6 from Takeshita 2007) *A complete set of representatives for the distinct orbits of Q is*

$$\{(i, \pi(i))|i = 0, 1, \ldots, L/\gcd(2q_2, L) - 1\} \tag{7.28}$$

Proof The set (7.28) is a complete set of representatives because the orbits are disjoint and in order to have a point belonging to the same orbit, they must be no closer than $L/\gcd(2q_2, L)$ in each coordinate, as results from Theorem 7.7. These must cover all representatives because the number of distinct orbits is $L/\gcd(2q_2, L)$. ■

The interleaver defined by $\pi(x) = 29x + 168x^2 \pmod{448}$ can be decomposed into $\epsilon = 448/\gcd(2 \cdot 168, 448) = 448/112 = 4$ disjoint orbits, shown in Fig. 7.4. We note that the placements of the points in plots in Fig. 7.4a, b are exactly the same. The regularity (linearity) of the interleaver gets clearly emphasized on the plot in Fig. 7.4b.

The next definition is for *local spread of a point p_x* in an interleaver code Q.

Definition 7.9 Let Q be an interleaver code generated by an arbitrary PP. The local spread of a point $p_x \in Q$ is

$$D_{p_x} = \min\{\delta_L(p_x, p_y) \mid \delta_L(p_x, p_y) \leq \sqrt{2L}, \sqrt{2L}\} \tag{7.29}$$

The following theorem gives an efficient computation procedure for the spread factor D of PP interleavers.

Theorem 7.10 (Theorem 7 from Takeshita 2007) *Let Q be an interleaver code generated by an arbitrary PP and let $\{p_x\}$ be a set of representatives for each orbit in Q. The spread factor Q can be computed by*

$$D = \min\left\{D_{p_y} \mid p_y \in \{p_x\}\right\}. \tag{7.30}$$

Proof All points are equivalent under translations in a particular orbit and a local spread cannot exceed the upper bound on D. ■

The randomness of an interleaver was assessed in Heegard and Wicker (1999) by a quantity named *dispersion*, denoted Γ. It is given by the number of distinct displacement vectors (Δ_x, Δ_y) (Heegard and Wicker 1999):

$$\Gamma = \left|\left\{(\Delta_x, \Delta_y) \in \mathbb{Z}^2 \middle| \Delta_x = j - i, \Delta_y = \pi(j) - \pi(i),\right.\right.$$
$$\left.\left. 0 \le i < j \le L - 1\right\}\right|. \tag{7.31}$$

The *normalized dispersion* is the value of Γ normalized to its maximum value, i.e.:

$$\gamma = \frac{2\Gamma}{L(L-1)} \tag{7.32}$$

The dispersion of an interleaver influences the multiplicities of the low weight code words; therefore a high dispersion is desirable. This desideratum was relieved in Trifina et al. (2006a, b), where interleavers with high dispersion have been proposed. However, this is not enough for a good performance, a good spread being also necessary.

In Takeshita (2007) the notion of randomness is replaced by degree of nonlinearity ζ (see Definition 7.4 and Eq. (7.22)). The nonlinearity metric ζ was shown to be inversely related to the degree of shift-invariance ϵ of an interleaver (see Eq. (7.21)). For QPP interleavers, the degree of nonlinearity ζ is computed in closed form as a function of the second-degree coefficient (see Eq. (7.22)), which gives a complete control of this parameter. When tail-biting convolutional codes are used as constituent codes, turbo codes using QPP interleavers become quasi-cyclic; for those codes, it is predicted that the multiplicity of many low-weight codewords is typically a multiple of the degree of shift invariance ϵ. Thus, for low multiplicities, the degree of nonlinearity ζ should be high.

It was shown in Sect. 7.2 that the maximization of the spread factor D is beneficial in the minimization of the number of low-weight codewords caused by self-terminating weight-2 input sequences or by short bursts of self-terminating input sequences. Thus, this parameter controls the effective free distance of the turbo code.

Taking into account that the minimum distance of a turbo code grows at most logarithmically with the interleaver length, the following metric is proposed in Takeshita (2007) to be maximized for a good interleaver:

$$\Omega = \ln(D) \cdot \zeta \tag{7.33}$$

The nonlinearity metric ζ does not capture the notion of orbits that are disjoint but there exists a linear curve that interpolates them. This can represent a problem. Consequently, in Takeshita (2007) the author proposed another nonlinearity metric $\zeta' \leq \zeta$ that fixes part of this problem. A QPP can be decomposed in two monomials, one corresponding to the first degree term $q_1 \cdot x \pmod{L}$ and the other to the second degree term $q_2 \cdot x^2 \pmod{L}$. In most cases, $\gcd(q_1, L) = 1$ for a valid QPP, which means that $q_1 \cdot x \pmod{L}$ is a LPP. When $\gcd(q_1, L) \neq 1$, a generalization is possible. Therefore, a QPP can be viewed as a LPP that is "affected" by $q_2 \cdot x^2 \pmod{L}$ at every position x. For example, the point at coordinate $x = 0$ gets affected by $q_2 \cdot 0^2 = 0$, the point at coordinate $x = 1$ gets disturbed by $q_2 \cdot 1^2 = q_2$, and so on. Considering Theorem 7.8, the periodicity of the disturbance is at most ζ. The *refined nonlinearity degree* ζ' simply measures how many distinct elements are in the set $\{q_2x^2 \pmod{L} \mid x = 0, 1, \ldots, \zeta - 1\}$. It is a very simple measure. From the values of ζ' in Table 7.3 we see that they do not grow as fast as ζ. With refined nonlinearity degree, the metric Ω becomes:

$$\Omega' = \ln(D) \cdot \zeta' \tag{7.34}$$

To minimize edge effects in the case of post-interleaver trellis termination, named uninterleaved dual termination in Takeshita (2007), the following metric is desirable to be maximized:

$$\mathcal{C} = \min_{x \in \mathbb{Z}_L} \delta\big((L-1, L-1), (x, \pi(x))\big), \tag{7.35}$$

where $\delta(p_i, p_j)$ is the Manhattan metric, given by relation (7.2).

The metric $\mathcal{C}$ is called the *corner merit* of an interleaver because geometrically it means avoiding points in the right upper corner of code interleaver Q.

Table 7.4 lists polynomials for which Ω' is maximized and with a spread factor D larger or equal to $\beta \cdot \mathrm{ub}_D(L)$; the associated interleavers will be called Ω'-QPP interleavers. The threshold β makes sure that the spread factor does not get too small for small block lengths. A reasonable threshold has been determined experimentally in Takeshita (2007). For the lengths given in Table 7.4, $\beta = 0.45$. For lengths greater than or equal to 2048, $\beta = 0.30$ is suggested in Takeshita (2007). In fact, as the block size increases, β is allowed to become smaller; this means that a maximization of the spread factor is considered less important for larger block lengths. When multiple polynomials with the same product merit Ω' exist, the one with the smallest coefficient q_2 and then q_1 is listed in Table 7.4.

Table 7.4 QPP interleavers with the best Ω' metric and $D \geq 0.45 \cdot \text{ub}_D(L)$ (Ω'-QPP) (Takeshita 2007) and the first term in the distance spectra of turbo codes of 1/3 nominal coding rate using these interleavers, generator matrix $G = [1, 15/13]$ and post-interleaver trellis termination

L	q_0	$\pi(x)$	D	ζ'	Ω'	C	$d_{min}/N_{d_{min}}/w_{d_{min}}$
40	14	$x + 10x^2$	4	2	2.77	16	12/24/96
80	72	$9x + 20x^2$	10	2	4.61	16	14/1/2
128	89	$7x + 16x^2$	8	3	6.24	21	12/1/2
160	115	$9x + 20x^2$	10	3	6.91	23	16/2/4
256	240	$15x + 32x^2$	16	3	8.32	30	16/1/2
320	304	$19x + 40x^2$	20	3	8.99	34	20/3/6
400	375	$7x + 40x^2$	16	5	13.86	39	16/1/2
408	273	$25x + 102x^2$	24	2	6.36	37	28/1/2
512	433	$15x + 32x^2$	16	4	11.09	45	20/1/2
640	549	$19x + 40x^2$	20	4	11.98	49	24/1/2
752	619	$23x + 94x^2$	26	3	9.77	51	25/1/3
800	786	$17x + 80x^2$	32	5	17.33	50	25/1/3
1024	992	$31x + 64x^2$	32	4	13.86	62	28/2/4

As in Table 7.2, to compare these interleavears with those found by methods presented in Sect. 7.5, we also give in Table 7.4 the first term in the distance spectra of turbo codes with these interleavers, 1/3 nominal coding rate, the generator matrix $G = [1, 15/13]$ and post-interleaver trellis termination.

7.4 A Method for Searching QPP Interleavers That Improve the Distance Spectrum of Turbo Codes Using the Spread Factor and Garrelo's Method

In Tarniceriu et al. (2009) two methods to obtain QPP interleavers with minimum distances superior to those given in Takeshita (2007) are proposed. The aim of these methods is to find QPP polynomials which lead to superior minimum distance and, concurrently, require less computation time than the exhaustive search over all QPP polynomials. The methods consider the spread factor, Ω' and corner merit metrics to reduce the search space for QPP interleavers. However, these methods only consider the minimum distance to enhance the performance of a QPP interleaver matched to the component codes of the turbo code.

In Trifina et al. (2011) a method for searching QPP interleavers which extends the second method in Tarniceriu et al. (2009) is presented. The search method of QPP interleavers for lengths up to 512, consists firstly in selecting polynomials with maximum D. In the second method in Tarniceriu et al. (2009), in the second step, QPPs with the highest minimum distance and lowest multiplicities were chosen, but in Trifina et al. (2011) QPPs with the best distance spectrum are chosen. The best spectrum of turbo codes is considered that which minimizes the truncated upper bound of frame error rate (TUB(FER)) for the independent Rayleigh fading channel with known channel state information (Eq. (2.25) in Chap. 2). SNR is considered high enough, so that TUB(FER) is a good approximation for real FER. In (2.25) more than one term in distance spectrum is considered (i.e. $M > 1$), because only the minimum distance and its multiplicities have proved to be insufficient in some cases, leading to polynomials with weaker performances. For example, for the interleaver length equal to 40, searching QPPs with the distance spectrum with only one term leads to the polynomial $\pi(x) = 19x + 30x^2$, for which the minimum distance is $d_{min} = 14$ and the multiplicities are $N_{d_{min}} = 2$ and $w_{d_{min}} = 4$. This polynomial, for 9 terms in distance spectrum, leads to TUB(FER)$\cdot 10^5 = 0.8106$. Searching QPPs with 9 terms in distance spectrum leads to the polynomial $\pi(x) = 13x + 30x^2$, for which TUB(FER)$\cdot 10^5 = 0.6539$, a value smaller than the previous one.

The complexity of the methods presented in Tarniceriu et al. (2009) can be reduced if in finding PP interleavers the equivalence between PPs is taken into account, thus avoiding to compute the distance spectrum more times for the same interleaver. This approach was used in Trifina and Tarniceriu (2013) for finding good CPP interleavers for LTE lengths up to 352.

7.5 Improved Method for Searching Interleavers in a Certain Class Using a Modified Version of Garrelo's Method

In Trifina et al. (2011) it was shown that there are lengths in the LTE standard for which the chosen QPP-based interleavers can be improved. The lengths considered in Trifina et al. (2011) are up to 512. The method used to find QPP interleavers consisted in minimizing the TUB(FER) metric in the class of interleavers with the largest spreading.

However, for medium and large lengths, the complexity of this method has to be reduced in order to find better QPPs in a reasonable time. In this section we present a method from Trifina and Tarniceriu (2014), which is an improvement of the method in Trifina et al. (2011), by reducing its complexity, allowing also to be applied for longer lengths.

The method consists in the calculation of TUB(FER) (or TUB(BER)), based on the distance spectrum calculated with Garello's algorithm, described in Sect. 2.4, for each $j = L, L-1, \ldots, 1$. Steps (1)–(5) calculate the complete distance spectrum and this algorithm is done in the C program given in Garello (C program) (2001), by the function named $gamma$. In the presented algorithm, some modifications of step (4) of $gamma$ function are operated, resulting in the function named $modified\_gamma$. This modification allows to stop the distance spectrum calculation for $j > 0$. In the algorithm of the improved method both functions $gamma$ and $modified\_gamma$ are used.

The search of interleavers based on the above mentioned method consists in:

- For the first set of interleavers, identical in terms of the distance spectrum, the complete distance spectrum is calculated with the original $gamma$ function and we denote $FER_{\min} = TUB(FER)$.
- For the following sets of interleavers we call the $modified\_gamma$ function, as follows:

(1) At each decrement of j, we compute $TUB_j(FER)$ with the spectrum computed for sequences $\boldsymbol{u}^{(j)}$.
   - If $TUB_j(FER) \leq FER_{\min}$, the calculation of the spectrum continues on the base of sequences $\boldsymbol{u}^{(j-1)}$.
   - If $TUB_j(FER) > FER_{\min}$, we check if there are next spectra possibly leading to lower $TUB_j(FER)$. Since the M distances in the spectrum are different, we choose the most optimistic case. Since for a smaller $TUB(FER)$ the distances should be as big as possible and the multiplicities as small as possible, the most optimistic case is when the real possible minimum distance is smaller only by 1 and its multiplicity is the smallest, i.e., 1.
   Thus in the first step, the temporary spectrum update is made as

$$d_{1\_temp} = d_1 - 1, d_{2\_temp} = d_1, \ldots, d_{M\_temp} = d_{M-1} \tag{7.36}$$

   The multiplicity for the first distance is set to the most favorable case, i.e. 1, and the other $M - 1$ multiplicities are updated similarly to the distances, i.e.

$$N_{1\_temp} = 1, N_{2\_temp} = N_1, \ldots, N_{M\_temp} = N_{M-1} \tag{7.37}$$

   For a $TUB_j(FER) > FER_{\min}$, the spectrum update continues until

$$d_{1\_temp} = d_1 - M, d_{2\_temp} = d_1 - (M - 1), \ldots,$$
$$d_{M\_temp} = d_1 - 1 \tag{7.38}$$

   with multiplicities

$$N_{1\_temp} = 1, N_{2\_temp} = 1, \ldots, N_{M\_temp} = 1 \tag{7.39}$$

   This is the last from the most optimistic cases of spectrum update.
   If, finally, $TUB_j(FER) > FER_{\min}$, we quit the loop for calculating the distance spectrum for the initial sequences $\boldsymbol{u}^{(j-1)}, \boldsymbol{u}^{(j-2)}, \ldots$ and so on. In this way the search time is reduced, especially if $TUB_j(FER) > FER_{\min}$, for j as close to L as possible.

(2) Continue similarly for the other sets of interleavers. Finally, we obtain the set of interleavers with the lowest $TUB(FER)$.

The method used in Trifina et al. (2011) to find QPPs with improved performances involved calculation of distance spectra using Garello's basic method (Garello et al. 2001). This was done for each QPP, as in the second method in Tarniceriu et al. (2009), or for a group of QPPs, leading to identical permutations, with the largest spreading factor D. This was done according to the condition provided in Trifina et al. (2011) and Zhao et al. (2010).

A reduction of the number of permutation polynomials, also including QPPs, for which the distance spectrum is calculated, has been given in Trifina and Tarniceriu (2013), where CPPs interleavers of small lengths have been found. One can note that for classical symmetric turbo codes the inverse interleaver results in the same distance spectrum. Thus, if a QPP admits an inverse QPP (Ryu and Takeshita 2006), different from the initial one or from one equivalent to it, then the distance spectrum was calculated only once for each group of such four QPPs. These are the groups of interleavers for which we apply step (1) in this section.

Furthermore, the complexity reduces especially if the QPP set with the lowest TUB(FER) is among the first in the search.

The method above was applied in Trifina and Tarniceriu (2014) to QPP interleavers able to outperform those chosen for the LTE standard, for lengths up to 1504. A model of independent Rayleigh fading wireless channel has been considered. Three classes of interleavers have been analyzed. They are described below.

For short lengths, maximizing the parameter D leads to better performance, therefore the first class examined was the one that maximizes the spread (denoted LS-QPP).

For longer lengths (greater than approximately 450), the simulation results did not show any improvement for the LS-QPP class. Therefore, it has been required to consider another class of QPP interleavers. Since the goal was to find interleavers better than those in LTE standard, the QPP class with the parameter as in the LTE standard (D_{LTE}-QPP) and the highest ζ' was considered. The condition to have the highest ζ' ensures a lower multiplicity in the distance spectrum and reduces the number of QPPs from the class. However, it can also affect the performance for smaller lengths.

The third class, consisting of all QPP interleavers, can be considered only for not too large lengths (up to 1008); it has been considered to verify if the metric TUB(FER) is sufficient to find the best interleavers.

Large tables with QPP interleavers found from the three classes were given in Trifina and Tarniceriu (2014). These results show that for lengths up to approximately 450, the best interleavers were found in the first class. For longer lengths, the second class contained the best ones.

Table 7.5 below displays QPP interleavers with minimum TUB(FER) from the class of largest spread, for the same lengths as in Tables 7.2 and 7.4, smaller than or equal to 408. The values of spread (D), nonlinearity degree (ζ), refined nonlinearity degree (ζ') and corner merit ($\mathcal{C}$) for the reported interleavers are also given in Table 7.5.

Table 7.6 displays QPP interleavers with minimum TUB(FER) from the class of QPPs with parameter D as in the LTE standard (D_{LTE}-QPP) and the highest ζ', for

Table 7.5 QPP interleavers with the largest spread and minimum TUB(FER) (LS-QPP-TUB(FER)min) (Trifina and Tarniceriu 2014) and the first term in the distance spectra of turbo codes with these interleavers, 1/3 nominal coding rate, the generator matrix $G = [1, 15/13]$ and post-interleaver trellis termination

L	$\pi(x)$	D	ζ	ζ'	$\mathcal{C}$	$d_{min}/N_{d_{min}}$ / $w_{d_{min}}$
40	$33x + 10x^2$	4	2	2	10	12/1/2
80	$11x + 20x^2$	10	2	2	14	19/3/5
128	$17x + 32x^2$	16	2	2	14	18/1/2
160	$99x + 40x^2$	16	2	2	18	19/1/1
256	$159x + 64x^2$	16	2	2	30	27/2/4
320	$21x + 80x^2$	20	2	2	28	25/1/3
400	$47x + 100x^2$	20	2	2	26	24/1/2
408	$229x + 102x^2$	24	2	2	30	27/2/4

Table 7.6 QPP interleavers with $D = D_{LTE}$ and minimum TUB(FER) (D_{LTE}-QPP-TUB(FER)min) (Trifina and Tarniceriu 2014) and the first term in the distance spectra of turbo codes with these interleavers, 1/3 nominal coding rate, the generator matrix $G = [1, 15/13]$ and post-interleaver trellis termination

L	$\pi(x)$	D	ζ'	Ω'	$\mathcal{C}$	$d_{min}/N_{d_{min}}$/ $w_{d_{min}}$
512	$289x + 192x^2$	32	3	10.40	30	27/1/1
640	$359x + 240x^2$	32	3	10.40	38	31/2/6
752	$165x + 94x^2$	26	3	9.77	46	26/1/2
800	$17x + 80x^2$	32	5	17.33	48	31/3/9
1024	$31x + 192x^2$	32	4	13.86	30	29/1/1

the same lengths as in Tables 7.2 and 7.4, greater than or equal to 512. The values of spread (D), refined nonlinearity degree (ζ'), Ω' metric and corner merit ($\mathcal{C}$) for the reported interleavers are also given in Table 7.6.

Comparing the interleavers of the same lengths from Tables 7.5 and 7.2 we observe that in all cases, with the method from this section, higher minimum distances (except the length of 160, for which the minimum distance is the same, but with smaller multiplicities) are obtained. It is interesting to observe that for the lengths of 256 and 320, we have obtained smaller values for ζ or ζ', but significantly higher minimum distances for the same maximum spread. The same effect can be observed comparing the interleavers of lengths 256 and 320 from Tables 7.5 and 7.4, even if the corner merit is maximized by the coefficient q_0 for the QPP in Table 7.4. This shows that, for the same spread, smaller values of the nonlinearity degree may lead to increasing the minimum distance. We have to mention that this effect is beneficial only for relatively short lengths.

Comparing the interleavers of lengths 128, 160 and 400 from Tables 7.5 and 7.4 we observe that a greater value of the spread (D) leads to a higher minimum distance. For length of 408 we observe that, for the same spread and the same refined nonlinearity degree, maximizing the corner merit by the coefficient q_0, for the QPP in Table 7.4, leads to a slightly higher minimum distance. However, this is not the case for the lengths of 40 and 80.

Comparing the interleavers of lengths 512 and 640 from Tables 7.6 and 7.2 we observe that, even if the spread and the refined nonlinearity degree values are the same, the interleavers from Table 7.6 have a better distance spectrum. For lengths of 752 and 1024, decreasing the spread and increasing the refined nonlinearity degree lead to a better distance spectrum for interleavers from Table 7.6.

Comparing the interleavers of lengths 512 and 640 from Tables 7.6 and 7.4, we observe that increasing the spread leads to higher minimum distance and a better distance spectrum for the interleavers from Table 7.6. The interleavers of lengths 752, 800 and 1024 from Tables 7.6 and 7.4 have the same values of Ω' metric (the same spread D and the same nonlinearity degree ζ'), but those from Table 7.6 have a better distance spectrum, considering the first three terms.

Tables 7.7 and 7.8 display some good QPP interleavers, with minimum TUB(FER), from the class of largest spread and from the class of QPPs with the parameter D as in the LTE standard (D_{LTE}-QPP) and the highest ζ', respectively. These good QPPs found have the simulated FER, for AWGN channel at high SNR, at least two times smaller compared to the simulated FER of LTE-QPP, as shown in Tables 7.7 and 7.8.

Finally, in Table 7.9, we give some CPPs better than some good QPPs or LPPs in terms of FER at high SNR for AWGN channels. The spread factor (D parameter), the refined nonlinearity degree (ζ') and the minimum distance with its multiplicity are also listed in Table 7.9. These QPPs/LPPs and CPPs were reported in Trifina and Tarniceriu (2017) and they were found using the method from this section applied for two different classes of interleavers, as follows:

- for lengths smaller than or equal to 448, by maximizing the spread factor and then by minimizing the TUB(FER) value;
- for lengths greater than or equal to 592, by maximizing the Ω' metric among the interleavers with spread factor greater than or equal to $0.45 \cdot \sqrt{2L}$ (Takeshita 2007) and then, by minimizing the TUB(FER) value.

From Table 7.9 we observe that CPPs of medium lengths can achieve a FER at high SNR of approximately 3 up to 9 times smaller than for QPPs. For these lengths, the spread factor of CPPs is always greater than or equal to that of QPPs, while the refined nonlinearity degree ofCPPs is always greater than that of QPPs. This fact togheter with smaller TUB(FER) values for these CPPs explains the FER performance differences (Takeshita 2007). For small lengths, the FER for CPPs is slightly smaller than the FER for good QPPPs or LPPs.

Table 7.7 LTE-QPP and LS-QPP with minimum TUB(FER) interleavers

L	SNR [dB]	LTE-QPP	D	$d_{min}/N_{d_{min}}$	FER ($\times 10^6$)	Good LS-QPP	D	$d_{min}/N_{d_{min}}$	FER ($\times 10^6$)
40	5.00	$3x + 10x^2$	4	11/1	11.77	$33x + 10x^2$	4	12/1	5.37
128	3.50	$15x + 32x^2$	16	16/1	3.32	$17x + 32x^2$	16	18/1	1.58
184	3.00	$57x + 46x^2$	12	16/1	7.75	$25x + 46x^2$	14	20/2	1.81
240	2.75	$29x + 60x^2$	16	24/2	4.21	$89x + 60x^2$	16	24/1	1.35
256	2.50	$15x + 32x^2$	16	16/1	6.93	$159x + 64x^2$	16	27/2	0.78
320	2.50	$21x + 120x^2$	20	20/1	3.01	$21x + 80x^2$	20	25/1	1.26
384	2.50	$23x + 48x^2$	24	22/1	1.17	$217x + 144x^2$	24	25/1	0.50
448	2.50	$29x + 168x^2$	28	22/105	18.00	$139x + 112x^2$	28	25/1	1.82

Table 7.8 LTE-QPP and D_{LTE}-QPP with minimum TUB(FER) interleavers

L	SNR [dB]	LTE-QPP	D	$d_{min}/N_{d_{min}}$	FER ($\times 10^6$)	Good D_{LTE}-QPP	D	$d_{min}/N_{d_{min}}$	FER ($\times 10^6$)
832	1.80	$25x + 52x^2$	26	23/1	1.55	$443x + 156x^2$	26	27/1	0.64
1248	1.60	$19x + 78x^2$	22	24/1	3.86	$19x + 234x^2$	22	26/1	1.84

Table 7.9 Some CPPs better than QPPs or LPPs

L	SNR [dB]	QPP or LPP	D	ζ'	d_{min}/ $N_{d_{min}}$	FER ($\times 10^6$) for QPP or LPP	CPP	D	ζ'	d_{min}/ $N_{d_{min}}$	FER ($\times 10^6$) for CPP
48	5.0	$11x$	8	1	15/1	1.5059	$13x + 6x^2 + 4x^3$	6	2	15/3	0.9258
120	4.0	$11x$	12	1	19/2	0.5223	$45x + 0x^2 + 8x^3$	12	5	20/4	0.2577
448	2.5	$139x + 112x^2$	28	2	25/1	1.8168	$251x + 56x^2 + 112x^3$	28	2	25/1	1.2802
592	2.0	$129x + 74x^2$	20	3	25/1	3.0400	$315x + 0x^2 + 74x^3$	20	5	26/1	0.6657
656	2.0	$21x + 246x^2$	22	3	29/1	1.9669	$185x + 164x^2 + 82x^3$	22	5	29/1	0.5617
688	2.2	$365x + 86x^2$	24	3	29/1	0.6749	$323x + 0x^2 + 258x^3$	24	5	28/1	0.1793
752	2.0	$165x + 94x^2$	26	3	26/1	1.8191	$541x + 188x^2 + 94x^3$	26	5	27/1	0.5096
816	2.2	$229x + 102x^2$	28	3	30/2	1.0478	$399x + 102x^2 + 34x^3$	28	8	29/1	0.3181
848	2.2	$185x + 318x^2$	28	3	29/1	0.8330	$157x + 212x^2 + 318x^3$	28	5	29/1	0.1390
912	1.8	$29x + 114x^2$	30	3	29/1	5.7172	$287x + 114x^2 + 114x^3$	30	4	36/1	0.9350
944	2.2	$265x + 118x^2$	32	3	32/1	1.0497	$179x + 0x^2 + 354x^3$	32	5	33/1	0.1557
976	2.3	$59x + 122x^2$	32	3	30/2	0.6804	$307x + 0x^2 + 122x^3$	32	5	33/1	0.0751

References

2001, http://www.tlc.polito.it/garello/turbodistance/turbodistance.html

2015, http://telecom.etti.tuiasi.ro/tti/papers/Text_files/Number_of_true_diff_LPPs_QPPs_CPPs_N_000002_100000.txt

E. Boutillon, D. Gnaedig, Maximum spread of D-dimensional multiple turbo codes. IEEE Trans. Commun. **53**(8), 1237–1242 (2005)

M. Breiling, A logarithmic upper bound on the minimum distance of turbo codes. IEEE Trans. Inf. Theory **50**(8), 1692–1710 (2004)

S. Crozier, New high-spread high-distance interleavers for turbo-codes, in *20th Bienn Symposium on Communications, Kingston, Ontario, Canada*, 28–31 May 2000, pp. 3–7

S. Crozier, P. Guinand, High-performance low-memory interleaver banks for turbo-codes, in *IEEE 54th Vehicular Technology Conference (VTC), Atlantic City, NJ, USA*, vol. 4, 7–11 October 2001, pp. 2394–2398

D. Divsalar, F. Pollara, Turbo codes for PCS applications, in *IEEE International Conference on Communications (ICC), Seattle, WA, USA*, 18–22 June 1995, pp. 54–59

S. Dolinar, D. Divsalar, Weight distribution of turbo codes using random and nonrandom permutations. TDA Progress Report (Jet Propulsion Laboratory, Pasadena, CA), 15 August 1995, pp. 42–122

G.D. Forney Jr., Geometrically uniform codes. IEEE Trans. Inf. Theory **37**(5), 1241–1260 (1991)

R. Garello, P. Pierleoni, S. Benedetto, Computing the free distance of turbo codes and serially concatenated codes with interleavers: algorithms and applications. IEEE J. Sel. Areas Commun. **19**(5), 800–812 (2001)

G.H. Hardy, E.M. Wright, *An Introduction to the Theory of Numbers*, 4th edn. (Oxford University Press, Oxford, 1975)

C. Heegard, S.B. Wicker, *Turbo Coding* (Kluwer Academic Publishers, Dordrecht, 1999)

C.Y. Lee, Some properties of nonbinary error-correcting codes. IRE Trans. Inf. Theory **4**(2), 77–82 (1958)

A. Perotti, S. Benedetto, A new upper bound on the minimum distance of turbo codes. IEEE Trans. Inf. Theory **50**(12), 2985–2997 (2004)

J. Ryu, O.Y. Takeshita, On quadratic inverses for quadratic permutation polynomials over integers rings. IEEE Trans. Inf. Theory **52**(3), 1254–1260 (2006)

O.Y. Takeshita, Permutation polynomial interleavers: an algebraic- geometric perspective. IEEE Trans. Inf. Theory **53**(6), 2116–2132 (2007)

O.Y. Takeshita, D.J. Costello Jr., New deterministic interleaver designs for turbo codes. IEEE Trans. Inf. Theory **46**(6), 1988–2006 (2000)

D. Tarniceriu, L. Trifina, V. Munteanu, About minimum distance for QPP interleavers. Ann. Telecommun. **64**(11–12), 745–751 (2009)

L. Trifina, D. Tarniceriu, Analysis of cubic permutation polynomials for turbo codes. Wirel. Pers. Commun. **69**(1), 1–22 (2013)

L. Trifina, D. Tarniceriu, Improved method for searching interleavers from a certain set using Garello's method with applications for the LTE standard. Ann. Telecommun. **69**(5–6), 251–272 (2014)

L. Trifina, D. Tarniceriu, On the equivalence of cubic permutation polynomial and ARP interleavers for turbo codes. IEEE Trans. Commun. **65**(2), 473–485 (2017)

L. Trifina, V. Munteanu, H. Balta, New types of interleavers based on the Welch-Costas permutation, in *Second European Conference on the Use of Modern Information and Communication Technologies (ECUMICT), Gent, Belgium*, 30–31 March 2006a, pp. 107–117

L. Trifina, V. Munteanu, D. Tarniceriu, Welch-Costas interleaver with cyclic shifts on groups of elements. Electron. Lett. **42**(24), 1413–1415 (2006b)

L. Trifina, D. Tarniceriu, V. Munteanu, Improved QPP interleavers for LTE standard, in *IEEE International Symposium on Signals, Circuits and Systems (ISSCS), Iasi, Romania*, June 30–July 1 2011, pp. 403–406

H. Zhao, P. Fan, V. Tarokh, On the equivalence of interleavers for turbo codes using quadratic permutation polynomials over integer rings. IEEE Commun. Lett. **14**(3), 236–238 (2010)

Printed by Printforce, the Netherlands